Medicinal Plants of Central Asia: Uzbekistan and Kyrgyzstan

Sasha W. Eisenman • David E. Zaurov
Lena Struwe
Editors

Medicinal Plants of Central Asia: Uzbekistan and Kyrgyzstan

Translations by David E. Zaurov, Sasha W. Eisenman,
Dilmurad A. Yunusov, and Venera Isaeva

Springer

Editors
Dr. Sasha W. Eisenman
College of Liberal Arts, School of Environmental Design
Department of Landscape Architecture and Horticulture
Temple University
580 Meetinghouse Road
Ambler, PA 19002
USA

Dr. David E. Zaurov
School of Environmental and Biological Sciences
Department of Plant Biology and Pathology
Rutgers University
20 Ag. Extension Way
New Brunswick, NJ 08901
USA

Dr. Lena Struwe
School of Environmental and Biological Sciences
Department of Plant Biology and Pathology
Rutgers University
59 Dudley Road
New Brunswick, NJ 08901
USA

ISBN 978-1-4614-3911-0 ISBN 978-1-4614-3912-7 (eBook)
DOI 10.1007/978-1-4614-3912-7
Springer New York Heidelberg Dordrecht London

Library of Congress Control Number: 2012939119

Printed on acid-free paper

Springer is part of Springer Science+Business Media (www.springer.com)

Preface

In ancient times, people used the gifts of nature found in their surrounding environments to treat their illnesses. Medicinal plants were of great significance, and the utilization of various plants in folk medicine has a very long history. As far back as 3000 BCE, herbs such as poppy, rhubarb, ginseng, etc., were well known. Hippocrates listed around 200 different medicinal herbs. In the first century, Dioscorides described about 400 medicinal plants, and the *Avesta*, the holy book of the Zoroastrians, included a thousand plants. In the eleventh century, Al-Beruni and Avicenna, two great scholars of Central Asia, made important contributions to the science of medicinal plants. Al-Beruni conceived a new area of science concerning medicinal herbs, now called pharmacognosy, and classified and described numerous plant species. In the year 1025, Avicenna gave the world *The Canon of Medicine*, where he described the herbs that were most widely researched and used in medical practice of the time.

Today, many of those plants are still used in medicine in Central Asia. Many centuries of herbal use has proven that plants contain substances that have healing power. Folk medicine has also shown that different parts of each plant often have different effects and, therefore, are used for different diseases, for example, roots for one type of disease and the aboveground parts for another. Similarly, leaves, flowers, fruits, and seeds may have different medicinal uses. Active compounds usually accumulate in large amounts in only certain parts of a plant (Wink 1999). The amounts of active substances in a plant, and consequently their physiological effect when taken as a medicine, significantly fluctuate depending on the season of the year, habitat, altitude, yearly climatic conditions, soil composition, and other factors (Evans 2002).

There are more than 20,000 plant species in the former Soviet Union. Of these, 4,500 grow in Uzbekistan and 4,100 in Kyrgyzstan (Komarov 1934; Pratov 1998; Umralina and Lazkov 2008). There are about 35,000–70,000 plants used in folk and scientific medicine worldwide (Hamilton 2004). As of 2004, at least 200,000 phytochemicals (excluding DNA-encoded proteins and peptides) have been characterized, but this is still thought to represent only a small percentage of phytochemicals that exist in nature (Raskin and Ripoll 2004). This further indicates the importance of drugs of herbal origin for folk and modern medicine. Currently, more than 400 wild and cultivated medicinal plants in Uzbekistan have been studied and described and more than 200 in Kyrgyzstan as well (Nikitina 1962). However, many medicinal plants found in Uzbekistan and Kyrgyzstan have not been thoroughly scientifically evaluated for their potential value in modern medicine.

Due to the increased interest in medicinal plants from different countries, the issue of preservation of the natural environment becomes important and, in particular, the conservation of medicinal plants in their original habitat. Habitat destruction and environmental pollution are factors that strongly affect medicinal plants in the wild. This complex issue is the subject of international agreements, which are united under the general concept of environmental preservation. For adequate conservation, it is important to identify the plant species that are most threatened due to over-collection in the wild. These species must receive the highest prioritization for preservation. It is important to bring the most utilized plants in medicine and veterinary science into cultivation with the goals of increasing the content of basic active compounds in the plants and providing a sustainable source of plant material. With the implementation of new agricultural practices, the industrial and medical sectors can be supplied with necessary amounts of high-quality product without depleting wild populations. Additional research is necessary to identify plants that have medicinal properties and to scientifically validate their pharmacological activities. It is important to conduct these efforts with the involvement of a wide circle of international researchers. Information exchange, job creation, and joint conferences will undoubtedly help researchers in their work and will also increase the conservation of the rich floras of Central Asian countries. A logical starting point for such systematic research would be the plants that were studied by our great ancestors and have traditionally been used in folk medicine in the different regions of Central Asia.

More than 200 of the most important medicinal plants of Central Asia are listed in this book, and it includes many whose medicinal uses and activities are being compiled for the first time. Most of the plants described grow wild in Central Asia, and some are endemic (e.g., *Vinca erecta* and *Ajuga turkestanica*). This book is aimed at scientists engaged in research on medicinal plants; physicians; as well as students of biology, pedagogy, agriculture, forestry, pharmacology, and medicine. This book is also a valuable reference for biodiversity conservation efforts and protection of rare and endangered species of the Central Asian flora.

We would like to warn our readers that conducting self-treatment with herbs and herbal preparations is dangerous. Medicinal plants can contain extremely strong physiologically active compounds and are often very poisonous. Without the proper recommendations of a medical doctor, no preparations of medicinal plants should be taken. The information in this book is not to be used to diagnose or treat any medical conditions.

Dr. Ravshanbek D. Kurbanov
Dr. Khasan Ch. Buriev
Dr. Djamin A. Akimaliev
Dr. Ilya Raskin

Acknowledgments

This collaborative project came into being as a result of the International Cooperative Biodiversity Groups (ICBG) Central Asia Project (Fogarty International Center grant No. U01TW006674-03). The ICBG program is managed by the Fogarty International Center of the National Institutes of Health (NIH) and is currently funded by the NIH, the National Science Foundation (NSF), the US Department of Agriculture (USDA) National Institute of Food and Agriculture, the Department of Energy (DOE) Office of Science, Biological and Environmental Research Program, and the National Oceanographic and Atmospheric Administration (NOAA) Oceans and Human Health Initiative. We would like to express our great appreciation to the many photographers who have allowed us to use their images in this book. SWE would like to thank Hillary, Eliyah, and Avishai for putting up with the long hours devoted to the production of this book. DZ would like to thank Dr. C. Reed Funk for his encouragement and support and Natalia Rogova for her assistance with gathering information on the plants used in Kyrgyzstan.

Contents

Contributors

Dr. Anarbek A. Akimaliev Biology and Soil Science Institute of the National Academy of Science of Kyrgyzstan, Bishkek, The Kyrgyz Republic

Dr. Djamin A. Akimaliev Kyrgyz Agricultural Research Institute, Bishkek, The Kyrgyz Republic

Dr. Igor V. Belolipov Tashkent State Agrarian University, Tashkent, The Republic of Uzbekistan

Dr. Khasan Ch. Buriev Tashkent State Agrarian University, Tashkent, The Republic of Uzbekistan

Dr. Sasha W. Eisenman College of Liberal Arts, School of Environmental Design, Department of Landscape Architecture and Horticulture, Temple University, Ambler, PA, USA

Dr. Ravshanbek D. Kurbanov The Specialized Center for Cardiology, Tashkent, The Republic of Uzbekistan

Dr. Anvar G. Kurmukov The Specialized Center for Cardiology, Tashkent, The Republic of Uzbekistan

Dr. Ilya Raskin Department of Plant Biology and Pathology, Rutgers University – School of Environmental and Biological Sciences, New Brunswick, NJ, USA

Dr. Ishenbay S. Sodombekov Kyrgyz Botanical Garden of the National Academy of Kyrgyzstan, Bishkek, The Kyrgyz Republic

Dr. David E. Zaurov School of Environmental and Biological Sciences, Department of Plant Biology and Pathology, Rutgers University, New Brunswick, NJ, USA

The Geography, Climate and Vegetation of Kyrgyzstan

1

Djamin A. Akimaliev, David E. Zaurov, and Sasha W. Eisenman

Kyrgyzstan is a mountainous country in the northeastern part of Central Asia. The Kyrgyz Republic shares borders to the south and southeast with Tajikistan and China, to the north and northwest with Kazakhstan, and with Uzbekistan to the west. The country covers 198,500 km^2 (76,621 sq miles) and has a population of approximately 5.3 million. Kyrgyzstan is divided into seven provinces (Fig. 1.1).

The highest point of elevation is in the Kakshaal-Too range, along the Chinese border, where Jengish Chokusu (Pik Pobedy) is the highest peak at 7,439 m (24,400 ft). The lowest point of elevation, 132 m (433 ft) above sea level, occurs along the Kara Darya River in the Fergana Valley. Other notable valleys are the low-montane Talas and Chui valleys, the mid-montane Issyk-Kul and Middle Naryn valleys, and the high-montane Ak-Say and Alai valleys. Ninety-four percent of the country is montane with the Tian Shan mountain system covering the major portion of the country. Lake Issyk-Kul, in the north western Tian Shan, is the largest lake in Kyrgyzstan and the second largest mountain lake in the world.

The principal river in Kyrgyzstan is the Naryn, which flows west through the Fergana Valley into Uzbekistan. There it meets another of Kyrgyzstan's major rivers, the Kara Darya. They merge to form the Syr Darya, which eventually flows into the Aral Sea. Heavy water usage for irrigation in Uzbekistan now causes the river to run dry before reaching the sea. The Chu River also briefly flows through Kyrgyzstan before entering Kazakhstan.

D.A. Akimaliev (✉)
Kyrgyz Agricultural Research Institute,
73/1, Timur Frunze St., Bishkek 720027, The Kyrgyz Republic
e-mail: krif@mail.kg

D.E. Zaurov
School of Environmental and Biological Sciences, Department of Plant Biology and Pathology, Rutgers University, 20 Ag. Extension Way, New Brunswick, NJ 08901, USA

S.W. Eisenman
College of Liberal Arts, School of Environmental Design, Department of Landscape Architecture and Horticulture, Temple University, 580 Meetinghouse Rd., Ambler, PA 19002, USA

Kyrgyzstan's climate is influenced by its position between the temperate and sub-tropical zones, its high elevation and its distance from oceans. These conditions cause intense sun radiation, lack of precipitation and a harsh continental climate. The mountain relief causes altitudinal zoning of climate parameters such as temperature and moisture. In July the average air temperature in the lowlands can range from 17 to 40°C (62.6–104°F), whereas at a higher elevation the temperature may be much cooler. During winters frosts may occur in all regions of Kyrgyzstan.

The southwestern Fergana Valley is dry-subtropical and hot in summer, with air temperatures reaching 40°C (104°F). The plains of southern and northern Kyrgyzstan have a hot desert or semi-desert climate and in these areas air temperatures can reach 35–40°C (95–104°F) during the summer months as well (Mamitov 1965). The northern foothills have a temperate climate and the climate in the Tian Shan mountain system varies from dry continental to polar, depending on elevation. The mountain regions have steppe, meadow-steppe, meadow, and high-mountain tundra climates and the highest areas are permanently snow covered (Ryazantseva 1965).

The yearly precipitation in Kyrgyzstan varies between 100 and 1,000 mm (3.9–39 in.) and is distributed unevenly throughout the country. The highest levels of precipitation (>900 mm; 35.4 in.) occur in the mid-belt of the southwestern slopes of the Fergana and Chatkal ranges, the high mountain areas of the northern slopes in the Kyrgyz Range, in the Kemin valley, and in the eastern Issyk-Kul area. The Talas and Chui valleys receive from 250 up to 500 mm (9.9–19.7 in.) precipitation and the valley and foothills in Fergana receive from 300 to 700 mm (11.8–27.6 in.) per year. Most of the internal and central areas of the Tian Shan system average 200–300 mm (7.9–11.8 in.) of rain annually and western Issyk-Kul and portions of Fergana may have less than 150 mm (5.9 in.) per year. On average, the foothills of the north and the eastern Issyk-Kul basin receive 15–20 cm (5.9–7.9 in.) of snow annually. The amount of snow fall in the high-altitude valleys of the Tian Shan is distributed very unevenly. The Ak-Shiyrak and Karakol valleys receive an

S. Eisenman et al. (eds.), *Medicinal Plants of Central Asia: Uzbekistan and Kyrgyzstan*,
DOI 10.1007/978-1-4614-3912-7_1, © The Editor 2013

Fig. 1.1 Kyrgyzstan with provincial boundaries

average of 3 cm (1.2 in.) of snow whereas, on average, the Karakudjur valley receives 9 cm (3.5 in.). The mid-altitude and high-altitude belts of the Fergana range can receive upwards of 150 cm (59 in.).

The flora of the Kyrgyzstan contains more than 4,100 species of vascular plants (Umralina and Lazkov 2008). Around 1,600 species have economic and/or useful value including species for fodder (450 species), for honey production (300 species) for medicinal use (200 species), for essential oils (62 species), and for food (50 species; Nikitina 1962). The largest portion of land used for agriculture is devoted to the cultivation of grain crops. Vegetables, oil crops and cotton are also grown to a lesser extent (UNDP 2007).

The distribution of the vegetation follows a pattern of elevation belts and is mainly influenced by land relief, climate, and soil zones. Twenty-two classes of ecosystems have been identified in Kyrgyzstan. The ecosystems are unevenly distributed throughout the country. Fourteen of the ecosystems occur in middle mountain zone (2,000–3,000 m), which occupies just 30% of the country's area. The Western and Central Tian Shan regions have 16 and the Alai has 13 ecosystems. In the Northern Tian Shan and Issyk-Kul regions 10 ecosystems can be found. The southern Kazakhstan biogeographic region has five of the ecosystems and the Fergana valley has the fewest with only three (Ministry of Environmental Protection 1998).

The ecosystems include deciduous and evergreen forests, shrublands, grasslands (savannahs, meadows and steppes), deserts, various wetlands and bodies of water. The river floodplains have shrubby forests (tugai) with *Rhamnus* spp., *Salix* spp., *Rosa* spp., etc. The valleys and foothills contain perennial herbs, ephemerals, and on stony soils, thorny herbs and succulents. In the mid-belt of the mountains, depending on precipitation levels, there are deserts, steppes, meadows and shrublands. The high elevation areas consist of glacial and subglacial areas as well as cryophylic steppes, alpine meadows and deserts. The majority of these deserts are *Artemisia* spp. dominated, fewer being *Salsola* spp. deserts, and a very few dominated by *Ephedra* spp. (Golovkova 1990).

In spring and in the beginning of the summer, *Astragalus* spp., *Crocus* spp., *Gagea* spp., *Iris kolpakowskiana*, *Ranunculus* spp. and *Tulipa* spp., as well as medicinal plants like *Betonica* spp., *Salvia* spp., *Thymus* spp., *Ziziphora* spp., etc. are found in the low- and middle mountain steppes. Meadows are less common than steppes, but they have a diverse floral composition including *Aconitum* spp., *Androsace ovczinnikovii*, *Aster alpinus*, *Cerastium* spp., *Codonopsis clematidea*, *Delphinium* spp., *Erigeron aurantiacus*, *Gentiana karelinii*, *Primula algida*, etc.

Only about 4.0% of Kyrgyzstan is covered with forests. Spruce and juniper forest account for a major portion of the forested area and over 350 herbaceous plant species can be found in the spruce forests. In the southern part of Kyrgyzstan the world's largest naturally occurring nut tree forests occupy about 608,500 ha (2,350 sq miles). These forests occur mainly in the Chatkal and Fergana ranges at an elevation of 1,000–2,200 m (3,280–7,218 ft). Many of the species in these forests are wild relatives of domesticated nut and fruit crops. These wild populations are important reservoirs of genetic diversity, which can be utilized in breeding programs to develop cultivars with cold tolerance, disease and insect resistance, and other important characteristics. The main forest species is *Juglans regia* (Persian walnut), which occupies about 40,000 ha (155 sq miles). Other wild fruits and nuts include *Prunus amygdalus* (almond) and *Pistacia vera* (pistachio), *Berberis oblonga*, *Cerasus mahaleb* and *C. tianschanica*, *Crataegus songorica* and *Cr. turkestanica*, *Malus kirghisorum* and *M. sieversii*, *Prunus sogdiana*, *Pyrus communis*, *P. korshinskyi*, and *P. regelii*.

Due to their extreme environment and climate, portions of the country have limited or no biodiversity. These areas account for around 45% of the country and consist of high altitude areas (above 3,500 m [11,483 ft]) of rock and glaciers, open areas of rock, gravel or clay, and deserts.

There are 65 plant species on the list of endangered species in the *Red Data Book of Kirghiz SSR* (1985). Sultanova et al. (1998) published a more up-to-date list with 386 species recommended for inclusion to the red book. At the present time there is a need for the establishment of organized medicinal plant farming and for the protection of endangered species. Many of the plants used in Kyrgyz folk medicine have not been studied using modern scientific techniques. Pharmacological studies are necessary to characterize the biological activity of the medicinal plants and their components. Folk medicine is an invaluable source of information on the properties and activities of medicinal plants and for discovery of novel medicines. Further study of the Kyrgyz ethno-medicine will help facilitate the identification of new medicinal plants, which may possibly serve as sources for new pharmaceuticals. Further expansion of botanical and floristic research is also necessary, including detailed mapping of all medicinal plant resources and determination of regions for cultivation of valuable and rare species.

Currently all ecosystems are subject to human influence. The overall biodiversity of Kyrgyzstan is threatened as a result of human disturbance. Over-grazing has degraded many of the plant communities and over-use has greatly reduced the overall size of forest ecosystems. Intensifying anthropogenic influence threatens the diversity of the natural resources of the country. Preservation and conservation of these unique natural resources is of extreme importance for future generations of Kyrgyz people.

The Geography, Climate and Vegetation of Uzbekistan

2

Igor V. Belolipov, David E. Zaurov, and Sasha W. Eisenman

Uzbekistan is a country in Central Asia that extends from the foothills of the Tian Shan and Pamir mountains in the east to just west of the Aral Sea. In the north Uzbekistan borders Kazakhstan, in the east and southeast Kyrgyzstan and Tajikistan, in the west Turkmenistan, and in the south Afghanistan. The country covers 447,400 km^2 (172,742 sq miles) and has a population of about 26 million. Uzbekistan is divided into 12 provinces and 1 autonomous republic (Fig. 2.1).

The highest point of elevation is in the Gissar mountain range at 4,643 m (15,233 ft), and the lowest point of elevation is the Sarykamysh depression at 20 m (ca. 65.6 ft) below sea level. About 80% of Uzbekistan's land consists of plains and deserts. The vast Kyzlkum desert lies in central Uzbekistan and is largely uninhabited except for mining towns.

There is a wide spectrum of natural environments from the hot sand and gypsum deserts of Kyzlkum to the eternal snows and glaciers of the Pamiro-Alai mountains. All valleys receive their water from glaciers in the Tian Shan and Pamiro-Alai mountains. Uzbekistan's two most important rivers, the Syr Darya and Amu Darya, flow from the Tian Shan and Pamiro-Alai mountain ranges to the Aral Sea.

The climate of Uzbekistan is continental with predominance towards harsh continental. It is characterized by low precipitation (70–100 mm [~2.75 to 3.94 in.] per year) in the plains of the northern-western part of the country and up to 1,200 mm (47.25 in.) of precipitation in mountainous regions. Over 70% of the precipitation falls in the autumn to spring period, with a maximum in March and April. Summers in Uzbekistan are long, dry, and hot, summer rains are very rare, and summer temperatures may reach 45°C (113°F). In the south the winter is mild, but sometimes with considerable frosts. In the northern regions winters are cold and temperatures may drop to −37°C (−35°F).

The flora of Uzbekistan contains more than 4,500 vascular plants in 650 genera, in 115 families (Chemonics International Inc. 2001). More than 4,000 species of algae and more than 2,000 species of fungi also occur in Uzbekistan (National Biodiversity Strategy Project Steering Committee 1998). The most species-rich plant families account for a large portion of the flora. These families include Asteraceae (600 species), Fabaceae (450 species), Poaceae (>250 species), Brassicaceae, Lamiaceae, Rosaceae, Boraginaceae and Apiaceae.

Agriculture and cultivated crops occupy considerable areas of irrigated and non-irrigated land. Some of the major crops are cotton, maize (corn), alfalfa, wheat, barley, sorghum, rice, mulberry for silkworm culture, vegetables, melons, fruit trees, and others. The natural vegetation of Uzbekistan is a very rich source of fodder (more than 1,700 species), medicinal plants (600 species) and plants with essential oils (>650 species), saponins (>100 species), and tannins (ca. 400 species).

The vegetation of Uzbekistan is divided into four main ecosystems. The main cause for ecosystem zonation is change in hydrothermal conditions. These zones form belts which are directly correlated to an increase in precipitation and elevation. As elevation increases there are changes in environmental conditions. Growing periods become shorter, temperature decreases and precipitation increases. Due to the increase in precipitation water is no longer a limiting factor above 2,500 m (~8,200 ft). Diverse soil conditions, in combination with the environmental conditions, result in a great diversity of vegetation. The local names "chul" (arid plain, desert), "adyr" (foothills), "tau" (mountains), and "yailau" (alpine zone) are widely used by the people of Uzbekistan and correspond to the zones produced by vertical changes in the landscape.

I.V. Belolipov (✉)
Tashkent State Agrarian University, Tashkent-140, Tashkent 100140, The Republic of Uzbekistan
e-mai: ivbelolipov@mail.ru

D.E. Zaurov
School of Environmental and Biological Sciences, Department of Plant Biology and Pathology, Rutgers University, 20 Ag. Extension Way, New Brunswick, NJ 08901, USA

S.W. Eisenman
College of Liberal Arts, School of Environmental Design, Department of Landscape Architecture and Horticulture, Temple University, 580 Meetinghouse Rd., Ambler, PA 19002, USA

S. Eisenman et al. (eds.), *Medicinal Plants of Central Asia: Uzbekistan and Kyrgyzstan*,
DOI 10.1007/978-1-4614-3912-7_2, © The Editor 2013

Fig. 2.1 Uzbekistan with provincial boundaries

The chul zone (arid plain, desert): The chul consists of the flat territory of Uzbekistan, which is usually considered desert. The chul continues up to 500–600 m (~1,640 to 1,970 ft) above sea level and has a dry period of 3–6 months. The climate of the chul zone is ultra-continental and is characterized by low precipitation of 70–208 mm (~2.75 to 8.2 in.) per year and humidity levels that drop to as low as 1–2%. The dry period in the chul zone lasts from May to October. Summer temperatures can reach 45°C (113°F) while winter temperatures often drop below –30°C (–22°F).

The chul zone occupies most of the Central Asian plain (Turan) and displays four soil types: salty chul, sandy chul, gypsum (stony) chul, and clay chul (National Biodiversity Strategy Project Steering Committee 1998). Portions of the salty chul ecosystem that have extremely high salt concentrations support no plant life. Areas of salty chul with lower salt content are dominated by *Artemisia halophila* and species in the Chenopodiaceae family such as *Halocnemum strobilaceum*, *Halostachys caspica*, *Haloxylon aphyllum*, *Salicornia herbacea*, *Salsola dendroides*, *Suaeda dendroides* and *S. microphylla.* Sandy chul is dominated by *Acanthophyllum korolkowi*, *Ammodendron conollyi*, *Astragalus villosissima*, *Calligonum aphyllum*, *Convolvulus hamadae*, *Ephedra strobilacea*, *Ferula foetida*, *Salsola arbuscula* and *S. richteri*. The gypsum chul is located in the hills of the southwestern and central Kyzylkum desert. *Artemisia* associations predominate in the gypsum chul zone. The most common association is *Artemisia diffusa* (less frequently *A. ferganensis*) with *Convolvulus hamadae* or with co-dominance of *Aellenia subaphylla*, *Anabasis eriopoda*, *Anabasis turkestanica* and *Salsola arbuscula*. The species *Calligonum junceum* and *Reaumuria turkestanica*, and others are commonly found in the gypsum chul and are characteristic for the area. *Nanophyton erinaceum* is less frequent and restricted mainly to the hills.

Where river valleys cut into the chul zone the increased humidity in the valleys facilitates the development of special mesophytic communities that are locally called "tugai". Common species that occur in these communities are *Alhagi persarum*, *Apocynum scabrum*, *Asparagus persicus*, *Clematis orientalis*, *Elaeagnus orientalis*, *Erianthus purpurascens*, *Glycyrrhiza glabra*, *Halimodendron halodendron*, *Hippophae rhamnoides*, *Karelinia caspia*, *Limonium otolepis*, *Lycium ruthenicum*, *Phragmites communis*, *Populus diversifolia* and *P. pruinosa*, and *Tamarix* spp.

The adyr zone (lowlands and foothills): The adyr zone is a broad belt at an elevation of around 500–1,500 m (1,640–4,921 ft). This band is found around all the mountains of Central Asia. It occupies the range between two contrasting ecological zones: the xerothermic chul (desert) and the mesothermic tau (mountain region). The soils of the adyr zone contain less salt and more humus than the chul soils and are classified as sierozem (Makhmudov 2001). Bedrock is often found exposed on the surface.

The annual precipitation is between 250 (9.8 in.) and 400 mm (15.7 in.) and rarely reaches 500 mm (19.7 in.). The mean monthly temperature for July is 25°C (77°F), which is 3–4°C lower than in the chul and 5–6°C higher than in the tau zone. The dry period lasts from June to September. Due to its location the adyr zone is exposed to the influence of both the hot desert along its lower edge, and the cooling effects of the mountains on its upper edge. This causes the lower section of the adyr zone to be closer to the environmental conditions of the chul and the upper section to be similar to the mountainous environment of the tau zone. Because of this gradient the adyr is divided into subzones: the lower adyr with rolling relief and the upper adyr with broken relief.

Typical species found in the lower adyr area are *Amygdalus spinosissima*, *Artemisia sogdiana*, *Carex pachystylis*, *Mediasia macrophylla*, *Phlomis thapsoides*, *Pistacia vera* and *Psoralea drupacea*. At altitudes of 1,200–1,500 m in the upper adyr zone, typical species are *Acanthophyllum gypsophiloides*, *Agropyron trichophorum*, *Astragalus eximius*, *Bunium persicum*, *Centaurea squarrosa*, *Cousinia pulchella*, *Onobrychis* spp., *Phlomis salicifolia* and *P. olgae*, *Potentilla soongarica*, *Scabiosa songarica* and *Ziziphora pamiroalaica*.

The tau zone (mid-mountain zone): The tau zone is a broad belt at an elevation of around 1,500–2,800 m (4,921–9,186 ft). The dominant soil of the tau zone is of the brown soil type. Precipitation in this zone exceeds more than 500 mm (19.7 in.) per year, with a dry period that lasts for 3 months from July to September. The growing period is in spring, summer, and autumn with a dormant period in the winter. The mean monthly temperature in July is 19°C (66°F).

In terms of economy, the tau zone is an important area for growing cereals and leguminous crops, for producing hay, and for use as pastures. The dominating wild and cultivated shrub and arboreal species (*Crataegus* spp., *Juglans regia*, *Malus* spp., *Prunus* spp., etc.) of the area provide the local population with fuel, building materials, and food. In the tau zone shrubs can be found in large groups or as individuals. The common species of shrubs are *Berberis oblonga*, *Cerasus tianshanica*, *Ephedra equisetina*, *Lonicera microphylla*, *Rosa kokanica* and *Spiraea hypericifolia*. Some of the woody species found in the tau zone are gymnosperms such as *Juniperus semiglobosa*, *J. seravschanica* and *J. turkestanica*, and broad-leaved deciduous trees such as *Acer turkestanicum*, *Betula tianschanica*, *Crataegus pontica* and *C. turkestanica*, *Juglans regia*, *Malus sieversii*, *Prunus sogdiana*, *Sorbus persica*, *Ziziphus jujuba* and others.

The yailau zone (high mountain zone): The yailau zone is the high-altitude, subalpine to alpine zone and extends from 2,800 to around 3,400 m (9,186–11,155 ft). This zone is characterized by environmental conditions that will not support the development of arboreal and shrub vegetation. The soil is mainly light brown and of the meadow-steppe type (Kaurichev 1989). Summer is short and rather warm, with sharp changes between day and night temperatures. Summer daytime temperature reaches up to 25°C (77°F), but can drop to 0°C (32°F) at night. In the winter the temperature may drop to −40°C (−40°F). Precipitation varies from 400 (15.7 in.) to more than 600 mm (23.6 in.) per year. This zone has stony taluses, glacial valleys, glacial cirques, and glacial tongues, and fields with heavy clay soils. There are many sheer rock formations in the southwestern Tian Shan and the western Pamiro-Alai.

In terms of economy, the yailau region is utilized as the main summer pasture. While the Karakul sheep graze mainly in the chul, the Merinos and fat-tailed breeds of sheep (including the Gissar breed) are pastured mainly in the yailau. Other agriculture is limited by low temperatures.

Tallgrass meadows are an important portion of the vegetation cover of the yailau. These meadows also contain *Polygonum bucharicum* and *P. hissaricum*. In western Tian Shan and the southern Pamiro-Alai there are Apiaceae-rich meadows with *Ferula tenuisecta* and *Prangos pabularia*. Meadows containing *Alopecurus*, *Artemisia*, and *Geranium* spp. are also common in the yailau zone. Shortgrass meadows, also known as alpine meadows, are found in small patches in the upper yailau. The high-altitude meadows are comprised of a diversity of grasses and other herbaceous genera such as *Gentiana*, *Oxytropis*, *Potentilla* and *Ranunculus*. Meadows of grasses such as *Alopecurus* spp., *Festuca* spp., *Poa alpina* and *Phleum alpinum*, and sedges such as *Carex* and *Kobresia* are also characteristic of the upper yailau zone.

The great extremes of elevation, temperature, precipitation, and soil types found in Uzbekistan provide a wide range of habitats, which support a great diversity of vegetation. Due to human activities many of the natural areas of the planet are being disturbed or destroyed. Conservation of natural environments and resources are of great importance for the future of mankind and the conservation of Uzbekistan's natural resources is no exception.

A Short History of Medicinal Plant Use in Central Asia

3

Anvar G. Kurmukov and Anarbek A. Akimaliev

Central Asia is a synthesis of many nations and many cultures. There is a long history of using and documenting medicinal plants in this region. Great contributions to the knowledge of medicinal plants were made by the Greeks as early as the seventh century BCE. In the sixth century BCE, Central Asia was part of the Persian Empire founded by Cyrus (Bobokhanova and Bekturgunava 1996). In the first half of the fourth century BCE, Alexander the Great helped expand the Greek empire into the regions of Bactria and Sogdiana (territories that included much of present-day Central Asia) and formed the Greco-Bactrian Kingdom. It is known that Chinese travelers visited Central Asia and surrounding regions since the establishment of the Great Silk Road. China has a long history of herbal medicine and undoubtedly had a great influence on the development of Central Asian herbology. The Zoroastrian holy book, the *Avesta*, written over a long period (the nineth century BCE – third century CE), is a valuable source of information about the social structure, medicine, and way of life in the ancient societies of Central Asia, Iran, and Azerbaijan. The book includes information concerning all aspects of life, including natural philosophies and medical views. In the *Avesta*, medicine is described as the art of keeping the body in health. There were several kinds of treatments: (1) surgery, (2) treatment with herbs, and (3) treatment with words. This source was expanded by other researchers and physicians up to the eighth century and by that time, contained information on 1,000 plants. The *Avesta* contains information about useful and unhealthy plants, herbs that were used as sedatives, anesthetics, narcotics, restoratives, tonics, antiseptics, antidotes, and as other types of remedies. In the fifth and sixth centuries CE a large Turkic kaganate was developed as result of the unification of diverse nomadic tribes. In the seventh century Arabs brought Islam to Central Asia and during this time the knowledge and science of medicinal plants grew greatly. In 1220 CE the Mongols, led by Genghis Kahn, invaded Central Asia and there is no doubt that this invasion had an influence on the culture of the local population.

The Central Asian scientists Abu Rayhan Muhammad ibn Ahmad Al-Beruni (973–1048) and Abu Ali ibn Sina (Avicenna; 980–1037) made considerable contributions to the knowledge of medicinal plants. Both were great scholars, and besides other sciences, studied pharmacognosy and pharmacology. The work *Kitab-al-Saidana* (*Materia Medica*) was written by Beruni towards the end of his life (1041–1048). It describes about 750 plants and contains information about the botanical characteristics of plants and their geographical locality. About 400 geographical place names from where the plants had been brought (Central Asia, Afghanistan, Iran, Arabia, Azerbaijan, Armenia, and others areas) are mentioned.

Abu Ali ibn Sina (Avicenna) is famous due to his works on philosophy and medicine. Being a doctor, he studied botany as well, and often used medicinal plants to treat his patients. His most important medical work is the *Al-Qanun fi al-Tibb* (*The Canon of Medicine*). The second volume of this work is dedicated to medicinal remedies used during his time. The book describes more than 800 pharmaceutical substances of vegetative, animal, and mineral origin. Besides remedies produced in Central Asia and other countries of the Near and Middle East, Avicenna described a number of drugs brought from India, China, Greece, Africa, Mediterranean islands, and other parts of the world. The book includes the practices of scientific medicine as well as the traditional folk medicine of the time. Many medicines (drugs) described by Avicenna have entered the pharmacopoeia and are still in use. The fifth volume of *The Canon of Medicine* represents his pharmacopoeia. It describes how to make and use different forms of drugs and complex medicinal formulations. In the chapter named "Necessity of complex drugs", Avicenna

A.G. Kurmukov (✉)
The Specialized Center for Cardiology, 4 Murtozoev St., Tashkent 100052, The Republic of Uzbekistan
e-mail: kurmukov_anvar@mail.ru

A.A. Akimaliev
Biology and Soil Science Institute of the National Academy of Science of Kyrgyzstan, 265 Chuy St., Bishkek 720071, The Kyrgyz Republic

S. Eisenman et al. (eds.), *Medicinal Plants of Central Asia: Uzbekistan and Kyrgyzstan*,
DOI 10.1007/978-1-4614-3912-7_3, © The Editor 2013

recommended making complex drug formulations in order to increase effects of a drug; to prevent side effects of one drug by another drugs composition; to strengthen the effect of the main drug by adding another one (synergy); to increase penetration of one drug into tissues with the help from another drug, or to slow down an effect of a drug by reducing absorbability caused by a second drug and this way elongate the effect of the first drug; and to use drugs for guiding delivery of the main active substance to a point (organ) of action. Carl Linnaeus later named *Avicennia*, a genus of tropical mangrove trees, in honor of Avicenna.

In the eleventh to twelfth centuries, Ismail al-Jurjani (Ismail ibn Muhammad al-Husayn Jurjani), wrote an encyclopedic work on medicine called *Zakhirah-i Khvarazm'Shahi* (*Treasure of Khorezm Shah*). Later, in the fourteenth century, Mansur ibn Ilyas (Mansur ibn Muhammad ibn Ahmad ibn Yusuf ibn Ilyas) published his work *Kifayah-i Mansuri* (*Mansur's Sufficient Book*), which was also known as *Kifayah-i Mujahidiyah* (*The Sufficient* [*Book*] *for Mujahid*). Besides being summaries of medical theory and practice, these works gave basic information about plant-based medicine. In the eighteenth century, Muhammad Husayn (Muhammad Husayn ibn Muhammad Hadi al-Aqili al-Alavi al-Khurasani al-Shirazi, also known as or Hakim Muhammad Hadikhan) described the therapeutic qualities of more than 2,000 plants, preparations of animal origin and minerals in his works *Majma al-javami va-zakha'ir al-Tarakib* (*The Assemblage of Generalities and Treasuries of Compounds*) and Makhzan-al-*Adviyah* (*The Storehouse of Medicaments*). These works were largely based on the earlier writings of his great uncle Alavi Khan and documented centuries of past achievements in the field of folk medicine, the practices of previous physicians, and his personal research.

In the past, the use of plants for medicine was not rigorously based in science. Modern scientific techniques have been used to prove the effectiveness of many plant remedies used in folk medicine and prescribed by ancient physicians. For instance, *Rauvolfia serpentina* has been used in Indian medicine for about 2,000 years, while Europeans discovered the value of this plant only in the middle of twentieth century (Balick and Cox 1996; Gupta 2002). Since ancient times, Africans have used *Strophanthus* seeds to make arrow poison and as a cardiac remedy, but only at the end of nineteenth century did *Strophanthus* enter the European pharmacopoeia (Norn and Kruse 2004). To the present day, these plants are considered irreplaceable cardiac remedies used to treat cardio-vascular diseases.

Many of the specific activities of plant remedies described by Avicenna have been confirmed by modern research conducted at the Institute of Chemistry of Vegetative Substances (ICVS) of the Academy of Sciences of the Republic of Uzbekistan. For example, according to Avicenna, the plant *Haplophyllum perforatum* has anti-inflammatory and sedative effects. The alkaloids perforine, evoksine, skimmianine, and others have been isolated from this plant. It has been found that at medium doses these alkaloids have sedative, and in higher doses sleep-inducing effects. Some of these alkaloids also have an anti-inflammatory action (Sadritdinov and Kurmukov 1980). Another example is *Khiltit* (the gum from *Ferula foetida*), which Avicenna noted could be used as a treatment for malignant and fatal tumors by cutting the tumor open and applying the gum. According to Avicenna this gum also strengthens the libido and stimulates menstruation. The esters of sesquiterpene alcohols, ferutinine, ferutin and others have been isolated from this species. A preparation from this plant, *Panoferol*, and also the individual compounds ferutin and ferutinine, have pronounced estrogenic action. *Panoferol* strengthens the libido and increases impregnation in sheep, pigs, and cattle. A mixture of ferutin and ferutinine (under the name *Tefestrol*) has been introduced to obstetric-gynecologic practice as an estrogenic preparation (Kurmukov and Akhmedkhodzhaeva 1994). There are many more examples of corresponding effects of various plants described by Avicenna and recent data gathered by modern pharmacologists (Sadritdinov and Kurmukov 1980).

In the twentieth century, research on Central Asian medicinal plants was especially productive, particularly in Uzbekistan. In 1943, the Laboratory of Chemistry of Alkaloids (headed by Professor S.Yu. Yunusov) was founded at the Institute of Chemistry in the Uzbek branch of the Academy of Sciences of the USSR. Due to the successful activity of the laboratory, Yunusov created the Institute of Chemistry of Plant Substances at the Academy of Sciences of Uzbek Soviet Socialist Republic in 1956. The Institute had laboratories devoted to distinct chemical groups including alkaloids, glycosides, fats, proteins, terpenes and acids, lignin, coumarins and phosphorous-containing organic compounds, as well as botany, laboratories of pharmacology and toxicology, phytotoxicology, experimental technology and physical and quantitative analysis and others.

The Institute's scientific directions consisted of a complex of investigations into plant substances. All plant parts collected during different growth periods and from different regions, were investigated. Applied laboratories had the task of studying the pharmacological activity of compounds; to determine the possibility of introduction into medical practice; to study the natural habitat of the medicinal plants; to organize long-term plant collecting; to maintain the safety of natural populations; and to organize the development of medicinal formulations and their production. The overall goal of the institute was to create medicinal products following a research pipeline, which included collecting of data on pharmacognosy, isolation of individual compounds, study of their pharmacological activity and creation of medicinal preparations up to the point of introduction into medical practice.

Researchers in the Laboratory of Alkaloid Chemistry isolated and studied many alkaloids, including a number of new alkaloids belonging to various chemical groups. Research on alkaloid chemistry was summarized in the monograph *Alkaloids* by Yunusov (1974, 1981). These newly isolated alkaloids were also studied by the pharmacology and toxicology labs. The Glycoside Chemistry Laboratory (headed by Professor N.K. Abubakirov) studied cardiac glycosides. This laboratory made significant contributions to the knowledge of the chemistry of triterpene glycosides. Among the studied compounds, glycosides with immunomodulatory, gonadotropic, and hypolipidemic activities were identified. Studies of *Astragalus* led to the isolation of methyl-steroids of the cycloartan series. Many species of *Allium* (onions) native to Central Asia were investigated and as a result more than 30 new compounds were isolated. One of the most important scientific directions of the laboratory during the past years has been investigations of phytoecdysteroids. This laboratory identified the structure of 25 of the 95 phytoecdysteroids described in the literature by 1980. Studies of *Amorpha fruticosa* led to the discovery of a new class of plant glycosides containing rotenone derivatives as the aglycone.

The Laboratory of Lactones, Coumarins, and Terpenoids (headed by Prof. G.P. Sidyakin) studied various plants for lactone content, particularly for the lactones leucomisine and austricine, which were isolated from *Artemisia leucodes*. Both lactones possess pronounced anti-inflammatory action. As a compound possessing pronounced angio-protective, hypolipidemic, hypo-cholesterolemic, and anti-inflammatory actions, leucomisine has passed medical tests and has been introduced into medical practice under the preparation name *Oligvon*. Since 1970, systematic studies of chemical compounds found in various species of the genus *Ferula*, which grows in the territory of Uzbekistan and adjacent republics, have been conducted. As a result, more than 50 species of *Ferula* have been investigated, from which more than 250 new terpenoids, coumarins, and esters have been isolated and their chemical structures determined. Natural esters of mono- and sesquiterpene alcohols with aromatic acids were discovered for the first time in this lab (Kurmukov and Akhmedkhodzhaeva 1994).

In 1957, the Pharmacology Laboratory was founded at the Institute (headed by associate prof. I.K. Kamilov). The initial activities of the laboratory were related to alkaloids. The findings were mainly summarized in the books *Pharmacology of Plant Alkaloids and Their use in Medicine* (Sadritdinov and Kurmukov 1980) and *Alkaloids and Herbal Preparations for Hypertensive Treatmen*t (Kurmukov and Zakirov 1992). During these years preparations containing the alkaloids vincanine (a preparation of vincanine hydrochloride, a strychnine-like analeptic), vincamine (a preparation of vincametrine, a stimulator of uterine smooth muscles), ervinine (a CNS analeptic with primary stimulating effect on the respiratory center) and others were introduced into medical practice. Later the alkaloid lappaconitine, in the preparation *Allapenin* developed by S.Yu. Yunusov and F.N. Dzhakhangirov and isolated from *Aconitum soongaricum*, was introduced into medical practice and was widely used as an antiarrhythmic drug. The same authors developed the compound preparation *Aklezin* from similar alkaloids and which was also used as an antiarrhythmic drug. Pharmacological investigations of alkaloids from *Peganum harmala* resulted in the introduction of an anticholinesterase preparation, *Desoxypeganine*, into medical practice (Tulyaganov et al. 1986). The rotenoid glycoside amorphine was isolated from the plant *Amorpha fruticosa* in the laboratory of chemistry of glycosides. Pharmacological studies revealed the hypolipidemic, hypocholesteremic, and angio-protecting actions of the preparation (Aizikov et al. 1984; Kurmukov et al. 1982, 1984a, b, 1986). After completion of clinical tests, the preparation *Glirofam* (containing amorphine), was introduced as a prophylaxis and treatment of atherosclerosis.

A series of studies on the pharmacology of phytoecdysteroids (ecdysterone, turkesterone, ciasterone, viticosterone) isolated from *Rhaponticum carthamoides, Ajuga turkestanica*, and various species of *Serratula* have been conducted. These compounds possess tonic and anabolic actions, and unlike the steranabols (nerobol) do not have androgenic effects. They increase exercise performance, accelerate rehabilitation of lost physical capabilities, and increase an organisms' ability to adapt to extreme environmental conditions (Kurmukov and Syrov 1976; Syrov and Kurmukov 1975a, b, c, 1976a, b, c, d, 1977, 1980; Kurmukov et al. 1980, 1982; Syrov 1984, 1994; Syrov et al. 1986; Saatov et al. 1994). The preparation *Ecdisten* was developed from ecdysterone, and is used in medical practice as a restorative, to improve memory, as a prophylaxis for and treatment of myocardial infarction, and especially for rehabilitation after cardiac infarction.

The Institute of Bioorganic Chemistry of the Academy of Sciences of Uzbekistan was founded by academician A.S. Sadikov. Besides various chemical laboratories, there is also a Laboratory of Pharmacology at this Institute (headed by Prof. S.Kh. Nasirov). In addition to natural compounds, the institute has studied medicinal plants, particularly alkaloids from the species *Anabasis aphylla* and *A. jaxartica*, *Ammodendron argenteum*, *Calligonum minimum*, *Colchicum kesselringii*, *Merendera raddeana* and others. Other plant compounds, including proanthocyanidins from the seeds of grapes, are studied at the institute as well (Pirniyazov et al. 2003).

Medicinal plants and their compounds are studied in the Pharmaceutical Institute of the Ministry of Health of The Rep. of Uzbekistan, especially in the subdepartments of Pharmacognosy (Prof. Kh.Kh. Khalmatov and his students), Pharmacology (Prof. Kh.U. Aliev) and Botany. Prof. Khalmatov and his associates published a series of books about the medicinal plants of Central Asia and Uzbekistan, and about their use in medicine. Similar studies are conducted in the subdepartments of the medical institutes and related laboratories of the scientific research institutes. As a result of the research on plant substances in the Laboratory of Experimental Cardiology of the Scientific Research Institute of Cardiology, now known as the Republican Specialized Center for Cardiology (headed by Prof. R.D. Kurbanov), the preparations *Oligvon*, *Glirofam*, *Ecdisten*, *Kavergal* and others were introduced into medical practice.

In Kyrgyzstan scientific studies of medicinal plants began in the pharmacology laboratory of the Institute of Regional Medicine of the Kyrgyzstan National Academy of Science in 1954. Later the name of the lab was changed to laboratory of pharmacognosy. The laboratory developed a tincture and the preparation *Foetidin* from the aboveground parts of *Thalictrum foetidum*, which was used to treat the first and second stages of hypertension. Later, Dr. P.K. Alimbaeva studied all species of the genus *Lagochilus* found in Kyrgyzstan. These studies showed that *Lagochilus platyacanthus* and *L. platycalyx* had the same effects on the cardiovascular system and blood coagulation as the species *L. inebrians*. Dr. B.N. Aronova conducted pharmacognostical studies of *Betonica foliosa*. As a result, a liquid extract of the aboveground parts of this species was introduced into medical practice as a treatment for uterine diseases.

The department of biopharmacology (headed by Academician Altimishev) was organized in 1969. This department included the laboratory of pharmacology and toxicology (led by Academician Altimishev), lab of resources (led by Dr. A.A. Akimaliev), and the lab of pharmacognosy (led by Dr. P.K. Alibaeva). The main scientific goals of the department were pharmacotoxicology studies and justification for the use of natural and synthesized physiologically active compounds. The preparation *Licorin* was introduced into medical practice to treat bronchial and lung diseases. The Ministry of Public Health Committee of the USSR (Pharmacology committee) permitted the use of the linament *Karagai* and *Hippophae rhamnoides* oils, in the preparation *Gippol*, which were developed by scientists from the department. The medicinal balsams (alcoholic plant extracts), including *Arashan*, *Uccurisky*, *Kobuctan* and *Sibir*, were developed and commercialized. *Arashan* was awarded a seal of quality by the USSR and a gold medal at an international exhibition in Leipzig in 1977.

With support from the Soviet Space Program, Drs. O.I. Gorelkina, E.P. Zotov and S.N. Khabibrakhmanov of the department of biopharmacology, developed and introduced special adaptogens such as *Gipkos*, *Giprex*, *Gipomin*, *Daugil*, etc. for use in the space program and in sports medicine. The preparation *Dipsacozide*, prepared from *Dipsacus azureus* roots, was developed and studied. Experiments showed that this preparation increased organisms' resistance to hypoxia and had hepatoprotective and antiatherosclerotic activities, which were proven after clinical studies. A non-alcoholic drink called *Omur*, based on the preparation *Dipsacozide*, was developed and recommended as a prophylactic for atherosclerosis. Additionally, the glycoside fraction, *Zongorozid*, was isolated from the roots of *Scabiosa songorica*. In experiments with animals the fraction significantly decreased arterial blood pressure and had sedative effects (Alimbaeva et al. 1986).

In recent years medicinal plants have been studied at the laboratory of biopharmacology (led by Dr. A.A. Akimaliev) at the Soil Biology Institute of the Kyrgyzstan National Academy of Science. Based on edible and medicinal plants, this lab developed the dietary supplement *Chabal*, which is recommended to people who have been exposed to radiation (such as atomic power station workers), as well as recommended to weak patients and athletes as a general tonic. *Chabal* has been approved by the Pharmacology and Pharmacopeia Committee of the Ministry of Public Health of the Kyrgyz Republic.

Many therapeutic syrups have been developed using medicinal plants from the flora of Kyrgyzstan. The syrup *Beykut* is used as a sedative and *Glitimal* is used as an expectorant and anti-inflammatory. The syrup *Akan* is used to prevent the development of stones in the urinary tract and bile pathways and is also recommended as a treatment for cholecystitis and hepatitis. All of these syrups were approved by the Pharmacology and Pharmacopeia Committee of the Ministry of Public Health of the Kyrgyz Republic. At the Medical Academy of Science, under the leadership of the Corresponding Academician of the National Academy of Science of the Kyrgyz Republic Professor A.Z. Zurdinov, a preparation *Immunaz*, with immunomodulatory properties, was developed from the leaves of *Padus grayana* and introduced into medical practice.

4 Phytochemistry of Medicinal Plants

Anvar G. Kurmukov

Plants contain organic as well as inorganic substances that can provide therapeutic effects. Different plants may possess a wide spectrum of effects due to the presence of various groups of chemical compounds and various microelements. A preparation obtained from one plant can simultaneously be an analgesic, sedative, cardiotonic, anti-inflammatory, and expectorant. Well-formed herbal preparations can be used protractedly when necessary, without injury to a patient, which is very important when treating chronic ailments. Medicinal plants are widely used as prophylaxis for, and treatment of, many diseases, including gastritis, stomach and duodenal ulcers, cholecystitis, colitis, enteritis, pyelonephritis, cystitis, atherosclerosis, cardiac insufficiency, and arrhythmia. They are also used for treatment of hypertensive and hypotensive neurocirculatory dystonia, neurosis and asthenia, menopausal disorders, and also to boost the body's immune system during times of disease, for rehabilitation of post-infarction conditions, as a tonic, and to increase adaptive capabilities of the organism.

Rational phytotherapy can promote recovery from dysbolism, normalize nervous system function, contribute to stabilization of blood pressure, improve coronary blood circulation and cerebral blood supply, help reduce insomnia and increase capacity for work. Herbal preparations promote excretion of toxic substances, help individuals to regain normal strength, increase energy metabolism and stop further disease progress during atherosclerosis and hypertension.

It is known that the effectiveness of medicinal plants and their pharmacotherapeutic action is due to their complex diversity of chemical compounds. Among these compounds are alkaloids, glycosides, lactones, tannins, proanthocyanidins, pigments, ecdysones, saponins and others.

Alkaloids – Alkaloids are nitrogen-containing organic bases. They are characterized by high pharmacological activity. In small doses, alkaloids represent valuable pharmaceutical substances such as lappaconitine, vincamine, reserpine, morphine, quinidine, strychnine, atropine, caffeine, ephedrine, nicotine and others. They form the main active ingredients of many medical products used for treatment of various diseases. Decoctions, infusions, extracts and others are made of alkaloid-containing plants.

Glycosides – Glycosides are organic compounds of vegetative origin, composed of a sugar component (glycoside, glycone) and a non-sugar component (aglycone, genin). The aglycone forms the main physiologically active part. Depending on their chemical nature and structure, glycosides are divided into cyanogenic glycosides (aglycones contain prussic acid), cardiac glycosides (aglycones are cardinolides and bufadienolides), saponins (aglycones are triterpene and steroid compounds), anthraglycosides (aglycones are derivatives of anthracene), phenolics (aglycones are coumarins, flavonoids, and others), and glycoalkaloids (aglycones are nitrogen-containing steroid compounds). Cardiac glycosides are used in medicine to treat cardiac disorders. They are toxic and have to be used under the supervision of a physician.

Saponins – Saponins are glycosides that make suds when shaken in water. The name comes from the Latin word "Sapo" meaning soap. Saponins are used as expectorants, diuretics, hypotensives and hypocholesterolemics. Saponins from *Aralia mandschurica*, *Echinopanax elatus*, *Eleutherococcus* spp., and *Panax* spp. have stimulating effects. Saponins cause hemolysis after intravenous introduction. Because of this, they are only introduced orally.

Anthraglycosides – Substances which belong to anthraglycosides look like red-orange crystals. Plant extracts containing anthraglycosides usually have a blood-red color. These compounds have purgative and choleretic actions.

Phenol compounds – Simple phenols, coumarins, chromones, lignan, tropolones, flavonoids and their glycosides, tannins, proanthocyanidins and others are in this group. This group of substances has the most diverse pharmacological activity. Among them there are substances that have antihypoxic, antioxidant, choleretic, cardio-, angio-, and hepatoprotecting and hemostatic actions.

A.G. Kurmukov (✉)
The Specialized Center for Cardiology,
4 Murtozoev St., Tashkent 100052, The Republic of Uzbekistan
e-mail: kurmukov_anvar@mail.ru

S. Eisenman et al. (eds.), *Medicinal Plants of Central Asia: Uzbekistan and Kyrgyzstan*,
DOI 10.1007/978-1-4614-3912-7_4, © The Editor 2013

Proanthocyanidins – These are polyphenol compounds, which possess pronounced antihypoxic, antioxidant and anti-inflammatory actions. They have vitamin-P activity.

Flavones and flavonoids – This group includes heterocyclic compounds, uneasily dissolved in water. Flavones and their derivatives have a yellow color, due to which they obtained their name (flavum = yellow). These compounds (rutin, quercetin, hesperidin, citrin and others) have the ability to decrease the permeability of vascular walls and fragility of capillary walls, have antispasmodic actions used for spasms of vessels and smooth-muscle organs, and are used to treat stomach and duodenal ulcers, and hepatitis.

Coumarins and furocoumarins – These compounds increase human and animal sensitivity to ultraviolet light and are used to treat vitiligo. Some have phyto-estrogenic action. When eaten by sheep and other animals, plants containing coumarins and furocoumarins have contraceptive action. Ingestion can cause fetal death in early pregnancy as well.

Tannins – Tannins promote inhibition of pathogenic microbial growth and reduce reproduction of viruses and bacteria. They also have astringent, tanning and hemostatic actions, and increase stability of capillary walls.

Organic acids – These acids are contained in plants in free form as well as in the form of salts and esters. Among them there are malic, citric, succinic, tartaric, oxalic, formic, acetic and other acids. They participate actively in metabolism, strengthen activity of salivary glands, and increase bile excretion and gastric juices. Organic acids are contained in lemons, apples, cranberries, currants, rosehips, sea-buckthorn berries, sorrel leaves, asparagus, greater celandine and other plants. Valeric and isovaleric acids (valerian, milfoil and others), and benzoic acid (in red whortleberry) have medicinal effects.

Esters of mono – and sesquiterpene alcohols with aromatic acids – These have estrogenic, hypolipidemic, and hypo-triglyceridemic activity and moderately increase blood pressure.

Fatty oils and fat-like substances – Fats and oils are esters of glycerin and higher fatty acids. In pure form, oils (castor, sea-buckthorn and others) are used as remedies or as solvents for pharmaceutical substances. Fatty oils are used in medicine to make ointments, liniments and emollients, and for skin care and therapeutic massage. Some oils have therapeutic action. For example castor oil is used as a purgative and sea-buckthorn is used internally to treat stomach and duodenal ulcers and externally for skin burns. Plant waxes, sterols and other substances are fat-like substances. Some of them are used in medicinal preparations. Unsaturated fatty acids (oleic, linoleic, linolenic, palmitic and other acids) prevent development of atherosclerosis. They are contained in plant seeds (almond, sunflower, flax and others), and fruits (olives and sea-buckthorn).

Mucilage – Consists of nitrogen-free compounds of various chemical compositions, mainly polysaccharides. They have coating and emollient actions, and can be found in *Althaea* roots and flax seeds.

Gums – Gums are polysaccharides. They are hardened fluids released out of damaged tree and shrub bark. They are used as emulsifying agents and also as adhesives. Apricot, cherry, plum and others are sources for gum.

Pectins, starch, and various sugars – Like mucilage and gums, these are related to carbohydrate groups and are used as additives in drug formulations.

All of the above mentioned groups of chemical compounds are the main active principles of the medicinal plants that are used today. However, only a very small percentage of the great diversity of plant-based compounds that exist in nature has been explored. Through the scientific process, new compounds having other effects are currently being revealed and will continue to be revealed in the future.

Milligram% (mg%) – A unit used to describe concentration. Milligrams of a specific substance contained in 100 ml of a solution or in 100 g of the analyzed material. This unit of measure is often used to describe vitamin content in plants and foods.

5 The Medicinal Plants of Uzbekistan and Kyrgyzstan

David E. Zaurov, Igor V. Belolipov, Anvar G. Kurmukov, Ishenbay S. Sodombekov, Anarbek A. Akimaliev, and Sasha W. Eisenman

D.E. Zaurov
School of Environmental and Biological Sciences, Department of Plant Biology and Pathology, Rutgers University, 20 Ag. Extension Way, New Brunswick, NJ 08901, USA

I.V. Belolipov
Tashkent State Agrarian University, Tashkent-140, Tashkent 100140, The Republic of Uzbekistan

A.G. Kurmukov
The Specialized Center for Cardiology, 4 Murtozoev St., Tashkent 100052, The Republic of Uzbekistan

I.S. Sodombekov
Kyrgyz Botanical Garden of the National Academy of Kyrgyzstan, 1A Akhunbabaev St. Bishkek 720064, The Kyrgyz Republic

A.A. Akimaliev
Biology and Soil Science Institute of the National Academy of Science of Kyrgyzstan, 265 Chuy St., Bishkek 720071, The Kyrgyz Republic

S.W. Eisenman
College of Liberal Arts, School of Environmental Design, Department of Landscape Architecture and Horticulture, Temple University, 580 Meetinghouse Rd., Ambler, PA 19002, USA

S. Eisenman et al. (eds.), *Medicinal Plants of Central Asia: Uzbekistan and Kyrgyzstan*,
DOI 10.1007/978-1-4614-3912-7_5, © The Editor 2013

Achillea asiatica Serg. – Asteraceae

Synonyms: *Achillea millefolium* var. *manshurica* Kitam., *Achillea setacea* ssp. *asiatica* (Serg.) Worosch.

English name: Chinese yarrow, Mongolian yarrow

Russian name: Тысячелистник азиатский (Tysyachelistnik aziatskiy)

Uzbek name: Unknown

Kyrgyz name: Азия каз тандайы (Aziya kaz tandayy)

Description: Herbaceous perennial with thin, branched rhizomes. Stems few or solitary, usually 25–50 cm tall, grayish with long, entangled, white hairs, often with short leafy branches in mid and upper leaf axils. Leaves bipinnatisect, usually oblong, green or grayish-green, more or less densely hairy; leaves of sterile shoots up to 25 cm long, long-petiolate; lower stem leaves 7–20 cm long, petiolate to subsessile; upper leaves sessile, usually 1–6 cm long. Inflorescences capitula arranged in loose, convex corymbs of unequal heights. Involucre cup-shaped; involucral bracts oviform, pale yellowish-green. Ray flower ligules pink, very rarely white. Fruits oblong, wedge-shaped achenes, truncated at the apex.

Other distinguishing features: Bases of mid-stem leaves partially clasping to auriculate. Ray flower ligules usually 1–3 mm long and 1.5–2.5 mm wide.

Phenology: Flowers in August and fruits in August and September.

Reproduction: By seeds and rhizomes.

Distribution: Ysyk-Kol, Osh, Jalal-Abad, and Chuy Provinces of Kyrgyzstan; not found in the flora of Uzbekistan.

Habitat: Found in forests, steppes, and abandoned fields.

Population status: Common, forming dense groups.

Traditional use: The aboveground parts are used as a hemostatic for bloody noses, bleeding gums, small wounds, abrasions, scratches, lung and uterine hemorrhages, and hemorrhoidal hemorrhages. It is used to treat inflammation, metropathy, and for gastrointestinal diseases, such as colitis and ulcers. It is also recommended for treating inflammation of the urinary tract (Plant Resources of the USSR 1993).

Documented effects: This species is used in the same manner as *Achillea millefolium*, and is anti-inflammatory, hemostatic, and antibacterial (Tolmachev 1976).

Phytochemistry: The plant contains alkaloids, flavonoids, sesquiterpene lactones, essential oils, vitamins C and K, resin, carotene, phytoncides, and bitter and astringent substances (Plant Resources of the USSR 1993; Glasl et al. 2001).

Achillea filipendulina Lam. – Asteraceae

Synonyms: *Achillea eupatorium* M. Bieb.

English name: Fern-leaf yarrow

Russian name: Тысячелистник таволголистный (Tysyachelistnik tavolgolistnyy)

Uzbek name: Dastarbosh

Kyrgyz name: Табылгы жалбырактуу каз тандай (Tabylgy zhalbyraktuu kaz tanday)

Description: Perennial herb. Stems erect, up to 60–80 cm tall, thick, striated, densely hairy, densely-leafy. Leaves pubescent, punctate glandular; basal leaves petiolate, oblanceolate, 10–20 cm long and 3–7 cm wide, pinnatipartite with acute segments; upper leaves pinnatifid with large, incised-dentate segments, sessile. Inflorescences capitula gathered into thick, unequally high, terminal corymbs. Ray flowers 1–4, yellow, trilobate. Disc flowers yellow with flattened corolla tube. Fruits oblong, wedge-shaped achenes, 2–2.25 cm long, grayish-black.

Other distinguishing features: Flowers have a specific pungent smell.

Phenology: Flowers in June-beginning of September, fruits in the end of August-September.

Reproduction: Reproduces abundantly by seeds and vegetatively by rhizomes.

Distribution: Widespread in Kyrgyzstan; Toshkent, Samarqand, Andijon, Farg'ona and Surxondaryo provinces of Uzbekistan.

Habitat: The adyr and tau zones. On stony, shallow-soiled, slopes with rocky debris, in mountain fissures, in valleys along rivers and brooks, in agricultural zones, and rarely along the banks of small irrigation canals.

Population status: Common, often found in large populations.

Traditional use: A decoction of the herb is used to treat gastric diseases, hemorrhoids, and as an abortifacient (Khalmatov 1964; Sadyrbekov et al. 2006a).

Documented effects: An extract of the inflorescences has anti-inflammatory activity and strongly inhibited expression of genes associated with inflammation processes (Dey et al. 2008).

Phytochemistry: The herb contains 0.07–0.26 % essential oil, alkaloid traces, asparagine, amino acids and nitrogen-containing substances. Plants growing in Uzbekistan have high variation in the amount of essential oils, which can vary from 0.04 % to 0.5 %. Around 3 % aldehydes and ketones and 0.5 % phenols are found in the oil composition. Flowering plants from Burchmulla village (Toshkent province, Uzbekistan) contained 0.2–0.27 % essential oil, which contained 10 % octylene, ~5 % pinene, 8 % camphene, 0.35 % $C_{10}H_{18}O$ alcohol, about 30 % borneol and formic, acetic and caprylic acid (Khakimov and Tsukervanik 1948; Khalmatov 1964). Essential oil extracted from plants growing in the Botanical Garden of the Institute of Phytochemistry, Karaganda, Kazakstan, consisted mainly of santolina alcohol (29 %), 1,8-cineol (19.1 %) and borneol (27.8 %; Sadyrbekov et al. 2006a). The sesquiterpene lactone leucomisine was isolated from the aboveground parts (Konovalov and Nesterova 2003).

Achillea setacea Waldst. & Kit. – Asteraceae

Synonyms: None
English name: Unknown
Russian name: Тысячелистник щетинистый (Tysyachelistnik shchetinistyy)
Uzbek name: Unknown
Kyrgyz name: Катуу туктуу каз тандай (Katuu tuktuu kaz tanday)
Description: Herbaceous perennial. Stems up to 80 cm tall, whitish hairy. Leaves linear-lanceolate, 3–10 cm long, up to 2 cm wide, bi- or tripinnatisect, lobes linear-lanceolate; basal and lower stem leaves petiolate; upper leaves sessile. Inflorescences capitula, densely arranged in convex, compound corymbs; involucres oblong-cylindrical; involucral bracts greenish-yellow. Ray flowers 4–5, white, slightly 3-lobed; disc flowers 10–20, yellow, 5-lobed. Fruits oblong achenes, 1.8–2 mm long, light brown.
Other distinguishing features: Capitula 2.5–3 mm across, with peduncles ca. 3 mm long.
Phenology: Flowers in April-June and fruits in July-August.
Reproduction: By seeds and rhizomes.
Distribution: Kungay Ala-Too and Terskey Ala-Too, Chuy valley, Kyrgyz Ala-Too and Alai mountain ranges of Kyrgyzstan; not found in the flora of Uzbekistan.
Habitat: Steppes, meadow-steppes, meadows, among shrubs, forests edges, in abandoned fields, and near roads.
Population status: Common.
Traditional use: Used in the same way as *Achillea millefolium* and *A. asiatica*. A decoction is used to treat internal and external bleeding and hemorrhoids (Plant Resources of the USSR 1993; Alimbaeva and Shambetov 1988).
Documented effects: The essential oil had antimicrobial effects against *Clostridium perfringens*, *Acinetobacter woffii*, and *Candida albicans* (Unlu et al. 2002). Sesquiterpenes isolated from this species exhibited anti-inflammatory activity in the croton oil ear test (Zitterl-Eglseer et al. 1991).
Phytochemistry: This plant contains essential oil, alkaloids, glycosides, tannins, resins, organic acids, vitamins C and K (Plant Resources of the USSR 1993). The aboveground parts contain sesquiterpenes (Zitterl-Eglseer et al. 1991). The essential oil, isolated from air-dried aerial parts, contained over 51 constituents with eucalyptol (1,8-cineole) being the major component (Unlu et al. 2002).

Aconitum karakolicum Rapaics. – Ranunculaceae

Synonyms: *Aconitum napellus* var. *turkestanicum* B. Fedtsch., *Aconitum soongaricum* Stapf. (some authors recognize this as a separate species), *Aconitum winkleri* Rapaics.

English name: Unknown

Russian name: Аконит каракольский (Akonit karakol'skiy)

Uzbek name: Karakool parpisi

Kyrgyz name: Исыккол уу коргошуну (Isykkol uu korgoshchunu)

Description: Herbaceous perennial with conical tuber-like roots. Stems up to 2 m tall, branched. Leaves appressed to stem, short-petiolate; blade circular, up to 10 cm long and 15 cm wide, palmatisect with 5 segments divided to the base; each segment pinnatifid with 2–3 linear lobes, lobes 1.5–3 mm wide. Inflorescence an dense apical raceme; pedicels with two bracteoles. Flowers irregular, with 5 petaloid sepals, dark-violet. Upper sepal hood-shaped, semispherical, with a small beak. Petals 2, each with a spur. Fruit a follicetum with 3–5 glabrous follicles.

Other distinguishing features: The roots form horizontal, chain-like rows. Distinguised from *Aconitum soongaricum* by having narrower leaf lobes and appressed pubescence on the inflorescence rachis and pedicels.

Phenology: Flowers in July-September and fruits in August-October.

Reproduction: By seeds.

Distribution: Ysyk-Kol province of Kyrgyzstan; not found in the flora of Uzbekistan.

Habitat: In meadows with diverse grass species and in spruce forests.

Population status: Common, forming dense groups.

Traditional use: In Kyrgyz folk medicine, an infusion of the tubers in fermented horse milk or water and ground tubers added to meat broth, are used to treat tuberculosis, radiculitis, and headaches. Tubers are also used to treat different types of cancer (Khalmatov et al. 1984).

Documented effects: An alcoholic tincture of the roots is applied externally to treat radiculitis, neuralgia, rheumatism, and as an analgesic. This tincture is a component of the preparation *Akofit*. An infusion of the tubers and the aboveground parts is used as a component of the preparation *Anginol*, which is used to treat sore throats. Because of the high toxicity the plant is not widely used in medicine (Khalmatov et al. 1984). Compounds isolated from the plant exhibited anti-tumor activity in vitro (Chodoeva et al. 2005).

Phytochemistry: The roots contain up to 2.35 % alkaloids and the aboveground parts up to 0.5 %. The roots contain starch and organic acids as well (Khalmatov et al. 1984). The alkaloids phenyl-β-naphthylamine, karakoline, neoline, delsosine, monticamine, songorine, napelline, acetylnapelline, isoboldine, karasamine and 1-benzoylkarasamine, etc. were found in the aboveground parts (Sultankhodzhaev et al. 1973; 1986; Sultankhodzhaev and Tadzhibaev 1976; Sultankhodzhaev 1993; Atta-ur-Rahman et al. 2005; Chodoeva et al. 2005).

▲**Achillea setacea Waldst. & Kit.** Photos: Andrei Lubchenko

◀**Achillea asiatica Serg.** Photo: Evgeny Davkaev

▲**Achillea filipendulina Lam.** Photos: Alim Gaziev

▼**Aconitum karakolicum Rapaics.** Photos: Alexander Naumenko

Aconitum leucostomum Worosch. – Ranunculaceae

Synonyms: None
English name: Unknown
Russian name: Аконит белоустый (Akonit beloustyy)
Uzbek name: Unknown
Kyrgyz name: Бурма кара, Ак темгилдуу, Үу коргошуну (Burma kara, Ak temgilduu, Uu korgoshchunu)
Description: Herbaceous perennial with rope-like roots. Stem 70–200 cm tall, erect. Leaves large, 10–20 cm long and 20–40 cm wide, reniform in outline, palmatisect with 5–11 lobes; basal leaves and lowers stem leaves long-petiolate. Inflorescence a dense, many-flowered raceme; pedicels with 2 bracteoles. Flowers irregular, with 5 sepals. Sepals petaloid, dark violet, interior almost white; upper sepal hood-shaped, beaked. Petals 2, each with a spur. Fruit a follicetum with 3 follicles, glabrous or glandular hairy.
Other distinguishing features: Interior of sepals almost white.
Phenology: Flowers in July and August and fruits in August and September.
Reproduction: By seeds.
Distribution: Ysyk-Kol and Chuy provinces of Kyrgyzstan; not found in the flora of Uzbekistan.
Habitat: On the edges of spruce and juniper forests, forest glades; found on northern slopes.
Population status: Common, forming dense groups.
Traditional use: Unknown.
Documented effects: The plant has antibacterial and antiarrhythmic activity. The preparation *Allapinin*, which is prepared from the aboveground parts and contains the hydrobromic salts of lappaconitine alkaloids, is used as an antiarrhythmic-class I (Gammerman et al. 1990). The alkaloid songorine was found to enhance excitatory synaptic transmission in rat hippocampus and may act as a non-competitive antagonist at the GABA(A) receptor (Zhao et al. 2003).
Phytochemistry: The plant contains high quantities of alkaloids: roots – 0.8–4.9 %, stems – 0.3–1, leaves – 0.6–3.9 %, and flowers – 1.3–4.5 %. Lappaconitine, lappaconidine, corydine, glaunidine, N-dimethyl colletine, and others have been isolated from the aboveground parts. The alkaloids mesaconitine, aksine, acsinatine, excelsine, lappaconitine, lappaconidine have been isolated from the root. Flavonoids, coumarins, saponins, and tannins are also found in the roots (Gammerman et al. 1990; Yue et al. 1996; Zhao et al. 2003).

Aconitum soongaricum Stapf. – Ranunculaceae

Synonyms: *Aconitum karakolicum* Rapaics. (some authors recognize this as a separate species).

English name: Unknown

Russian name: Аконит джунгарский (Akonit dzhungarskiy)

Uzbek name: Zhoongar parpisi

Kyrgyz name: Жунгар уу коргошуну (Zhungar uu korgoshchunu)

Description: Herbaceous perennial with tuber-like roots. Stems 70–130 cm tall, simple or branched. Leaf blades circular-cordate in outline, 5–9 cm long, 8–12 cm wide, palmatisect with 5 segments divided to the base; segments pinnatifid with 2 or 3 linear lobes, lobes 3–5 mm wide. Inflorescence an apical raceme. Flowers irregular. Sepals 5, petaloid, violet; upper sepal hood-shaped, convex, with a long beak. Petals 2, each with a spur. Fruit a follicetum with 3 follicles. Seeds 4–5 mm long.

Other distinguishing features: Forms conical, horizontally segmented (chain-like) roots. Distinguised from *Aconitum karakolicum* by having wider leaf segments and glabrous inflorescence rachis and pedicels.

Phenology: Flowers in June-July, fruits in July-August.

Reproduction: By seeds.

Distribution: Ysyk-Kol province of Kyrgyzstan; not found in the flora of Uzbekistan.

Habitat: In meadows with diverse grass species.

Population status: Common, forming dense groups.

Traditional use: Prepared in the same way as *Aconitum karakolicum*. Used in Kyrgyz folk medicine to treat tuberculosis, radiculitis, and headaches, and also to treat different types of cancer (Khalmatov et al. 1984).

Documented effects: An infusion of the tubers is a component in the preparations *Acofit* (*Radiculin*), which is used to treat radiculitis, neuritis, and rheumatism, and *Ehinor* (*Anginol*), which is used to treat tonsillitis and malignant tumors. The coumarin fraction has antitumor properties. Because of high toxicity the plant is not widely used in medicine (Tolmachev 1976). Alkaloids isolated from the plant have antiarrhythmic and CNS-stimulating activities (Salimov et al. 2004).

Phytochemistry: Underground parts contain carbohydrates, starch, organic acids, 1.23–3.4 %, alkaloids, and coumarins up to 0.3 %. Aboveground parts contain 0.56–0.7 % alkaloids and vitamin C. The inflorescence contains flavonoids, and seeds contain up to 32 % fatty oil (Plant Resources of the USSR 1985; Salimov et al. 2004).

Aconitum talassicum Popov – Ranunculaceae

Synonyms: None

English name: Monkshood

Russian name: Аконит таласский (Akonit talasskiy)

Uzbek name: Ok parpi

Kyrgyz name: Талас уу коргошуну, Кара барпы (Talas uu korgoshchunu, Kara barpy)

Description: Herbaceous perennial with narrow-conical roots. Stems up to 1.5 m tall. Leaf blades circular-pentagonal, 6–11 cm long, 7–16 cm wide, palmatisect nearly to the base, with 3–5 wedge-shaped, narrow segments; each segment divided into 2–3 sharply toothed lobes; lobes broadly lanceolate. Inflorescence an apical raceme. Flowers irregular. Sepals 5, petaloid, light-blue to blue; upper sepal hood-shaped with beak; lateral sepals obovate. Petals 2, each with a spur. Fruit a follicetum with 3 follicles.

Other distinguishing features: Forms segmented (chain-like), horizontal roots. Leaves not as finely dissected as *Aconitum karakolicum* and *A. soongaricum*.

Phenology: Flowers in June-August and fruits in August-September.

Reproduction: By seeds.

Distribution: Talas province of Kyrgyzstan; Toshkent and Samarqand provinces of Uzbekistan.

Habitat: In sub-alpine zones, in wet meadows in river valleys, and among junipers. Endemic to the Tian Shan and Pamiro-Alai mountains.

Population status: Common, forming dense groups.

Traditional use: An infusion of the roots is used to treat rheumatism and malaria (Khalmatov 1964). In veterinary medicine an infusion is used for flesh wounds and skin ulcers (Aldashev 1979).

Documented effects: The alkaloid talatizamine has effects similar to those of curare as well as ganglio-blocking actions (Khamdamov 1972).

Phytochemistry: The aboveground parts contain 1.01 % total alkaloids and roots contain 1.92–3.63 % total alkaloids (Khalmatov 1964). Talatizamine, talatizine, talatizidine, isotalatizine, condelphine, and others compounds were isolated from the total alkaloids (Yunusov et al. 1954; Yunusov 1981; Nishanov et al. 1991; Yue et al. 1994).

Acroptilon repens (L.) DC. – Asteraceae

Synonyms: *Acroptilon picris* (Pall.) C.A. Mey., *Centaurea repens* L.

English name: Russian knapweed

Russian name: Горчак ползучий (Gorchak polzuchiy)

Uzbek name: Kakra

Kyrgyz name: Сойлоочу кекире (Soyloochu kekire)

Description: Herbaceous rhizomatous, perennial. Stems 20–60 cm tall, straight, arachnoid-hairy, with sessile glands. Leaves coriaceous, grayish-green, oblong, sessile; basal and lower leaves oblong, 4–15 cm long; upper leaves oblong, linear or linear-lanceolate, 1–7 cm long. Inflorescences oval capitulas, from 8 to 65, arranged in panicles. Disk flowers 1–1.5 cm long, dark pink. Ray flowers absent. Fruits obovate achenes, 2–4 mm long, light in color, 8–30 in each capitula. Pappus white, short-pinnate.

Other distinguishing features: Staminal filaments are free and smooth. Basal and lower leaves often withering by flowering time.

Phenology: Flowers in June-July, fruits in July-August.

Reproduction: Reproduces abundantly by seeds and rhizomes.

Distribution: All of Uzbekistan and Kyrgyzstan.

Habitat: The chul, adyr, and tau zones. Grows on stony and clay-soiled slopes, and in abandoned fields.

Population status: Common, forms large populations.

Traditional use: A water infusion of the herb is used to treat malaria, epilepsy, and other diseases. The root of the herb is used as an emetic (Khalmatov 1964). In the folk medicine of Central Asia, Azerbaijan, and Crimea, a water infusion of the plant is used to treat malaria, and in Azerbaijan for treatment of epilepsy. Because this plant is poisonous, internal use of this species must be done with caution (Makhlayuk 1992).

Documented effects: Severe poisoning in farm animals occurs when animals are fed hay containing small amounts of the herb, but toxicity has only been observed when plants are in flower; plants mowed before flowering do not seem to be poisonous (Ogolevitz 1951). The plant causes a nervous system disease and neural cell necrosis when consumed by horses. Repin, a sesquiterpene lactone isolated from the plant, showed high toxicity to chicken embryo sensory neurons (Stevens et al. 1990). Volatile oil isolated from the aboveground parts strongly inhibited the growth of the bacteria *Staphylococcus saprophyticus* and *Staphylococcus epidermidis* (Norouzi-Arasi et al. 2006).

Phytochemistry: The plant contains traces of saponins, tannins and bitter substances, 4 % glycoalkaloids, 0.06 % essential oils and sesquiterpene lactones (Ogolevitz 1951; Stevens et al. 1990). The main constituent of volatile oil isolated from the aboveground parts was caryophyllene oxide (36.6 %; Norouzi-Arasi et al. 2006).

◄Aconitum leucostomum Worosch.
Photos: *left*: Vladimir Epiktetov; *center* and *right*: Denis Mirin

Aconitum soongaricum Stapf. ►
Photos: Vladimir Epiktetov

▼**Aconitum talassicum Popov**
Photos: Evgeny Davkaev

Acroptilon repens (L.) DC. ▼►
Photos: *right*: Evgeny Davkaev; *below* and *lower right*: Alim Gaziev

Agrimonia asiatica Juz. – Rosaceae

Synonyms: *Agrimonia eupatoria* ssp. *asiatica* (Juzepczuk) Skalický

English name: Agrimony

Russian name: Репейничек азиатский (Repeynichek aziatskiy)

Uzbek name: Sariq choiy

Kyrgyz name: Азия уйгакчасы (Aziya uygakchasy)

Description: Herbaceous perennial, 30–130 cm tall, with a short, thick rhizome. Stem densely hairy with very dense, stiff, horizontal hairs and fewer shorter, softer hairs. Leaves odd-pinnate, stipulate, hairy, with few small yellow glands; leaflets with large-dentate margins. Inflorescence a spike-shaped raceme, reaching 40 cm during fruiting. Flowers 10–12 mm in diameter, with short pedicels, petals yellow, twice as long as sepals. Fruits achenes enclosed in the hypanthium. Hypanthium 6–9 mm long and almost as wide, with rows of prickles towards the top.

Other distinguishing features: Wounded roots exude a fluid that quickly turns black.

Phenology: Flowers in May-June, fruits in July-August.

Reproduction: By seeds.

Distribution: All of Uzbekistan; Chuy, Osh, Talas, Jalal-Abad, and Batken provinces of Kyrgyzstan.

Habitat: The chul, adyr, and tau zones. Plains, walnut forests, fields, along small canals, along roads, in bushy thickets, and shaded areas of orchards.

Population status: Common, usually found as individual plants.

Traditional use: A decoction of the underground parts and dried stems and leaves is used in case of gastrointestinal diseases, as an astringent, to treat rheumatism, intestinal infections, fever, edema, as diuretic, and as a mouth wash. A decoction of the flowers is used to treat hemorrhoids, body rashes, and as a hemostatic (Akopov 1981).

Documented effects: An infusion and liquid extract showed hemostatic effects (Khalmatov 1964). An aqueous extraction of the aboveground plant parts increases diuresis, and it has been shown that an infusion and liquid extracts have hemostatic actions (Akopov 1981). An aqueous extraction of the aboveground parts of *Agrimonia eupatoria* inhibited hepatitis B surface antigen production in vitro (Kwon et al. 2005).

Phytochemistry: The plant contains ursolic acid (Ibragimov and Khazanovich 1972). Above and underground parts contain tannins, flavonol glycosides, B-vitamins, saponins, and trace alkaloids (Akopov 1981).

Ajuga turkestanica (Regel) Briq. – Lamiaceae

Synonyms: None
English name: Unknown
Russian name: Живучка туркестанская (Zhivuchka turkestanskaya)
Uzbek name: Kapalak kunmas
Kyrgyz name: Unknown
Description: Subshrub with a robust root. Stems 10–50 cm tall, pale brown, reddish or whitish, upper portions covered with fine, soft hairs. Leaves opposite, oblong-elliptic or obovate, 4.5–6 cm long, 1.4–1.8 cm wide, often soft-hairy, nearly sessile, margins usually entire. Flowers solitary, axillary. Calyx campanulate, hairy, with 5 narrowly lanceolate lobes. Corolla 2-lipped, bright pink-purple, with dark veins, rarely white, 2.5–4 cm long; upper lip very short; lower lip large, trilobite, the center lobe clawed and with 2 lobules; stamens 4. Fruits oblong nutlets, 7 mm long, olive-brown.
Other distinguishing features: Corolla tube nearly twice as long as calyx.
Phenology: Flowers in May- June, fruits by the end of May.
Reproduction: By seeds.
Distribution: South Pamiro-Alai: Surxondaryo Province of Uzbekistan; absent in Kyrgyzstan.
Habitat: The adyr and tau zones. On clay-soiled and stony slopes in areas with mixed soil types and areas with gypsum and red sandstone.
Population status: Usually found in populations of 10–300 individuals, and in greater numbers in herbaceous *Artemisia*-grass complexes.
Traditional use: Plants in the genus *Ajuga* are used medicinally to treat weight deficiency, reduced hair growth, ulcers, burns, and to heal wounds. They are also used as a restorative for weakened people (Iordanov et al. 1970; Ikan and Ravid 1971; Kovaleva 1971; Zavrazhanov et al. 1972).
Documented effects: Biological activity of ethanolic extracts of the aboveground parts is due to the presence of phytoecdysteroid compounds. The phytoecdysones ecdysterone, turkesterone, and cyasterone have anabolic activity. In contrast to the stero-anabolics (nerobol), androgenic action is absent in studied phytoecdysones. In animals, these compounds have a tonic action and increase resistance to various stress factors (Syrov et al. 1975a, b; Syrov and Kurmukov 1975b, 1976b, c, d; Aizikov et al. 1978; Syrov et al. 1986; Mamatkhanov et al. 1998). During animal tests, ecdysterone decreased the area of necrosis after experimental myocardial infarction, decreased intracellular enzyme release into the blood and accelerated enzyme reduction until normal (Ermishina et al. 1982; Kurmukov and Ermishina 1986, 1991). Ecdysterone, under the preparation name *Ecdysten*, has successfully passed clinical tests in several clinics in Russia and Uzbekistan and has been allowed for medical use in the treatment of cardiac infarction, rehabilitation of post-infarction conditions, and to treat fetal hypoxic hypotropia (Iskanderova and Sharipova 1992; Kurmukov and Ermishina 1991; Kurmukov and Kurmukova 1992; Kurmukova 2000a, b; Kurmukova and Kurbanov 1999). The preparation *Ayustan*, which contains phytoecdysones, is also used in medicine.
Phytochemistry: This species contains the following oxysteroid compounds and phytoecdysones: turkesterone, ecdysterone, cyasterone, and others (Saatov et al. 1977; Usmanov et al. 1975, 1977; Mamatkhanov et al. 1998; Abdukadirov et al. 2005).

Alhagi pseudalhagi (M. Bieb.) Desv. – Fabaceae

Synonyms: *Alhagi camelorum* Fisch. ex DC., *Alhagi maurorum* Medic., *Alhagi persarum* Boiss. & Buhse, *Hedysarum alhagi* L., *Hedysarum pseudalhagi* M. Bieb.

English name: Camel's thorn

Russian name: Янтак ложный, Верблюжья колючка ложная (Yantak lozhnyy, Verblyuzh'ya kolyuchka lozhnaya)

Uzbek name: Yontok

Kyrgyz name: Жантак (Zhantak)

Description: Herbaceous perennial with a deep root system. Stems green, 50–120 cm tall, much branched, with upward curving thorns. Leaves alternate, simple, obovate, 7–20 mm long, margins entire, apex rounded. Inflorescences axillary racemes with 3–8 flowers; inflorescence rachis ending with a spine. Calyx campanulate, with or without 5 teeth. Corolla papilionaceous, 8–9 mm long, pink to brownish-red. Fruit a moniliform legume, 1–3 cm long, curved or straight with 1–5 seeds. Seeds small, glabrous, kidney-shaped, brown.

Other distinguishing features: Stamens 10, nine of the filaments fused. Banner petal obovate, keel blunt, equal in length to banner, wings shorter than keel.

Phenology: Flowers in May-September, fruits in August-October.

Reproduction: By seeds and rhizomes.

Distribution: Almost all provinces of Uzbekistan; Chuy, Osh, and Jalal-Abad provinces of Kyrgyzstan.

Habitat: The chul, adyr and tau zones. In abandoned fields, as a weed in fields, and along irrigation networks.

Population status: Common.

Traditional use: Used in Iran and other eastern countries as a laxative and antipyretic (Khalmatov 1964). An infusion of the roots is used to treat liver diseases and stomach and duodenal ulcers as well as diuretic. A galenic preparation of the aboveground parts is used to treat colitis, gastritis, stomach ulcers, dysentery, cervical erosion, to heal wounds, to treat inflammation of the ear, nose and throat, as a choleretic, to quench thirst, to reduce sweating and as an antipyretic, anti-inflammatory, and cough remedy (Karimov and Shomakhmudov 1993).

Documented effects: A dry extract from the aboveground parts, as well as the total proanthocyanidins have antihypoxic, antioxidant, angioprotective, and hypocholesterimic abilities (Aizikov et al. 1986). An ethanolic extract of the aboveground plant parts had antiulcerogenic activity in rats (Amani et al. 2006), and a methanolic extract exhibited an antidiarrheal effect (Atta and Mouneir 2004).

Phytochemistry: The herb contains 3.9–8.2 % tannins, up to 0.2 % coumarins, up to 1,000 mg%,[1] vitamin C, about 0.8 % essential oils, as well as up to 3.4 % flavonoids. The roots contain alkaloids (0.17–0.19 %), glycosides, resins (up to 5.67 %), pigments and sugars. Proanthocyanidins have been isolated from this species (Karimov and Shomakhmudov 1993). The flavonoids catechin, epigallocatechin, gallocatechin, leucodelphinidin, quercetin, rutin, etc. and the flavanone glycosides alhagitin and alhagidin have also been isolated from the plant (Singh et al. 1999; Awaad Amani et al. 2006).

[1] Milligram% (mg%) – A unit used to describe concentration. Milligrams of a specific substance contained in 100 ml of a solution or in 100 g of the analyzed material. Often used to describe vitamin content in plants and foods.

Allium karataviense Regel – Alliaceae

Synonyms: None
English name: Unknown
Russian name: Лук каратавский (Luk karatavskiy)
Uzbek name: Chuchka kuloq
Kyrgyz name: Кара Тоо пиязы (Kara Too piyazy)
Description: Herbaceous perennial with large, spherical bulb. Bulb 2–6 cm in diameter, with a blackish or grayish paper-like coat. Stem short, 10–25 cm tall, sometimes half buried in the soil, stems shorter than leaves. Leaves lanceolate, oblong, (3–)5–15 cm wide with smooth margins. Inflorescence a dense, many-flowered, spherical umbel. Pedicels equal in length, 3–4 times longer than perianths, lacking bracts. Flowers with 6 tepals. Tepals 5–7 mm long, linear, apex rounded, pink-violet with a dark vein. Stamens 6. Fruit an obovate capsule, 8 mm wide.
Other distinguishing features: Ovary with a rough surface. Leaves broader than those of related species.
Phenology: Flowers in April-May, fruits in May-June.
Reproduction: By seeds and vegetatively by lateral bulblets.
Distribution: Toshkent and Farg'ona provinces of Uzbekistan; Chuy and Osh provinces of Kyrgyzstan.
Habitat: The tau zone. On limestone taluses.
Population status: Common.
Traditional use: A decoction of the bulb is used in folk medicine to treat lung diseases and shortness of breath (Khalmatov 1964).
Documented effects: No data.
Phytochemistry: Most *Allium* species contain essential oils, volatile organic compounds, flavonol glycosides, phenols, vitamins, ascorbic acid, mineral salts and microelements (Khalmatov 1964), as well as steroidal saponins and sapogenins (Mimaki et al. 1999).

▲**Agrimonia asiatica Juz.**
Photos: *far left* and *left*: Andrei Lubchenko; right and far right: Evgeny Davkaev

◄**Ajuga turkestanica (Regel) Briq.** Photos: *far left*: Authors; *left*: Diana Cheng

▼**Alhagi pseudalhagi (M. Bieb.) Desv.** Photos: *left*: Evgeny Davkaev; *right*: Alim Gaziev

▼**Allium karataviense Regel**
Photos: *left*: Alim Gaziev; *middle* and *right*: Evgeny Davkaev

Allium suvorovii Regel – Alliaceae

Synonyms: None
English name: Unknown
Russian name: Лук Суворова (Luk Suvorova)
Uzbek name: Yowoiy piyoz
Kyrgyz name: Суворов пиязы (Suvorov piyazy)
Description: Herbaceous perennial plant to 1 m tall, with a spherical bulb. Bulb 2–3 cm in diameter, covered with grayish, cracked, almost coriaceous coat that sheathes the base of the stem. Stem 30–100 cm tall. Leaves 2–6, belt-like, much shorter than stem, 5–20 mm wide, margins rough. Inflorescence a dense, many flowered, semispherical or spherical umbel. Pedicels equal in length, 2–5 times longer than perianths, lacking bracts. Flowers with 6 tepals. Tepals 6, ~4 mm long, linear, apex rounded, pink-violet with a darker vein. Stamens 6. Fruit a capsule, broadly-ovate, 5 mm wide.
Other distinguishing features: It differs from closely related species by having a smooth ovary.
Phenology: Flowers in May, fruits in June.
Reproduction: By seeds and vegetatively by lateral bulblets.
Distribution: Toshkent, Farg'ona, Samarqand, Qashqadaryo, and Surxondaryo provinces of Uzbekistan; Chuy and Osh provinces of Kyrgyzstan.
Habitat: The adyr zone. Grows in shallow soil on foothills, as a weed along canals, along the edges of plowed fields, and in orchards and cemetaries. Found in places inaccessible for pasturing and mowing.
Population status: Rare. Found sporadically as individual plants and in small populations; listed in the Red Book of Rare and Endangered Species of Uzbekistan.
Traditional use: The bulbs pickled in wine vinegar, are used to treat hemoptysis and to treat incipient tuberculosis. Also used as a phytoncidal remedy to treat various skin diseases, especially eczema and psoriasis (Khalmatov 1964).
Documented effects: No data.
Phytochemistry: See *Allium karataviense*. The bulbs of *A. suvorovii* contain various carbohydrates (Khodzhaeva and Turakhozhaev 1992; Khodzhaeva 1994); the seeds contain the carbohydrate stachyose (Khodzhaeva and Kondratenko 1984).

Allochrusa gypsophiloides (Regel) Schischk. – Caryophyllaceae

Synonyms: *Acanthophyllum gypsophiloides* Regel.

English name: Turkestan soaproot

Russian name: Аллохруза качимовидная, Колючелистник качимовидный, Мыльный корень (Allokhruza kachimovidnaya, Kolyuchelistnik kachimovidnyy, Myl'nyy koren')

Uzbek name: Beh, Etmak, Kachimsimon etmak

Kyrgyz name: Качимдай кок тикен (Kachimday kok tiken)

Description: Herbaceous perennial, 30–80 cm tall, with a strong taproot reaching 6 m deep. Stems thin, branched, short-pubescent or glabrous. Leaves opposite, linear or linear-lanceolate, acute, 1–2.5 cm long, glabrous, sessile. Inflorescence paniculiform, loose, branched. Flowers pale-pink, with long pedicels. Fruit a capsule with 1–2 seeds. Seeds rough, flattened, light-brown.

Other distinguishing features: Capsule obovate or spherical, ca. 2 mm long.

Phenology: Flowers in June-July, fruits in July-August.

Reproduction: By seeds.

Distribution: Toshkent, Jizzax, Samarqand, Qashqadaryo, and Surxondaryo provinces of Uzbekistan; in the Chatkal, Talas, and Pskem ranges in Kyrgyzstan.

Habitat: The adyr and tau zones. Stony slopes with rocky debris.

Population status: Rare, listed in the Red Book of Rare and Endangered Species of Uzbekistan.

Traditional use: Roots are used to treat coughs and applied externally to heal wounds. A decoction of the root is recommended as an expectorant for bronchitis (Khalmatov 1964). An infusion of the roots is used as a choleretic, diuretic, and laxative. The root is brewed in a tea and drunk to treat gastrointestinal, skin and venereal diseases, spleen, liver and kidney diseases, as well as metabolism dysfunction. An infusion of the aboveground parts is used as an expectorant and laxative (Khodzhimatov 1989).

Documented effects: Saponins from this species increase the secretory activity of glands. Pure saponin from this species is used in veterinary medicine to prepare vaccines against anthrax and brucellosis (Khodzhimatov 1989). Treatment with a saponin extracted from the roots antagonized the narcotic effect of chloral hydrate, potentiated the convulsive effect of strychnine, decreased the convulsive and toxic effect of Corazole, and increased dieresis in mice (Polievtsev and Sultanov 1971).

Phytochemistry: An important source of saponins. Roots contain up to 30 % saponins with a hemolytic index of 1:1000 or 1:2860 and aboveground parts of the plant have saponins with an index of 1:240 (Khalmatov 1964; Yukhananov et al. 1972). The roots contain 10–30 % triterpene saponins from which the glycosides gypsogenin and acanthophyllosides B, C and D were isolated (Putieva et al. 1970, 1975, 1979). The aboveground parts contain polysaccharides as well as many saponins (Arifkhodzhaev and Kondratenko 1983; Khodzhimatov 1989).

Althaea nudiflora Lindl. – Malvaceae

Synonyms: *Alcea leucantha* Fisch., *Alcea nudiflora* (Lindl.) Boiss.

English name: Naked-flowered hollyhock

Russian name: Алтей голоцветный (Altey golotsvetnyy)

Uzbek name: Oq gulhairy

Kyrgyz name: Туксуз гулдуу гулкайыр (Tuksuz gulduu gulkayyr)

Description: Herbaceous biennial or perennial, to 1.5–2 m high, stellate hairy. Stems cylindrical. Leaves simple, long-petiolate; blade 5–7-lobed with coarse-dentate margins, rough with crowded, stiff, stellate hairs on both sides. Inflorescences terminal, racemiform. Calyx with 5 triangular-lanceolate lobes, densely stellate hairy. Corolla white, 5–8 cm in diameter; petals 5, obovate. Stamens fused into a column. Fruits wheel-shaped schizocarps. Seeds 3–4 mm long, kidney-shaped, red-brown.

Other distinguishing features: Foliaceous bracts absent. Style with numerous branches.

Phenology: Flowers in June-August, fruits in July-September.

Reproduction: By seeds.

Distribution: Toshkent, Andijon, Namangan, Farg'ona, Samarqand, Qashqadaryo and Surxondaryo provinces of Uzbekistan; Chuy, Ysyk-Kol, Talas, Osh and Jalal-Abad provinces of Kyrgyzstan.

Habitat: The adyr and tau zones. Shallow soil and stony slopes.

Population status: Common.

Traditional use: An infusion of the dried flowers is given to children to treat diarrhea and sialorrhea. A decoction of the roots and seeds is recommended as a hemostatic for post-natal bleeding. A plaster of the flower and leaf powder is used to treat tumors (Khalmatov 1964). The fresh stem is cut and applied to cuts on the skin. The roots and seeds are made into a tea to treat dysuria (Sezik et al. 2004).

Documented effects: None.

Phytochemistry: The plant contains mucilage. Leaves contain 165–176 mg% of vitamin C (Khalmatov 1964). The roots contain lipids with cyclopropenoid fatty acids (Sagdullaev et al. 2001) and the flowers contain kaempferol glycosides (Pakudina et al. 1970).

Althaea officinalis L. – Malvaceae

Synonyms: *Althaea kragujevacensis* Pančić ex Diklić & Stevan., *Althaea micrantha* Borbás, *Althaea sublobata* Stokes, *Althaea taurinensis* DC., *Althaea vulgaris* Bubani, *Malva althaea* E.H.L. Krause, *Malva maritima* Salisb., *Malva officinalis* (L.) Schimp. & Spenn.

English name: Common marshmallow

Russian name: Алтей лекарственный (Altey lekarstvennyy)

Uzbek name: Dorivor gulhairi

Kyrgyz name: Дары гулкан (Dary gulkan)

Description: Herbaceous perennial. Stems single or multiple, 40–150 cm tall, tomentose. Basal leaves 3–5-lobed; cauline leaves with rounded or cordate base and acuminate apex, margins coarsely serrate. Flowers up to 3 cm across, clustered in leaf axils. Epicalyx with 8–12 segments. Corolla pale-pink with 5 petals. Fruit a disc-shaped schizocarp, with 15–25 small, laterally flattened mericarps. Seeds kidney-shaped.

Other distinguishing features: Stamens many, staminal filaments connate and forming a tube.

Phenology: Flowers in June-September, fruits in June-October.

Reproduction: By seeds.

Distribution: All provinces of Kyrgyzstan; Toshkent, Farg'ona and Samarqand provinces of Uzbekistan.

Habitat: In places with a high water-table, and along rivers and canals.

Population status: Common, forming dense groups.

Traditional use: The roots, flowers and leaves are used as an anti-inflammatory and to treat flu, sore throat, hepatitis, and urinary incontinence. They are also used to treat kidney stones, cystitis, prostate tumors, chronic prostatitis, and joint pain (Kurochkin 1998).

Documented effects: The plant is used internally to treat eczema, psoriasis, dermatitis and to normalize metabolism. In combination with other preparations, this species is used to treat gastritis, stomach and duodenal ulcers, enterocolitis, food poisoning, dysentery, kidney inflammation, and urinary incontinence. The preparation *Mucaltin*, which is prepared from the herb, is used as an expectorant to treat bronchitis and pneumonia (Kurochkin 1998). A methanolic extract and a decoction of the roots inhibited a variety of bacteria known to cause periodontal disease (Iauk et al. 2003). An extract of the root has been shown to be a potent inhibitor of calcium mobilization associated with UVB-induced pigmentation of skin (Kobayashi et al. 2002a).

Phytochemistry: Roots consist of 35 % mucilage substances, about 37 % starch, 10 % sucrose, betaine, flavonoids, coumarins, phenolic acids, and fatty oil. Aboveground parts contain mucilage, carbohydrates (glucose and sucrose), essential oils, vitamin C, and carotene. Seeds contain up to 12 % fatty oil, 1 % phospholipids and pectin (Khalmatov 1964; Tolmachev 1976; Capek et al. 1987; Gudej 1991).

◄**Allium suvorovii Regel**
Photos: Alim Gaziev

▲**Allochrusa gypsophiloides (Regel) Schischk.**
Photos: Evgeny Davkaev

▲**Althaea nudiflora Lindl.** Photos: Evgeny Davkaev

▼**Althaea officinalis L.**
Photos: *left*: Andrei Lubchenko; *center* and *right*: Vadim Prokhorov

Amaranthus retroflexus L. – Amaranthaceae

Synonyms: some authors consider *Amaranthus tricolor* L. a synonym.

English name: Redroot amaranth

Russian name: Щирица запрокинутая (Shchiritsa zaprokinutaya)

Uzbek name: Gultojihuroz, Eshakshura

Kyrgyz name: Кайрылган амарант (Kayrylgan amarant)

Description: Herbaceous annual with a taproot. Stems 20–100 cm tall, pubescent. Leaves ovate-rhomboid, apex obtuse. Inflorescence a dense panicle; bracts lanceolate. Flowers unisexual. Pistillate flowers with 5 tepals and 3 stigmas. Staminate flowers at top of inflorescence; tepals 5, stamens 3–5. Seeds shiny black, lenticular, contained in circumscissile utricles.

Other distinguishing features: Pistillate tepals membranaceous with emarginate or obtuse apices.

Phenology: Flowers in May-June, fruits in August.

Reproduction: By seeds.

Distribution: Agricultural zones in all provinces of Kyrgyzstan and Uzbekistan.

Habitat: In vegetable gardens, orchards, waste places, and along the edges of fields.

Population status: Common, found in dense groups.

Traditional use: A water infusion of the aboveground parts is used to treat colitis, intestinal colic, and as a laxative for constipation, as well as a hemostatic to treat hemoptysis, and menstrual and hemorrhoid hemorrhages. A water extract of the dried plant collected during flowering stage is used as an antiprotist and antibacterial. A decoction of the roots is used to treat guinea worm and jaundice. Young stems are used as a source of vitamins. Leaves are used as a diuretic and a decoction of the leaves is used to treat headaches (Zolotnitskaya 1965; Makhlayuk 1967).

Documented effects: An antimicrobial peptide was isolated from the seeds and effectively inhibited the growth of multiple fungi species (Lipkin et al. 2005).

Phytochemistry: Roots contain the betacyanins amaranthin and isobetanin. Leaves contain nitrogenous compounds, 0.96 % betaine and fatty oils, which contain the following fatty acids: miristic, palmitic, stearic, linoleic and linolenic acid. Seeds contain 4.3–7 % fatty oil with the following fatty acids: palmitic (18.9 %), stearic (1.9 %), oleic (51.5 %), linoleic (27.9 %; Plant Resources of the USSR 1985).

Anagallis arvensis L. – Myrsinaceae (formerly in Primulaceae)

Synonyms: *Anagallis latifolia* L., *Anagallis phoenicea* Scop.

English name: Scarlet pimpernel

Russian name: Очный цвет пашенный (Ochnyy tsvet pashennyy)

Uzbek name: Savun ut, Savunak

Kyrgyz name: Кызгылт анагаллис (Kyzgylt anagallis)

Description: Herbaceous annual or biennial with multiple branches. Stems quadrangular, glabrous, 10–25 cm long. Leaves opposite, sessile, ovate to elongate-ovate, with black dots on abaxial surface. Flowers brick-red, individual, with long pedicels. Fruit a spherical capsule, opening by a small cover. Seeds small, oval, 3-sided, black, many in each capsule.

Other distinguishing features: Differs from related species by having a brick-red corolla.

Phenology: Flowers and fruits in April-September.

Reproduction: Only by seeds.

Distribution: Toshkent, Farg'ona, Samarqand and Surxondaryo provinces of Uzbekistan; Chuy, Talas and Osh provinces of Kyrgyzstan.

Habitat: The chul, adyr and tau zones. Along banks of small canals, in river valleys, along roads, in orchards, in fields, and on loess slopes.

Population status: Common, often found in small populations.

Traditional use: A decoction of the herb is used to treat shortness of breath, tuberculosis, gynecological disorders, rabies, and as a diuretic in cases of edema, and is used externally for washing wounds (Khalmatov 1964). The essence from blooming plants is used in homeopathy (Ogolevitz 1951).

Documented effects: Plants collected in the Toshkent region contained saponins with a hemolytic index of 1:2230 (Khalmatov 1964). An aqueous extract of the plant showed significant antifungal activity against isolates of *Microsporum canis, Trichophyton mentagrophytes*, and *Trichophyton violaceum* (Ali-Shtayeh and Abu Ghdeib 1999). Saponins isolated from the plant exhibited strong molluscicidal activity when tested against *Biomphalaria glabrata* and *Oncomelania quadrasi* (Abdel Gawad et al. 2000). A triterpene saponin isolated from the plant inhibited the replication of herpes simplex virus type 1 and poliovirus type 2 in vitro (Amoros et al. 1987). The plant is noted as being poisonous and extracts of the plant have been shown to be highly toxic to rats (Ogolevitz 1951; Al-Sultan et al. 2003).

Phytochemistry: The herb contains the glucoside cyclamine, saponoids and other terpenoid saponins as well as the enzyme primveraza (Ogolevitz 1951; Amoros et al. 1987). The aboveground plant parts contain flavonoids (kaempferol, quercetin), phenylcarbonic acids (caffeic, ferulic, etc.), anthocyans and fatty oil (MedicineLib.ru 2008).

Anagallis foemina Mill. – Myrsinaceae (formerly in Primulaceae)

Synonyms: *Anagallis arvensis* f. *coerulea* (Schreb.) Arechav., *Anagallis arvensis* var. *coerulea* (Schreb.) Gren. & Godr., *Anagallis arvensis* ssp. *foemina* (Mill.) Schinz & Thell., *Anagallis coerulea* Schreb.

English name: Blue pimpernel

Russian name: Очный цвет голубой (Ochnyy tsvet goluboy)

Uzbek name: Savun ut, Savunak

Kyrgyz name: Когултур анагаллис (Kogultur anagallis)

Description: The botanical description of this plant is very similar to that of *Anagallis arvensis*. This species differs by having a blue corolla. Some botanists consider *Anagallis coerulea* a form or variety of *Anagallis arvensis*.

Other distinguishing features: Differs from related species by having a blue corolla with dentate lobes and no glands.

Phenology: Flowers and fruits in April-September.

Reproduction: Only by seeds.

Distribution: Surxondaryo province of Uzbekistan; Chuy, Osh and Jalal-Abad provinces of Kyrgyzstan.

Habitat: The chul, adyr and tau zones. Along banks of small canals, river valleys, along the roads, in orchards, fields, and on loess slopes.

Population status: Common, found in small populations.

Traditional use: Same as *Anagallis arvensis*.

Documented effects: No data.

Phytochemistry: Similar to *Anagallis arvensis*.

Anchusa azurea Mill. – Boraginaceae

Synonyms: *Anchusa italica* Retz.

English name: Italian bugloss, large blue alkanet

Russian name: Анхуза итальянская (Ankhuza ital'yanskaya)

Uzbek name: Hukuz tili

Kyrgyz name: Италия анхузасы (Italiya ankhuzasy)

Description: Perennial herb to 1.5 m tall, with multiple, thick, conjoined taproots. Entire plant densely covered with bristly hairs. Stem usually single, sometimes branching, erect. Basal leaves in a rosette, oblanceolate, 10–30 cm long, petiolate; upper leaves alternate, oblong or lanceolate, sessile. Inflorescences terminal, bracteate, helicoid racemiform. Calyx lobes linear, divided nearly to the base. Corolla bright blue, funnelform, 10–15 mm in diameter, 5-lobed, with 1–1.5 cm long, bristly-hairy pedicels. Fruits gray nutlets, erect, 5–8 long mm long, 3–5 mm wide.

Other distinguishing features: Stamens inserted at the top of corolla tube. Fruits 3-sided.

Phenology: Flowers in the end of April-July, fruits in May-August.

Reproduction: Only by seeds.

Distribution: Toshkent, Farg'ona, Andijon, Samarqand, and Surxondaryo provinces of Uzbekistan; Chuy, Osh and Jalal-Abad provinces of Kyrgyzstan.

Habitat: The chul, adyr and tau zones. In fields, waste grounds, orchards, wheat fields, and oases.

Population status: Common, found as single individuals.

Traditional use: A decoction of the flowers is used to treat chest aches, neurasthenia, and asthma, and is used as a laxative, febrifuge, and cough remedy as well. An ointment, prepared by boiling the roots in cow fat, is used as hemostatic and to heal wounds (Khalmatov 1964). In Iraq, a decoction of the flowers is used as a sedative, analgesic, sudorific, and diuretic (Al-douri 2000).

Documented effects: Saponins from this species have a hemolytic index of 1:2800–1:20000 (Khalmatov 1964). An extract of the aboveground parts showed significant antibacterial effect against *Pseudomonas aeruginosa* (Fazly Bazzaz and Haririzadeh 2003).

Phytochemistry: All plant tissue contain saponins. The roots contain dyes (alkanin and anchusin), anchusa acid, resins, and waxes (Khalmatov 1964). Oil extracted from the seeds contains γ- and α-linolenic acid as well as stearidonic acid (Guil-Guerrero et al. 2001).

▲**Amaranthus retroflexus L.** Photos: Maxim Kucherov

▼**Anagallis foemina Mill.** Photos: *top*: Alim Gaziev; *bottom*: Andrei Lubchenko

▲**Anagallis arvensis L.** Photos: Sasha Eisenman

▼**Anchusa azurea Mill.** Photos: *top left*: Evgeny Davkaev; *lower left* and *right*: Alim Gaziev

Artemisia absinthium L. – Asteraceae

Synonyms: None

English name: Common wormwood, absinthium, armoise absinthe

Russian name: Полынь горькая (Polyn' gor'kaya)

Uzbek name: Erman, Achik erman

Kyrgyz name: Эрман шыбак (Erman shybak)

Description: Herbaceous perennial with a short taproot. Stems up to 1.5 m tall, sometimes with short, lateral, vegetative stems. Basal leaves bi- or tripinnatisect, long-petiolate; cauline leaves alternate; lower cauline leaves short-petiolate, bipinnatisect; upper cauline leaves small, almost sessile. Inflorescences many-flowered capitula with 40–70 flowers, ca. 3 mm in diameter, globose, nodding, in narrow to broadly pyramidal panicles; involucral bracts linear; receptacle convex, densely hairy. Disc flowers yellow; ray flowers absent. Fruits oblong to wedge-shaped achenes, about 1 mm long.

Other distinguishing features: The whole plant is gray-silver due to short, appressed hairs.

Phenology: Flowers in July-August, fruits in August-September.

Reproduction: By seeds.

Distribution: Almost all provinces of Kyrgyzstan and Uzbekistan.

Habitat: From valleys to the mid-belt of mountains. On slopes, in meadows, along rivers, and near cultivated and in abandoned fields.

Population status: Common, in some places forming dense groups.

Traditional use: Leaves, which are collected before and at the beginning of flowering, are used in a decoction as a carminative, a vermifuge, and to treat dyspepsia, loss of appetite, insomnia, diseases of the liver, stomach, spleen, and gall bladder, fever, hemorrhoids, malaria, and intestinal ulcers, as well as to heal wounds (Khalmatov et al. 1984).

Documented effects: This species is used to make preparations to treat cases of gastritis with low stomach acidity, to increase appetite, and as a choleretic. In combination with other medicines, preparations are also used to treat chronic diseases of the pancreas, stomach, and intestinal tract. Because of the presence of azulene, this species is used to treat allergic reactions of the skin (Kurochkin 1998). Ethyl acetate and chloroform extracts of the whole plant inhibited a variety of microorganisms (Erdogrul 2002).

Phytochemistry: The herb contains 0.17–2 % essential oil which contains sesquiterpene lactones (absinthin, anabsinthin and artabasin), flavonoids (artemetin), tannins, organic acids, vitamin C and carotene (Khalmatov et al. 1984). The roots were found to contain many lignans (Greger and Hofer 1980).

Artemisia annua L. – Asteraceae

Synonyms: *Artemisia chamomilla* C. Winkl.

English name: Sweet sagewort, sweet wormwood, sweet annie, chinese wormwood

Russian name: Полынь однолетняя (Polyn' odnoletnyaya)

Uzbek name: Burgan

Kyrgyz name: Бир жылдык шыбак (Bir zhyldyk shybak)

Description: Herbaceous annual. Stems often single, 15–200 cm tall, erect. Lower leaves up to 7 cm long and wide, ovate in outline, bi- or tripinnatisect, petiolate; cauline leaves bipinnatisect, triangular to broadly ovate, becoming simpler and smaller towards top of stem. Inflorescences globose capitula with ca. 30 flowers, in a leafy, open panicle; involucral bracts linear. Disc flowers pale- or greenish-yellow; ray flowers absent. Fruits flat achenes, 0.5–1 mm long.

Other distinguishing features: Plant has a sweet aroma.

Phenology: Flowers in July and August, fruits in August and September.

Reproduction: By seeds.

Distribution: All provinces of Kyrgyzstan and Uzbekistan.

Habitat: From valleys to the mid-belt of mountains. In agricultural zones, near canals, in orchards, and vegetable gardens.

Population status: Common, forming dense groups.

Traditional use: Leaves are collected in spring and aboveground parts in autumn. The juice from fresh leaves is used to treat skin diseases (scabies, abscesses, bacterial and fungal diseases, etc.). The dried leaves are used to prepare an ointment that is used to treat eczema. A decoction of the aboveground parts is used to increase appetite. Traditional doctors use an infusion of the herb to treat rheumatism and skin diseases (Khodzhimatov 1989).

Documented effects: An extract of this species inhibits the development of anthrax, by causing loss of pathogenic ability and killing bacterial cells (Khodzhimatov 1989). This plant species is the source of artemisinin, which, in combination with other drugs, is used as a highly effective treatment for malaria worldwide (World Health Organization 2006). Artemisinin has also been shown to cause apoptosis in human cancer cells (Singh and Lai 2004).

Phytochemistry: The aboveground parts contain lactones (artemisinin and arteannuin), coumarins (scopoletin), 0.12–0.65 % essential oil, tannins, alkaloids, resins, sugars, and vitamin C. Maximum essential oil content was observed during the flowering period. Plants collected near Bishkek (Kyrgyzstan) contained 0.21 % essential oil and 2.44 % tannins (Khalmatov 1964; Khodzhimatov 1989).

Artemisia dracunculus L. – Asteraceae

Synonyms: *Artemisia aromatica* A. Nelson, *Artemisia dracunculina* S. Watson, *Artemisia dracunculoides* Pursh, *Artemisia dracunculoides* ssp. *dracunculina* (S. Watson) H. M. Hall & Clements, *Artemisia glauca* Pallas ex Willdenow, *Oligosporus dracunculus* (L.) Poljak.

English name: Russian tarragon, wild tarragon, estragon, silky wormwood

Russian name: Полынь Эстрагон (Polyn' estragon)

Uzbek name: Sherolgin

Kyrgyz name: Шыраалжын шыбак (Shyraalzhyn shybak)

Description: Herbaceous, rhizomatous perennial to 50–120 cm tall, with a woody caudex and fibrous roots. Stems numerous, erect, green, yellowish or reddish brown, partially woody, glabrous. Leaves alternate, 5–8 cm long, linear-lanceolate, usually entire; lower leaves often irregularly lobed or trilobate, mostly glabrous. Inflorescences globose to ovate capitula arranged in panicles. Disk flowers pale-yellow, only peripheral flowers fertile; ray flowers absent. Fruits oblong achenes, ca.1 mm long, brown.

Other distinguishing features: Receptacle where flowers are attached is naked (lacking chaff, scales, hairs, etc.). The plant has a unique smell.

Phenology: Flowers in June, fruits in September.

Reproduction: By seeds and rhizomes.

Distribution: All of Uzbekistan and Kyrgyzstan.

Habitat: The tau zone. On soft and rocky mountain sides. Often planted in vegetable gardens for use as a culinary herb.

Population status: Common, often found in dense groups.

Traditional use: It is used to treat edema and scurvy, dyspepsia, to improve appetite, and as a carminative. A powder of the plant is used to treat oral diseases. Tarragon from Uzbekistan has been noted to have anti-helminthic action. Leaves are also recommended as a good source of carotene (Khalmatov 1964).

Documented effects: After clinical tests, use of the liquid extract of tarragon was recommended to treat patients with chronical low-acid gastritis (Khalmatov et al. 1984). Essential oil isolated from the aboveground parts of *Artemisia dracunculus* "Piemontese" exhibited strong antifungal activity when tested against *Candida albicans, C. lusitaniae*, *C. glabrata*, and *C. tropicalis,* and weak antimicrobial effects against *Xanthomonas maltophilia* and *Proteus mirabilis* (Curini et al. 2006). An ethanolic extract of the plant significantly reduced hyperglycemia in mice with chemically induced insulin deficiency and diabetes, and reduced hyperglycemia in genetically diabetic mice (Logendra et al. 2006; Ribnicky et al. 2006).

Phytochemistry: The flowering herb contains 0.1–0.7 % essential oils, 41.8 mg% (for absolute dry weight) carotene, 190 mg% vitamin C and alkaloid traces (Khalmatov et al. 1984). The essential oils of Central Asian plants contain 65–85 % d-sabinene, about 10 % myrcene, 5 % sesquiterpene fractions, about 0.5 % methoxy-cinnaroic aldehyde, and 7–15 % resins. Central Asian tarragon oil is substantially different from Western European tarragon oil because it doesn't contain methyl-chavicol (Khalmatov 1964). The herb contains flavonoids, alkamides, and coumarins (Mallabaev et al. 1971, 1970; Mallabaev and Sidyakin 1976; Hofer et al. 1986; Bohm and Stuessy 2001; Saadali et al. 2001; Logendra et al. 2006).

Artemisia leucodes Schrenk – Asteraceae

Synonyms: *Seriphidium leucodes* (Schrenk) Poljak.
English name: Unknown
Russian name: Полынь беловатая (Polyn' belovataya)
Uzbek name: Oq shuvoq
Kyrgyz name: Ак шыбак (Ak shybak)
Description: Herbaceous annual or biennial, 30–90 cm tall, covered with long, white, raised hairs. Stems single or multiple, erect, branched. Lower stem leaves petiolate, tripartite-pinnatisect, 3–7 cm long; cauline leaves sessile, tripartite. Inflorescences 3–5-flowered capitula in panicles. Disk flowers yellow, punctate glandular; ray flowers absent. Fruits obovate achenes, 2–2.25 mm long, olive-colored.
Other distinguishing features: Leaves produce a strong smell of camphor when rubbed.
Phenology: Flowers in September, fruits in October.
Reproduction: Only by seeds.
Distribution: Karakalpakstan autonomous republic, Toshkent, Andijon, Farg'ona, Samarqand, Surxondaryo, and Buxoro provinces of Uzbekistan; Chuy and Osh provinces of Kyrgyzstan.
Habitat: The chul, adyr, and tau zones. On sandy soil, clay-soiled slopes with rocky debris, and in areas with soils containing a wide diversity of minerals.
Population status: Uncommon, found as small populations in *Artemisia*-ephemeral communities.
Traditional use: Unknown.
Documented effects: The lactone leucomisine has strong anti-inflammatory action which is due to its antagonism of the main inflammation mediators: histamine, serotonin (5-hydroxytryptamine), and prostaglandin F2ά and E2 (Kurmukov 1987). It reduces the volume of atherosclerotic aorta involvement in (tested) rabbits with hypercholesteremic atherosclerosis (method of Anichkov and Holatova), reduces aorta wall permeability, has angioprotective action, and has medicinal effects on experimental myocarditis (Kurmokov and Rasulova 1983; Aizikov et al. 1991; Azizov et al. 1992; Kurmukova et al. 1997a, b; Kurmukov et al. 1991a, b; Prokhorova et al. 1992a). Ascorbic acid strengthens the effects of leucomisine (Kurmukova and Aizikov 1997). *Oligvon*, a preparation containing leucomisine, is used to prevent and treat artherosclerosis. The lactone austricine also has combined angioprotective and hypolipidemic activity (Prokhorova et al. 1993; Aizikov et al. 1993a, b). A total lactone extract increased the intensity of bile production and increased the concentration of cholesterol in the bile of normal rats, as well as in rats with chemically induced hepatitis (Tursunova et al. 2002).
Phytochemistry: Leaves and inflorescences contain up to 1 % essential oils, which consist of up to 90 % levorotatory camphor. The lactones leucomisine and austricine are obtained from the aboveground plant parts, as well as the sesquiterpenoids matricarin, anhydroaustricine, parishin B, parishin C, artelin, and artelein (Ribalko 1978; Tursunova et al. 2002). The seeds contain lipids with epoxy-, monohydroxy-, and dihydroxyacids (Ul'chenko and Glushenkova 2001).

▲**Artemisia absinthium L.** Photos: Andrei Lubchenko

◀**Artemisia dracunculus L.**
Photos: Sasha Eisenman

▼**Artemisia annua L.**
Photos: *left*: Kristian Peters; *right*: Sasha Eisenman

▼**Artemisia leucodes Schrenk**
Photo: Authors

Artemisia scoparia Waldst. & Kit. – Asteraceae

Synonyms: *Oligosporius scoparia* (Waldst. & Kit.) Less.
English name: Redstem wormwood
Russian name: Полынь метёльчатая (Polyn' metyol'chataya)
Uzbek name: Kizilburgan
Kyrgyz name: Шыпыргы шыбак (Shypyrgy shybak)
Description: Herbaceous annual or biennial plant with a thin vertical root. Stems single or few, 30–90 cm tall. Basal leaves petiolate, bi- or tripinnatisect, segments linear-lanceolate, apex acute; middle cauline leaves smaller, sessile, segments narrow, linear; upper cauline leaves deeply tri-lobed or entire. Inflorescences small capitula with 10–12 flowers, in a wide, nodding panicle; involucral bracts brownish or pink-violet. Disc flowers yellow; ray flowers absent. Fruits achenes, ca. 0.6 mm long, ovate, flat with narrow ribs.
Other distinguishing features: Stems and leaves sparsely hairy or glabrous. Capitula subglobose.
Phenology: Flowers in July, fruits in August-September.
Propagation: By seeds.
Distribution: Agricultural lands in all provinces of Kyrgyzstan and Uzbekistan.
Habitat: Near and in cultivated and abandoned fields.
Population status: Common, often forming dense groups.
Traditional use: Used to treat respiratory disease and rheumatism, and used as a diuretic (Gammerman et al. 1990). A tincture (2.5–10 %) of the plant is used to treat radiculitis. An infusion of the plant is used as a vermifuge, and to treat epilepsy and irregularities in the menstruation cycle (Khodzhimatov 1989).
Documented effects: Aboveground parts are collected during the flowering stage for use as raw material. The plant is slightly toxic. An infusion of the plants has diuretic properties. Its essential oil has laxative properties and is included in the preparation *Artemisol*, which has antispasmodic action, increases the solubility of salts in urine and promotes the passage of kidney stones (Maksudov 1964). Intravenous administration of a hydro-methanolic extract of the plant produced hypotensive and bradycardiac effects. Studies indicate that the plant contains Ca^{++} channel-blocking constituents (Gilani et al. 1994). The essential oil exhibited considerable inhibitory effects against a number of different bacteria (Cha et al. 2005).
Phytochemistry: Aboveground parts contain 4.35–5.57 % resins, with 1.08–1.37 % resins in roots and 7.91 % in flowers. The whole plant contains organic acids (citric, malic, oxalic, acetic, propionic, and valerianic) and tannins (3.61–4.74 % in aboveground parts and 2–2.5 % in the roots). The aboveground part contains essential oil of which the maximum accumulation (0.96 %) happens during the flowering stage (Khodzhimatov 1989). The major components of the essential oils are camphor, 1,8-cineole, and β-caryophyllene (Cha et al. 2005).

Artemisia viridis Willd. – Asteraceae

Synonyms: *Artemisia rupestris* ssp. *viridis* (Willd.) Ameljczenko, *Artemisia rupestris* var. *viridis* (Besser) A. DC.

English name: Wormwood

Russian name: Полынь зелёная (Polyn' zelyonaya)

Uzbek name: Unknown

Kyrgyz name: Мамыр шыбак (Mamyr shybak)

Description: Herbaceous perennial. Stems 6–20 cm tall; vegetative stems short, densely leafy; flower-bearing stems wide, reddish-violet, rarely green, hairy. Leaves 1.5–5 cm long, 1–2.5 cm wide, wrinkled, hairy below; basal leaves with wide petioles, bipinnatisect, lower segments entire, mid- and upper segments with 3–5 narrow lobes. Inflorescences multiflorous capitula with ca. 70–80 flowers, arranged in racemes or spikes. Disc flowers reddish-brown; ray flowers absent. Fruits achenes, oblong-oviform, striated.

Other distinguishing features: Outer involucral bracts linear; internal involucral bracts triangular or elliptic.

Phenology: Flowers in July, fruits in September.

Reproduction: By seeds.

Distribution: High mountain regions of all provinces in Kyrgyzstan; not found in the flora of Uzbekistan.

Habitat: In high mountain steppes.

Population status: Common, forming dense groups.

Traditional use: An infusion of the herb is used to treat stomach ulcers, and diseases of the kidneys, liver, and bile ducts (Nanaeva 1960; Isakov 1969).

Documented effects: No data.

Phytochemistry: Aboveground parts contain essential oil, phenols, ketones, flavonoids, alkaloids, and coumarins (Plant Resources of the USSR 1993).

Artemisia vulgaris L. – Asteraceae

Synonyms: *Artemisia coarctata* Forselles, *Artemisia opulenta* Pampanini.

English name: Common mugwort, felon-herb, green-ginger, armoise vulgaire

Russian name: Полынь обыкновенная (Polyn' obyknovennaya)

Uzbek name: Oddiy erman

Kyrgyz name: Кадимки куурай (Kadimki kuuray)

Description: Herbaceous perennial. Stems 30–180 cm tall. Basal leaves petiolate, 3–15 cm long, 1.5–12 cm wide, pinnatifid or pinnatisect, segments lanceolate or linear; cauline leaves sessile, entire or pinnatisect. Inflorescences capitula in compact racemiform or paniculiform clusters; involucral bracts hairy; outer bracts oblong; inner bracts elliptical. Disc flower corollas brownish. Fruits ellipsoid achenes, grayish-brown, glabrous.

Other distinguishing features: Leaves densely white-haired beneath, usually glabrous above.

Phenology: Flowers in July, fruits in September.

Reproduction: By seeds.

Distribution: In agricultural lands in all provinces of Kyrgyzstan and Uzbekistan.

Habitat: From valleys to mid-belt of mountains. Near canals and in vegetable gardens, orchards and waste grounds.

Population status: Common, forming dense groups.

Traditional use: Leafy tops and roots are used for medicinal purposes. Leaves are collected from plants during the flowering stage and the roots are collected in the autumn. Avicenna used the herb in baths to treat kidney stones and uterine ulcers, and to induce menstruation. A decoction of the herb is used to treat sinus colds (Khalmatov et al. 1984), nervous diseases, epilepsy, and neurasthenia, and is also used as an anticonvulsant. The aboveground parts are used to treat poisoning, inflammation of the gastrointestinal tract, tuberculosis, and to increase the appetite. It is also used externally as a lotion to treat ulcers and persistent wounds (Maznev 2004).

Documented effects: Data suggest that aqueous and chloroform extracts from leaves of *A. vulgaris* have antihypertensive actions (Tigno et al. 2000). Essential oils showed a broad spectrum of antimicrobial activity (Blagojevic et al. 2006). Two flavonoids, eriodictyol and apigenin, found in *A. vulgaris*, exhibited estrogenic effects in vitro (Lee et al. 1998).

Phytochemistry: The herb contains essential oils (contains cineol, thujone, borneol, and aldehydes), flavonoids, alkaloids, carotene, and ascorbic acid (Khalmatov et al. 1984; Lee et al. 1998).

Arum korolkowii Regel – Araceae

Synonyms: *Arum elongatum* Steven, *Biarum sewertzowii* Regel.

English name: Korolkov's arum

Russian name: Аронник Королькова (Aronnik Korol'kova)

Uzbek name: Kuchala, Chayon ut

Kyrgyz name: Корольков аруму (Korol'kov arumu)

Description: Perennial herb, 30–50 cm high, with a flat-spherical tuber that is 3–4 cm in diameter. Base of leaf petiole sheathing, petiole short to twice as long as the blade. Leaf blade cordate, acuminate (spear-shaped) or triangular. Inflorescence a spadix; peduncle longer than leaf petioles, 50–60 cm long with reddish stripes; spathe exterior green, white inside, elongate-lanceolate, narrow-cylindrical, almost 2 times longer than spadix, apex acute. Fruits red berries.

Other distinguishing features: Fruits are densely clustered on spadix.

Phenology: Flowers and fruits in May-June.

Reproduction: By seeds.

Distribution: All of Uzbekistan; Jalal-Abad province of Kyrgyzstan.

Habitat: The adyr and tau zones. Shady, wet places, in gorges, among rocks.

Population status: Not common, found as single individuals.

Traditional use: The powdered tuber is used to treat scorpion and poisonous snake bites and is mixed with honey to treat fungal skin diseases and white spots on the skin of the neck. Bread made with tuber powder and sesame oil is prescribed (to be eaten) to treat hemorrhoids (Khalmatov 1964).

Documented effects: All parts of the fresh plant are poisonous due the presence of saponins (Khalmatov 1964).

Phytochemistry: Tubers contain poisonous saponins, which produce hydrocyanic acid as a result of hydrolysis, alkaloids (possibly volatile cicutine), lipids, pectic substances, fructosans and 28–30 starch. A carotenoid, lycopene, was found in the fruits (Khalmatov 1964; Chernenko et al. 2000).

▲Artemisia scoparia Waldst. & Kit. Photos: Andrei Lubchenko

▲Artemisia viridis Willd.
Photos: Vladimir Epiktetov

►Artemisia vulgaris L.
Photos: Dmitri Oreshkin

▼Arum korolkowii Regel
Photos: *left*: Evgeny Davkaev
right: Alexander Naumenko

Asparagus persicus Baker – Asparagaceae

Synonyms: *Asparagus inderiensis* Blume ex Ledeb., *A. ledebourii* Mishchenko.

English name: Persian asparagus

Russian name: Спаржа персидская (Sparzha persidskaya)

Uzbek name: Tomirdori

Kyrgyz name: Персия спаржасы (Persiya sparzhasy)

Description: Perennial herb. Stems 60–120 cm high, smooth, glabrous, branched; branch angles at 90° or obtuse to the stem. Cladodes 1–8 per cluster, usually 1.5–2 cm long and unequal in length, glabrous, smooth; upper and middle leaves scale-like with a sharp spur. Flowers arise from the stems and branches; female flowers 3 mm long, semispherical, campanulate, greenish-white; male flowers campanulate, 5–6 mm long. Fruit a red berry, spherical, 6–7 mm wide; on a long pedicel up to 2 cm in length.

Other distinguishing features: Stems often winding, curling, or trailing.

Phenology: Flowers in May, fruits in June.

Reproduction: By seeds and rhizomes.

Distribution: Toshkent, Farg'ona, and Samarqand provinces of Uzbekistan; Naryn, Ysyk-Kol and Chuy provinces of Kyrgyzstan.

Habitat: The tau zone. Among tall grasses and in the tree-shrub belt of mountains.

Population status: Uncommon, found as single individuals.

Traditional use: Of the wild species of *Asparagus* found in Uzbekistan, this is the only species used in folk medicine. In some regions of Toshkent province (Uzbekistan) it is used to treat numerous diseases (Khalmatov 1964).

Documented effects: No data.

Phytochemistry: All the species of *Asparagus* found in Uzbekistan contain alkaloids, essential oils, vitamins, asparagine, saponins, steroid sapogenins and related substances. The seeds contain fatty oils (Khalmatov 1964; Tairov 1969).

Astragalus sieversianus Pall. – Fabaceae

Synonyms: *Lithoon sieversianum* (Pall.) Nevski.

English name: Unknown

Russian name: Астрагал Сиверса (Astragal Siversa)

Uzbek name: Pakhtak

Kyrgyz name: Тулку куурай (Tulku kuuray)

Description: Herbaceous perennial. Stems 60–150 cm tall, up to 2 cm in diameter, densely hairy. Leaves alternate, pinnate, 15–30 cm long, long-stipulate; leaflets in 8–12 pairs, from narrow-ovate to elliptic, densely hairy on undersides, margins entire. Inflorescence axillary racemes with 3–9 flowers. Calyx tubular, densely hairy. Corolla papilionaceous, pale-yellow. Fruits ovate-spherical legumes, 15–20 mm long, densely covered with long, entangled hairs. Seeds kidney-shaped, brown.

Other distinguishing features: Stamens 10, nine filaments are fused. Legumes ovate-spherical, very hairy.

Phenology: Flowers in May-June, fruits in July-August.

Reproduction: By seeds.

Distribution: Farg'ona, Toshkent, Samarqand, Navoiy, Qashqadaryo, and Surxondaryo provinces of Uzbekistan; Chuy, Talas, Osh and Jalal-Abad provinces of Kyrgyzstan.

Habitat: The adyr and tau zones. Stony slopes in the mountain-steppe belt and lower belt of juniper stands.

Population status: Common.

Traditional use: A decoction of the fruits is drunk to remove kidney and bladder stones. The seeds, taken internally, are recommended to treat hernias in children and are smoked to treat syphilis (Khalmatov 1964).

Documented effects: In experiments, the ethanol extracts of flowers showed high antioxidant, lipid-reducing, and anti-aggregating activities. Preparations of the plant have sedative, antibacterial, anti-inflammatory, and bile-stimulating actions. Intravenous introduction of the alkaloid smirnovine in narcotized animals, at the dose of 2 mg/kg, reduced blood pressure by 32–56 % for a short time and excited breathing, which is apparently due to ganglio-blocking actions (Sadritdinov and Kurmukov 1980). Saponins extracted from the roots protected the liver from induced chemical injury in mice (Zhang et al. 1992).

Phytochemistry: The aboveground parts contain saponins, up to 0.1 % alkaloids (especially smirnovine), coumarins, tannins, flavonoids (0.9 % in stems, up to 4.9 % in leaves), vitamins C, E and P, and carotene. The roots contain triterpenoids, alkaloids, coumarins, and saponins (Khalmatov 1964; Svechnikova et al. 1983; Gan et al. 1986a, b).

Atraphaxis spinosa L. – Polygonaceae

Synonyms: *Tragopyrum spinosum* (L.) C. Presl.
English name: Goat's wheat, manna
Russian name: Курчавка колючая (Kurchavka kolyuchaya)
Uzbek name: Tuya singren
Kyrgyz name: Боз караган (Boz karagan)
Description: Woody shrub, 30–100 cm tall. Branches long, slender, apex leafless, spine-tipped. Leaves alternate, sessile or short-petiolate, 3–7 mm long, 2–5 mm wide, elliptic to ovate, coriaceous, glabrous, margins entire. Ocreae cylindric, 1–3 mm long, membranous, brown at base. Inflorescences 2–6-flowered clusters, occurring in leaf axils of current year's branchlets; pedicels ~5 mm. Tepals 4, pink. Fruits lenticularly compressed nutlets, light brown, smooth, shiny.
Other distinguishing features: Stamens 6, styles 2.
Phenology: Flowers and fruits in May-June.
Reproduction: By seeds.
Distribution: Karakalpakstan autonomous republic, Buxoro province, and probably other provinces of Uzbekistan; Naryn and Ysyk-Kol provinces of Kyrgyzstan.
Habitat: The tau zone. Stony slopes in lower mountain areas.
Population status: Common.
Traditional use: A decoction from the leaves and flowers are used in folk medicine to treat fever (Khalmatov 1964).
Documented effects: In animal studies, the total alkaloids isolated from leaves increased blood pressure (Khalmatov 1964).
Phytochemistry: The leaves contain alkaloids and tannins (Khalmatov 1964). Flavonoids has also been isolated from the plant (Chumbalov et al. 1970, 1971).

Berberis integerrima Bunge – Berberidaceae

Synonyms: Some consider this species synonymous with *Berberis oblonga* (Regel) Schneid.
English name: Unknown
Russian name: Барбарис цельнокрайний (Barbaris tsel'nokrayniy)
Uzbek name: Kizil zirk
Kyrgyz name: Бёру карагат (Byoru karagat)
Description: Branched shrub, up to 4 m. Oldest branches gray, young branches reddish brown, glabrous. Branches armed with 3- to 5-fid spines; spines straight, ca. 3 cm long. Leaves clustered on short shoots, petiolate, 3–3.5 cm long, 1.5–1.7 cm wide, coriaceous, obovate or elongate, margins mostly entire. Inflorescences racemiform, 6–10 cm long, axillary. Flowers ca, 1 cm in diameter; pedicel ca. 1 cm long. Sepals similar to 6 yellow petals. Style very short. Fruit an elongated berry, purple-red, gray-glaucous, 7–8 mm long. Seeds elongated, dark brown.
Other distinguishing features: 10–12 berries per raceme. Differs from *Berberis nummularia*, which has red fruits when fully ripe.
Phenology: Flowers in May-June, fruits in July-August.
Reproduction: By seeds.
Distribution: Toshkent, Farg'ona, Qashqadaryo, and Surxondaryo provinces of Uzbekistan; all of Kyrgyzstan.
Habitat: The tau zone. Usually grows as a single shrub along mountain river banks, sometimes growing in small populations at 1,500–1,600 m elevation.
Population status: In Uzbekistan the prevalence of this species is decreasing due to human disturbance of its natural habitat.
Traditional use: Fruits are used as an antipyretic, to relieve thirst and as a spice (Khalmatov 1964). In northern Tajikistan the roots are used to treat wounds, bone fractures, rheumatism, radiculitis, heart pain, and stomach aches. A decoction of the leaves is used to treat kidney stones. A tea made with the flowers is used to treat lung tuberculosis, chest pains, and headaches. An infusion of the fruits is used to treat constipation and wounds (Khodzhimatov 1989).
Documented effects: In experiments with animals, the alkaloid berberine lowers blood pressure, has minor ganglion blocking action, stimulates contractility of smooth muscles of the uterine horns and intestines, depresses central nervous system, prevents tumors, and has a pronounced choleretic action (Supek 1946; Selivanova 1954; Shvarev and Tsetlin 1972; Idzumi and Conti 1962; Conti 1962). Berbamine, tetrandrine, and hydroxyacanthine had similar hypotensive effects, but only tetrandrine exhibited anti-inflammatory, analgesic, and analeptic effects (Naidovich et al. 1976). In medical practice, a preparation with berberine is prescribed to treat chronic cholecystitis. Berberine has antitumor and bacteriostatic activity, increases phagocytic activity of leucocytes, and prevents animal death from septicemia. It also is effective for patients with initial pulpitis. In vitro, berberine has bactericidal action against *Vibrios cholera*e (Turova et al. 1984). An extract prepared from the dried berries protected rat hepatocytes against induced cytotoxicity in vitro. In vivo, pretreatment and treatment of animals with the extract protected the liver against induced injuries (Jamshidzadeh and Niknahad 2006).
Phytochemistry: The plant contains many alkaloids (including berberine, columbamine, jatrorrhizine and oxyacanthine) and organic acids. Leaves from plants at the fruit-bearing stage from the Chon-Kemin valley in Kyrgyzstan, contained 0.18 % total alkaloids and the young shoots contained 1.5 % total alkaloids. Berberine, berbamine, berbamunine, isoboldine, isocorydine, isotetrandrine, oxyacanthine, magnoflorine, palmatine, talicmidine, reticuline and others were isolated from the total alkaloids (Karimov et al. 1977; Yunusov 1981; Karimov et al. 1993a, b; Khamidov et al. 1996).

Berberis oblonga (Regel) Schneid. – Berberidaceae

Synonyms: Some consider this species synonymous with *Beberis integerrima* Bunge.

English name: Unknown

Russian name: Барбарис продолговатый (Barbaris prodolgovatyy)

Uzbek name: Zirk, Kora zirk, Kora qand

Kyrgyz name: Созунку бёру карагат (Sozunku byoru karagat)

Description: Branched shrub, up to 4 m tall. Older branches dark, bark with long, shallow cracks; younger branches reddish-brown, often grayish with simple or 3-branched spines, spines ca. 1.5 cm long. Leaves clustered on short shoots in groups of 5–7, up to 6 cm long and 3 cm wide, wide-elliptic or obovate, narrow cuneate, glabrous, margins usually entire, occasionally with short, spiny teeth. Inflorescences usually racemiform, 3–4.5 cm long, with 10–30 flowers in each cluster. Flowers up to 1 cm in diameter with 6 yellow petals. Fruits ellipsoid berries, up to 1 cm long, 6 mm wide, black-purple, gray-glaucous. Seeds 2, rarely 1, dark brown.

Other distinguishing features: Young plants have 5–11-branched spines. The bark on branches and roots is bright-yellow inside.

Phenology: Flowers in May, fruits in July-August.

Reproduction: By seeds.

Distribution: Toshkent, Namangan, Andijon, Farg'ona, Samarqand, Qashqadaryo, and Surxondaryo provinces of Uzbekistan; all of Kyrgyzstan.

Habitat: The tau zone. Rocky slopes of medium and, sometimes, lower mountain zones.

Population status: Common, sometimes forming dense groups.

Traditional use: In folk medicine, the fruits are used as a heart tonic, to treat neurasthenia, as an antipyretic, to relieve thirst and as an antidiarrheal remedy. The root decoction is used to treat rheumatism, fever, eye diseases, and as oral rinse for wounds of the mouth (Khalmatov 1964). The residue from a dehydrated water extract of the root is eaten, mixed with hot water and drunk, or applied to a cloth and applied externally, to treat jaundice, stomach aches, back pain and arthralgia (Sezik et al. 2004; Pak 2005).

Documented effects: Giving an infusion of the plant to laboratory animals resulted in cardiotonic action and a mild decrease of blood pressure. In experiments with dogs the preparation decreased blood coagulability (Ibragimov and Dzhumabaev 1971; Dzhumabaev et al. 1970; Dzhumabaev 1972).

Phytochemistry: Roots from plants collected in Kyrgyzstan (Arslanbob) at the end of the growing season contained 4.5 % total alkaloids. Young shoots collected from flowering plants in Uzbekistan (Chingan) contained 1 % alkaloids, and the leaves contained 0.01 % total alkaloids. Berberine, berbamine, berbamunine, glaucine, isocorydine, columbamine, magnoflorine, oblongine, oxyacanthine, palmatine, thalicmidine, and others have been isolated from the total alkaloids (Karimov et al. 1975, 1976, 1977; Yunusov 1981; Khamidov et al. 2003).

◄**Asparagus persicus Baker**
Photos: Alim Gaziev

▼**Astragalus sieversianus Pall.**
Photos: Evgeny Davkaev

▼**Berberis integerrima Bunge** Photos: *top* and *lower left*: Vladimir Epiktetov; *lower right*: Alim Gaziev

▼**Atraphaxis spinosa L.**
Photos: Avinoam Danin

Betonica foliosa Rupr.– Lamiaceae

Synonyms: *Stachys betoniciflora* Rupr., *Stachys betonicifolia* Regel, *Stachys foliosa* Regel.

English name: Unknown

Russian name: Буквица олиственная, Чистец буквицецветный (Bukvitsa olistvennaya, Chistets bukvitsetsvetnyy)

Uzbek name: Tog kudusi

Kyrgyz name: Жалбырактуу бетоника (Zhalbyraktuu betonika)

Description: Herbaceous perennial with short rhizomes. Stems 60–100 cm tall, 4-sided, densely hairy below, scattered hairy above. Leaves opposite; lower leaves petiolate, 13–15 cm long, 4–5 cm wide, obovate, bases oblique, margins crenate; upper leaves sessile, ovate-lanceolate, 5–6 cm long, 2–3 cm wide, margins serrate; terminal leaves lanceolate, entire. Inflorescences 10–12-flowered verticillasters, in terminal, compact spikes. Flowers sessile. Calyx 10–15 mm long, campanulate with lanceolate teeth. Corolla 2-lipped, lilac. Fruits dark-brown nutlets.

Other distinguishing features: Fruits 3-sided with longitudinal grooves.

Phenology: Flowers in June-August, fruits in August and September.

Propagation: By seeds.

Distribution: All provinces of Kyrgyzstan; Toshkent province of Uzbekistan.

Habitat: Among shrubs and juniper stands, in steppes, forests, and in high mountain meadows.

Population status: Common, forming dense groups.

Traditional use: The aboveground parts are used to treat hysteria, hypertension, epilepsy, fainting, gout, jaundice, and rheumatism. A tea made from the herb is used to treat gastrointestinal pain, hemoptysis, respiratory disease, inflammation of the kidneys, and bladder, and is also used as a sedative. An infusion of the roots is used as a laxative (Khodzhimatov 1989).

Documented effects: A tincture and liquid extract of this species is used in obstetric-gynecological practices as a treatment to increase uterine muscle tonus, increase uterine contractions, and as a hemostatic (Tolmachev 1976). Preparations of the plant have anti-inflammatory, anti-asthmatic, antiseptic, analgetic, hemostatic, and choleretic properties. The preparations are used as expectorants, to decrease blood pressure, to increase metabolism, to improve blood circulation and to improve digestion (Plant Resources of the USSR 1991).

Phytochemistry: Plants collected in Talas-Alatau (Kyrgyzstan) contained ~54 mg/% vitamin C (in the leaves). The aboveground parts contained 1.54 % flavonoids, alkaloids (up to 0.49 % stachydrine), 1 % iridoids, 3.11 % resins, 0.12 % essential oil, ~49.5 mg/% vitamin C, 2 % organic acids, 1.02 % calcium salts, 3.98 % sugars, phenolcarbonic acids, and vitamin k_1 (Khalmatov et al. 1984).

Bidens tripartita L. – Asteraceae

Synonyms: *Bidens comosa* (Gray) Wieg., *Bidens orientalis* Velen.

English name: Threelobe beggarticks

Russian name: Череда трехраздельная (Chereda trekhrazdel'naya)

Uzbek name: Eteetkanak, Karakeez

Kyrgyz name: Уч болуктуу ит уйчан (Uch boluktuu it uychan)

Description: Herbaceous annual with a taproot. Stems 20–110 cm tall. Leaves opposite, lower and middle tripartite; upper leaves unlobed, lanceolate. Inflorescences capitula, single or in groups of 2–3; involucral bracts ovate or lanceolate-ovate, internal bracts shorter. Flowers yellow, usually only disc type. Fruits dark brown achenes, flattened with 4 edges, often with retrorsely barbed awns.

Other distinguishing features: 1–5 ray flowers occasionally present. Fruits usually not tuberculate.

Phenology: Flowers in June-September, fruits in July-October.

Reproduction: By seeds.

Distribution: All provinces of Kyrgyzstan and Uzbekistan.

Habitat: In wet places near ponds, lakes, bogs and canals.

Population status: Common, sometimes forming dense groups.

Traditional use: The top parts of the plant, with leaves and immature capitula, are collected for use in traditional medicine. They are used to treat respiratory diseases, scrofula, scurvy, scabies, bacterial and fungal skin diseases, poor digestion, toothaches, blood diseases (including anemia), arteriosclerosis, anthrax, and tuberculosis, and also to regulate the metabolism (Maznev 2004).

Documented effects: Used as an antipyretic, as a diuretic for treatment of urogenital diseases, and as a diaphoretic and anti-inflammatory (Kurochkin 1998). A bath infused with the herb is used to treat diathesis in children. An infusion of the herb is drank to induce sweating and to treat common colds (Grinkevitch 1991). Although the content of flavonoids in the flower heads was found to be half of that found in the herb, an extract from flowers had nearly 2 times higher antioxidant activity (Wolniak et al. 2007).

Phytochemistry: The herb contains flavonoids, coumarins, ascorbic acid, carotene, tannins, mucilage, γ-lactones, and traces of essential oils (Serbin et al. 1972 a, b, c, 1975; Khalmatov et al. 1984; Maznev 2004; Wolniak et al. 2007).

Biebersteinia multifida DC. – Biebersteiniaceae

Synonyms: None
English name: Unknown
Russian name: Биберштейния многораздельная (Bibershteyniya mnogorazdel'naya)
Uzbek name: Kontepar
Kyrgyz name: Кёп балуктуу биеберштения (Kyop baluktuu biyebershteniya)
Description: Herbaceous perennial with a thick, tuberiform root. Stem 30–60 cm tall, sturdy. Plant woolly-hairy and glandular. Leaves alternate, stipulate, short-petiolate, 10–20 cm long, 2–8 cm wide, tripinnatisect, both sides spreading-hairy. Inflorescences racemiform. Flowers orange-yellow. Sepals 5. Petals 5. Stamens 10; filaments glabrous, connate at the base, forming a ring. Styles 5, connate; stigma capitate. Fruit a schizocarp. Seeds very wrinkled, coriaceous.
Other distinguishing features: Root turns pink when fractured.
Phenology: Flowers and fruits in April-June.
Reproduction: By seeds.
Distribution: Karakalpakstan autonomous republic, Toshkent, Samarqand, Buxoro, Qashqadaryo, and Surxondaryo provinces of Uzbekistan; Chuy, Talas and Osh provinces of Kyrgyzstan.
Habitat: The tau zone. Gypsum soil on low mountains. Shallow-soiled and stony slopes of lower and medium mountain zones.
Population status: Uncommon.
Traditional use: A decoction of the root is used as a hemostatic for post-natal bleeding and to treat gastric diseases (Khalmatov 1964). In Iran an ointment made of the powdered root mixed with tallow is used to treat musculoskeletal disorders and bone fractures (Farsam et al. 2000).
Documented effects: An extract of the root had anti-inflammatory effects on induced rat paw edema and analgesic effects in tests with rats (Farsam et al. 2000). In parenteral toxicity tests in mice, an extract of the total alkaloids was classified as a moderately toxic agent. Dermal acute toxicity tests showed no sign of toxicity (Ostad et al. 2003).
Phytochemistry: The aboveground parts have essential oils. The roots contain tannins, carbohydrates, and saponins. 0.08 % total alkaloids were obtained from the extracts of tops of plants collected at the Usturt Plateau during budding stage and vasicinone was isolated (Yunusov 1981). The roots and aboveground parts contain polysaccharides (Arifkhodzhaev et al. 1985; Arifkhodzhaev and Rakhimov 1986, 1993) as well as the flavones luteolin 7-glucoside and 7-rutinoside (Omurkamzinova et al. 1991).

Bunium chaerophylloides (Regel & Schmalh.) Drude – Apiaceae

Synonyms: *Buniella chaerophylloides* (Regel & Schmalh.) Schischk., *Carum chaerophylloides* Regel & Schmalh., *Carum confusum* O. Fedtsch., *Carum sogdianum* Lipsky.

English name: Unknown

Russian name: Буниум бутеневый (Bunium butenevyy)

Uzbek name: Qarga oyeq

Kyrgyz name: Бутен зиреси (Buten ziresi)

Description: Herbaceous, glabrous perennial, with a spherical tuber up to 2 cm in diameter. Stem 50–70 cm high, narrowly striated, cylindrical, hollow. Basal leaves long-petiolate, blade triangular-oval, tri-pinnatisect, with lanceolate lobes; upper leaves alternate, sessile or with short membranous, sheathing petioles. Inflorescences compound umbels with 10–16 radials, flat-topped. Calyx toothless. Petals broadly obovate, white. Fruits oblong-linear schizocarps, 4–4.5 cm long, sometimes curved.

Other distinguishing features: Bractlets lacking. The fruits are similar to those of *Bunium persicum*, but do not smell when crushed.

Phenology: Flowers in April- May, fruits in June.

Reproduction: By seeds.

Distribution: Toshkent, Farg'ona, Samarqand, Qashqadaryo, and Surxondaryo provinces of Uzbekistan; Osh province of Kyrgyzstan.

Habitat: The tau zone. Mountainous slopes.

Population status: Common.

Traditional use: A powder of roasted and ground fruits mixed with honey is used in folk medicine to dissolve renal and cystic stones and to treat skin diseases (for white spots on the skin). The powdered tuber is applied to mouth injuries and for reddening of the tongue (Khalmatov 1964).

Documented effects: No data.

Phytochemistry: Fruits contain essential oil and roots contain around 24 % starch (Khalmatov 1964).

▲**Betonica foliosa Rupr.** Photos: Alexander Naumenko

▲**Biebersteinia multifida DC.** Photos: *top right*: Vladimir Epiktetov; *bottom*: Evgeny Davkaev

▼**Bunium chaerophylloides (Regel & Schmalh.) Drude** Photo: Komiljon Tojibaev

►**Bidens tripartita L.**
Photos: Dmitri Oreshkin

Bunium persicum (Boiss.) B. Fedtsch. – Apiaceae

Synonyms: *Carum persicum* Boiss., *Carum heterophyllum* Regel & Schmalh.

English name: Black cumin, wild cumin

Russian name: Буниум персидский (Bunium persidskiy)

Uzbek name: Zira

Kyrgyz name: Персия зиреси (Persiya ziresi)

Description: Herbaceous perennial with an irregular spherical tuber. Stem 40–60 cm high, striated, pale green, glabrous, branching from the middle to upper portion. Basal leaves with long petioles, blades wide-triangular, bi- or tripinnatisect; cauline leaves alternate, bipinnatisect with highly dissected sections, sessile, sheathing. Inflorescence a compound umbel with 15–20 rays. Petals white, ca. 1 mm long. Fruits oblong schizocarps, 3–4 mm long, dark brown, ridged, shorter than the pedicel.

Other distinguishing features: Involucel with 2–5 linear bractlets. The ripe fruits have a very specific smell, unique to *Bunium persicum.*

Phenology: Flowers in June, fruits in July.

Reproduction: Only by seeds.

Distribution: Samarqand, Qashqadaryo, and Surxondaryo provinces of Uzbekistan; Osh province of Kyrgyzstan.

Habitat: The adyr zone. Slopes of loess foothills.

Population status: Due to mass collection of seeds, it now occurs infrequently.

Traditional use: Seeds are used to prevent stomach aches and to eliminate spleen tumors. An infusion of the fruits in vinegar is used as a hemostatic to stop nose bleeds. Roasted fruits are recommended as a diuretic (Khalmatov 1964). The fruits are taken to increase appetite, and to treat kidney stones and liver diseases. A decoction and infusion is used as a diaphoretic and vermifuge, as well as to improve digestion. Roasted fruits are used to treat bladder incontinence and obesity. The fresh tubers are used to improve digestion (Khodzhimatov 1989).

Documented effects: Modern research has shown that a decoction of the fruits stimulates gastric secretion, favors creation of complete gastric fluids with a higher acid index of pepsin and pepsinogen and also has evident cholagogic, anti-inflammatory, and antispasmodic actions. It improves disinfectant and secretory functions of the liver. Decoctions of *Bunium persicum* are recommended to treat chronic hypo- and anacidity gastritis, chronic colitis and cholecystitis. Roasted fruits have a diuretic action (Khalmatov and Kosimov 1994). Extracts of the aboveground plant parts exhibited significant antibacterial effects against *Bacillus subtilis*, as well as antifungal activity and low cytotoxicity (Sardari et al. 1998; Fazly Bazzaz and Haririzadeh 2003). The essential oils has antioxidant and anti-inflammatory properties (Ur-Rehman et al. 1991; Jassbi et al. 2005).

Phytochemistry: Seeds contain up to 3 % essential oils (carene, cymol, terpinolene, carvone, linalool, carvacrol) and 13.6 % oils and proteins (Khalmatov and Kosimov 1994). Essential oil collected from plants in Iran contained mostly monoterpenes and phenylpropanoids, such as α-pinene, p-cymene, limonene, γ-terpinene, cuminaldehyde, cuminyl alcohol, myristicin, and dillapiole (Jassbi et al. 2005).

Campanula glomerata L. – Campanulaceae

Synonyms: None
English name: Clustered bell flower, Dane's blood
Russian name: Колокольчик скученный (Kolokol'chik skuchennyy)
Uzbek name: Kungrok gul
Kyrgyz name: Топтолгон конгуроо гул (Toptolgon konguroo gul)
Description: Perennial herb, gray-hairy. Stems 25–70 cm high, erect, slightly angled. Leaves alternate, simple, slightly toothed; lower leaves oblanceolate, apex acute, long-petiolate; upper leaves ovate to narrowly triangular, sessile, sometimes clasping. Flowers sessile in terminal compact clusters or few in the upper leaf axils. Sepals narrowly lanceolate. Corolla campanulate, 5-lobed with ovate-triangular acute lobes, lilac to blue-violet, 1–3 cm long, hairy outside. Fruit a capsule opening by lateral pores.
Other distinguishing features: Stems slightly angled.
Phenology: Flowers and fruits in June-August.
Reproduction: By seeds.
Distribution: Toshkent, Samarqand, and Farg'ona provinces of Uzbekistan; all of Kyrgyzstan.
Habitat: The tau zone. Shallow-soiled and stony slopes.
Population status: Uncommon, found as single individuals.
Traditional use: An infusion and decoction of the leaves are used as an oral rinse, to treat sore throat and hoarseness and are applied externally as a lotion for erysipelatous inflammations and taken internally to treat headache. A decoction of the flowering herbs is also used to treat hydrophobia, for bathing children with a fear of water and for treating people who have seizures (Khalmatov 1964).
Documented effects: An ethanolic extract of the roots has hypolipidemic and antioxidant properties (Eliseeva 2005).
Phytochemistry: The plant contains small amounts of alkaloids. The leaves contain up to 1,000 mg% of vitamin C (Khalmatov 1964).

Capparis spinosa L. – Capparidaceae

Synonyms: *Capparis herbacea* Willd.

English name: Caper bush

Russian name: Каперцы колючие (Kapertsy kolyuchiye)

Uzbek name: Kovul

Kyrgyz name: Тикендуу конуз баш (Tikenduu konuz bash)

Description: Herbaceous perennial plant with thick roots. Stems numerous, decumbent, up to 2.5 m long, woody towards the base, glabrous. Leaves alternate, ovate, obovate, or round, 3.5–6 cm long, glabrous, short-petiolate with stipular spines. Flowers single, 5–8 cm in diameter, white or sometimes cream to pinkish in color, with long pedicels. Fruits fleshy, berry-like capsules, round to obovate, 2.5–5 cm long, green, glabrous, smooth. Seeds 3–3.5 mm long, round-elliptic or kidney-shaped, brown.

Other distinguishing features: Flowers have strong aroma similar to honey.

Phenology: Flowers in May-June, fruits in July-August.

Reproduction: By seeds.

Distribution: All of Uzbekistan; Chuy, Talas, Osh and Jalal-Abad provinces of Kyrgyzstan.

Habitat: The chul and adyr zones. A weed and ruderal; found on hills, among unirrigated winter wheat fields, along roads, in dry river-beds and on stony slopes of the lower mountains.

Population status: Common, forms large populations.

Traditional use: One of the oldest folk remedies and used to treat a variety of diseases. Avicenna recommended caper bush as an analgesic and vermifuge, for healing wounds, and to treat asthma and gastrointestinal diseases. A decoction of the roots is used to treat hepatitis, and the root bark is smoked to treat syphilis. Juice from the flowers are used as a treatment for scrofula. A decoction of the fruit is used for hemorrhoids and toothaches, and to strengthen the gums (Akopov 1981).

Documented effects: 25 % root extract in 96 % ethanol and 25 % root decoction accelerate blood coagulation (Akopov 1981). A tincture of the root increased the number of thrombocytes in blood. In experiments with guinea pigs, treatment with a decoction of the roots caused desensitization to animal and plant allergens. The fresh juice from the fruits was clinically tested and recommended for the treatment of exophthalmic goiters (Khodzhimatov 1989). A methanol extract of the flower buds exhibited strong antioxidant activity (Germano et al. 2002). p-Methoxy benzoic acid isolated from an extract of this plant species was found to possess significant activity against induced hepatotoxicity in vivo and in vitro (Gadgoli and Mishra 1999).

Phytochemistry: Flowers and buds contain rutin, quercetin, vitamin C, saponins, pigments and glycosides. Seeds contain 25–35 % semi-drying oils, 25 % oleic and 33 % linoleic acids. The aboveground plant parts contain 0.32 % rutin and quercetin, up to 100 mg% vitamin C, stachydrine, and thioglycocides. Fruits contain up to 36 % sugar, 25–35%mg vitamin C, flavonoids, and thioglycosides. Roots contain 1.2 % alkaloids (stachydrine), 0.44 % flavonoids, 4.5 % sugars, coumarins, and other substances (Khalmatov and Kosimov 1994).

Capsella bursa-pastoris (L.) Medik. – Brassicaceae

Synonyms: *Bursa bursa-pastoris* (L.) Britton, *Bursa pastoris* Weber ex F.H. Wigg., *Capsella hyrcana* Grossh., *Crucifera capsella* E.H.L. Krause, *Iberis bursa-pastoris* (L.) Crantz, *Thlapsi bursa-pastoris* L.

English name: Shepherd's purse

Russian name: Пастушья сумка обыкновенная (Pastush'ya sumka obyknovennaya)

Uzbek name: Ochambiti, zhag-zhag

Kyrgyz name: Койчу баштык (Koychu bashtyk)

Description: Herbaceous annual. Stems 5–60 cm tall. Basal leaves in a rosette, petiolate, oblanceolate entire to pinnatipartite with triangular lobes; cauline leaves elongate, upper leaves almost linear with sagittate bases. Inflorescence an apical raceme. Flowers small, pedicellate with 4 white petals. Fruit a silicle, triangular to heart-shaped. Seeds small, oval, slightly flattened, yellow-brownish.

Other distinguishing features: Stamens 6 (4 long, 2 short).

Phenology: Flowers in April-June, fruits in June-July.

Reproduction: By seeds.

Distribution: All provinces of Kyrgyzstan and Uzbekistan.

Habitat: From the foothills to high mountains, on waste grounds, abandoned fields, near houses, along roads and canals.

Population status: Common, forming dense groups.

Traditional use: The aboveground parts are used in Kyrgyz folk medicine to treat uterine bleeding, malignant ulcers, stomach cancer, dysentery, gastritis, tuberculosis, and venereal diseases. In Chinese medicine the roots are used to treat dysentery and eye diseases, and in Tibet they are used as an antiemetic (Plant Resources of the USSR 1986).

Documented effects: The herb strengthens the tonus of uterine muscles and narrows the peripheral veins (Maznev 2004). The peptides shepherin I and shepherin II, isolated from the roots, exhibited antimicrobial activity against gram-negative bacteria and fungi (Park et al. 2000).

Phytochemistry: The herb contains rhamnoglucosides (including hyssopin), choline, acetylcholine, tyramine, inosine, tannins, bursic, fumaric, malic, tartaric and citric acids, vitamins A, B, C and K, saponins, phytoncides, and essential oils (Kurochkin 1998).

▲**Campanula glomerata L.** Photos: *left* and *center*: Alexander Naumenko; *right*: Ilya Raskin

▲**Capparis spinosa L.** Photos: *left* and *center*: John B. Taft; *right*: Alexander Naumenko

▼**Capsella bursa-pastoris (L.) Medik.** Photos: Sergey Appolonov

▼**Bunium persicum (Boiss.)** B. Fedtsch. Photo: Authors

Carum carvi L. – Apiaceae

Synonyms: *Carum gracile* Lindl., *Carum rosellum* Woronow.

English name: Caraway

Russian name: Тмин обыкновенный (Tmin obyknovennyy)

Uzbek name: Korazira

Kyrgyz name: Кадимки карум (Kadimki karum)

Description: Herbaceous biennial, occasionally annual or perennial. Stems 30–90 cm tall. Leaves bi- or tripinnatisect; basal leaves long-petiolate, segments linear-lanceolate; cauline leaves short-petiolate. Inflorescence a compound umbel with 8–16 rays. Flowers small with 5 petals, white or pink. Fruit a 2-seeded schizocarp, brown, 3–5 mm long, 1–2 mm wide, sides flattened.

Other distinguishing features: Fruits have a distinct aroma.

Phenology: Flowers in June, fruits in July.

Reproduction: By seeds.

Distribution: All provinces of Kyrgyzstan; Toshkent, Samarqand and Surxondaryo provinces of Uzbekistan.

Habitat: From valleys to high mountains. Found in meadows, along canals and river floodplains, near bogs and in forest glades.

Population status: Common, forming dense groups.

Traditional use: Fruits, which are collected in July and August, are used for medicinal purposes. They are used as a sedative, expectorant, diuretic, and is included in a preparation used as a carminative, laxative, sedative, and to increase appetite (Turova and Sapozhnikova 1984).

Documented effects: Caraway strengthens the appetite, promotes digestion, reduces spasms in smooth muscles (intestinal, uterine, and urethral), increases diuresis, and promotes expelling of phlegm and sputum (Turova and Sapozhnikova 1984). The essential oil isolated from the fruits exhibited antibacterial activity against a variety of gram-positive and gram-negative bacteria (Iacobellis et al. 2005).

Phytochemistry: Fruits contain 3–7 % essential oil, 14–22 % fatty oil, and tannins. The essential oil contains limonene, carvacrol, carvone, and other compounds. The flavonoids quercetin, camphorol, isorhamnetin and polyenes were isolated from the aboveground parts and flowers that were collected during the flowering stage (Khalmatov et al. 1984; Turova and Sapozhnikova 1984; Iacobellis et al. 2005).

Centaurea depressa M. Bieb. – Asteraceae

Synonyms: *Cyanus depressus* (M. Bieb.) Soják.

English name: Low cornflower, Iranian knapweed

Russian name: Василек придавленный (Vasilek pridavlennyy)

Uzbek name: Butakuz

Kyrgyz name: Жагалак кёп башы (Zhagalak kyop bashy)

Description: Herbaceous annual plant, to 15–60 cm tall. Stems multiple, abundantly branched, gray-tomentose, foliaceous from the base. Basal and lower leaves simple, petiolate, oblong, 5–10 cm long, entire to pinnatifid, gray-tomentose; upper leaves sessile, linear-lanceolate, entire. Inflorescences peduculate capitula, arranged solitarily; involucral bracts coriaceous, silvery fimbriate along the edges. Ray flowers bright blue or blue-violet; disc flowers violet. Fruits obovate achenes, mostly smooth, shiny, with pappus.

Other distinguishing features: Outer pappi of stiff bristles, unequal, up to 8 mm long.

Phenology: Flowers in May-June, fruits in June-July.

Reproduction: Only by seeds.

Distribution: All of Uzbekistan; Chuy, Talas, Osh and Jalal-Abad provinces of Kyrgyzstan.

Habitat: The adyr and tau zones. Clay-soiled slopes with rocky debris. Often occurs as a weed in wheat fields.

Population status: Common, found in small populations.

Traditional use: A decoction of the flowers is used for melancholy, neurasthenia, eye diseases, and as a cholagogue for hepatitis (Khalmatov 1964).

Documented effects: An extract of the aboveground parts had antibacterial effects against *Bacillus subtilis*, *Escherichia coli*, *Proteus mirabilis*, and *Pseudomonas aeruginos*a (Fazly Bazzaz and Haririzadeh 2003; Arif et al. 2004). A hexane extract of the plant showed antifungal activity against *Candida krusei* (Karamenderes et al. 2006).

Phytochemistry: The herb contains flavonoids, phenolic compounds and small amounts of alkaloids (Khalmatov 1964; Bandyukova et al. 1969; Hosseinimehr et al. 2007). The main components of the essential oil isolated from plants in Iran, were piperitone and elemol (Esmaeili et al. 2005).

Ceratocephala testiculata (Crantz) Roth – Ranunculaceae

Synonyms: *Ceratocephala orthoceras* DC., *Ceratocephala reflexa* Steven, *Ranunculus testiculatus* Crantz. (In some references the invalid genus name *Ceratocephalus* is used for this species).

English name: Bur buttercup, tubercled crowsfoot, curveseed butterwort

Russian name: Рогоглавник яичковидный, Рогоглавник прямрогий (Rogoglavnik yaichkovidnyy, Rogoglavnik pryamorogiy)

Uzbek name: Uchma, Kuitikan

Kyrgyz name: Unknown

Description: Herbaceous annual. Stems erect or suberect, tomentose. Basal leaves petiolate, blades broadly spatulate in outline, 1–3-dissected, segments linear. Flowers solitary, terminal. Sepals 5, 2–8 mm long, spreading, tomentose. Petals 5, yellow. Fruits achenes in ovoid clusters. Achenes 4.5–6.5 mm long, with a straight, sharp beak.

Other distinguishing features: Early flowering ephemeral. Sepals persistent in fruit.

Phenology: Flowers in March-April, fruits in April-May.

Reproduction: By seeds.

Distribution: Widespread throughout Uzbekistan and Kyrgyzstan.

Habitat: The chul, adyr, and tau zones. Clayey and sandy soils, pebbly, dry slopes, and very salty areas.

Population status: Common, weedy.

Traditional use: The plant is used to treat wounds, injuries, eczema, and other skin diseases (Khalmatov 1964).

Documented effects: The plant is used as a treatment for pyoderma and furunculosis. An ointment made of the dried herb is used to treat sores. Oil extracts of the fresh herb can accelerate reduction of inflammatory edema and stimulate steady increase in tissue granulation and wound epithelization (Khalmatov 1964). Studies showed that plants growing in Utah (USA) were toxic to sheep with a minimum lethal dosage of 11 g (wet weight) of green plant material/kg. Signs of poisoning are weakness, depression, diarrhea, labored breathing, anorexia, and occasional fever (Gusinin 1962; Nachman and Olsen 1983).

Phytochemistry: Contains anemonin, uronic acids, resins, carotene, and sugars (Khalmatov 1964). Analyses of plants collected in Utah (USA), revealed that the "early flower" stage contained the highest concentration of the toxic compound ranunculin (Nachman and Olsen 1983).

Cichorium intybus L. – Asteraceae

Synonyms: None
English name: Chicory
Russian name: Цикорий обыкновенный (Tsikoriy obyknovennyy)
Uzbek name: Sachratki
Kyrgyz name: Кадимки дарчын (Kadimki darchyn)

Description: Herbaceous perennial with a large taproot. Stems 15–120 cm tall. Basal leaves long-elliptic, 10–28 cm long, 2–6 cm wide, pinnatifid to pinnatisect, base tapering to the petiole; lower cauline leaves oblong-ovate to broadly lanceolate, large dentate; upper cauline leaves small, linear to lanceolate, margins almost entire. Inflorescences capitula, axillary, in groups of 1–3; involucral bracts in 2 rows. Flowers only ligulate; ligules blue, 5-toothed. Fruits 3–6-sided achenes, 2–3 mm long, brown; pappus of very short scales.

Other distinguishing features: Anthers bluish, connate around style. Sap milky.

Phenology: Flowers in June-July, fruits in July-August.

Reproduction: By seeds.

Distribution: All provinces of Kyrgyzstan; Karakalpakstan autonomous republic, Toshkent, Samarqand, Farg'ona, Buxoro and Surxondaryo provinces of Uzbekistan.

Habitat: On waste grounds, near roads, in fallow meadows and dry, stoney or clayey waterways, in vegetable gardens and in cultivated fields.

Population status: Common, forming dense groups.

Traditional use: The roots and inflorescences are used to prepare folk medicines. The roots are used to increase the appetite and improve digestion. Inflorescences are used to treat inflammation of the stomach lining, diseases of the large and small intestines, gall bladder, and kidneys, as well as to treat kidney and gallstones. The inflorescences are also used swelling due to various heart conditions (Ladigina and Morozova 1987; Nogaller et al. 1987).

Documented effects: According to experiments, an infusion of the inflorescences has sedative effects on the central nervous system and strengthens the heart function (Akopov 1990). A decoction of the aboveground parts has diuretic, astringent, and antimicrobial effects. Liquid extracts from the roots reduced the blood sugar content in people with early stage diabetes (Khalmatov et al. 1984). Extracts of the plant have been shown to affect cholesterol uptake, tumor development, prevent immunotoxicity induced by ethanol, and have anti-inflammatory properties (Schmidt et al. 2007). Experiments with mice and rats showed that an extract of the root, rich in sesquiterpene lactones, significantly reduced inflammation, by down-regulating pro-inflammation gene expression and reducing nitric oxide production (Ripoll et al. 2007).

Phytochemistry: Roots contain up to 65 % inulin, the glycoside intibin, alkaloids, organic acids, and vitamin B and C. The flowers contain the glycoside cichoriin, coumarins, flavonoids, and tannins. The plant sap contains lactucin, lactucopicrin, and taraxasterol (Khalmatov et al. 1984; Schmidt et al. 2007).

▲**Carum carvi L.** Photos: *left* and *center*: Sergey Appolonov; *right* : Dmitri Oreshkin

▲**Centaurea depressa M. Bieb.** Photos: *left*: Alim Gaziev; *right*: Evgeny Davkaev

◄**Ceratocephala testiculata (Crantz) Roth** Photos: Maxim Kucherov

▼**Cichorium intybus L.** Photos: *top*: Evgeny Davkaev; *bottom*: Sasha Eisenman

Clematis orientalis L. – Ranunculaceae

Synonyms: *Clematis grata* Wall., *Clematis incisodentata* Rich., *Clematis orveniae* Harvey & Sonder, *Clematis petersiana* Klotzsch, *Clematis thunbergii* Steud., *Clematis triloba* Thunb., *Clematis viridiflora* Bertol., *Viticella orientalis* (L.) W.A. Weber.

English name: Oriental virginsbower

Russian name: Ломонос восточный, Клематис восточный (Lomonos vostochnyy, Klematis vostochnyy)

Uzbek name: Ilan chup

Kyrgyz name: Чыгыш жебелгеси (Chygysh zhebelgesi)

Description: Perennial, semi-woody climbing vine. Stems 2–8 m long, sometimes reddish, glabrous or densely short-hairy. Leaves pinnately compound; leaflets (3–)5–7, lanceolate to ovate, slightly lobed, entire or coarsely dentate, with short appressed hairs or nearly glabrous. Flowers pedicellate, solitary or in axillary cymes. Sepals 4, greenish-yellow, recurved, often hairy. Petals absent. Fruits hairy achenes, 2 mm long with a long beak (3–8 cm).

Other distinguishing features: Plant climbs using tendril-like petioles and leaf-rachises. Staminal filaments hairy towards base.

Phenology: Flowers in June-September, fruits in July-October.

Reproduction: By seeds.

Distribution: Widespread throughout all of Uzbekistan and Kyrgyzstan.

Habitat: The chul, adyr, and tau zones. Along river banks and irrigation canals, along fences and among bushes.

Population status: Common.

Traditional use: The fresh herb is used as a source for antivenom to treat snake bites. A powder and decoction has strong insecticidal properties (Khalmatov 1964). In Central Asia the plant is used to treat tuberculosis (Ogolevitz 1951). Preparations are applied externally to treat chronic eczema with itching. In Chinese medicine preparations are used as a sedative, analgesic, diuretic, diaphoretic, to treat cystitis and as an anti-inflammatory to treat rheumatism, gout, and chronic gonorrhea. Other species, particularly *Clematis hexapetala*, are used as an antivenom remedy to treat snake bites and as an analgesic (Ibragimov and Ibragimova 1960).

Documented effects: Extract of the leaves have strong bactericidal and fungicidal actions, possibly due to the presence of anemonin. The fresh herb is considered poisonous, probably because of anemonin, which disappears after drying (Ogolevitz 1951). In experiments, an extract of the herb had antibacterial actions on gram-positive microbes (Khalmatov 1964).

Phytochemistry: The aboveground parts contain the alkaloid clematine, green resin with melissic acid, myricyl alcohol, and caulosapogenin glycoside. The roots contain alkaloids (Khalmatov 1964).

Cnicus benedictus L. – Asteraceae

Synonyms: *Centaurea benedicta* (L.) L., *Hierapicra benedicta* (L.) Kuntze.

English name: Blessed thistle

Russian name: Кникус благословенный, Волчец кудрявый (Knikus blagoslovennyy, Volchets kudryavyy)

Uzbek name: Saryq gul, Kushkunmas

Kyrgyz name: Тармал кникус (Tarmal knikus)

Description: Herbaceous annual, 5–70 cm high with taproot. Stems prostrate to erect, usually branching, slightly striated, often reddish, loosely hairy. Basal leaves elongate, pinnatipartite, up to 20 cm long, margins spiny-toothed, base of leaf tapering to winged petiole; stem leaves alternate, sessile; upper leaves simple, up to 5.5 cm long, sinuate with small spiny teeth. Inflorescences wide-ovate capitula, each solitary at the ends of branches; involucral bracts pinnate with spiny tips. Disc flowers yellow; ray flowers few, very slender. Fruits achenes 6–10 mm long, yellow-brown, with 20 ribs and pappi consisting of 2 rows of awns.

Other distinguishing features: Leaves glandular and slightly or densely hairy. Ray flowers sterile, 3-lobed. Achenes slightly curved.

Phenology: Flowers in May-June, fruits in June-July.

Reproduction: Only by seeds.

Distribution: Toshkent, Samarqand, and Surxondaryo provinces of Uzbekistan; all of Kyrgyzstan.

Habitat: The chul and adyr zones. A weed along roads and in waste places.

Population status: Common, does not grow in large populations.

Traditional use: The plant has a long history of use in folk medicine. Preparations were used as a cancer remedy. A decoction of the upper plant parts (capitula and leaves) is used to treat constipation, jaundice, liver diseases, hypochondria, respiratory tract catarrh, intermittent fever, gastrointestinal atonia, gout, ulcers, kidney diseases, urination disorders, and indigestion and is also used as an emetic (Ogolevitz 1951; Khalmatov et al. 1984; Khodzhimatov 1989).

Documented effects: In modern medicine, infusions and extracts of the plant are used to stimulate the appetite and improve digestion (Khalmatov et al. 1984). The herb increases the flow of gastric juices, which relieves dyspepsia, indigestion, and headaches associated with liver congestion. Extracts of the plants, including cnicin, essential oil, and polyacetylenes, have antibiotic properties. The compound cnicin has been shown to have antitumor, antimicrobial, cytotoxic, and anti-inflammatory activities. The lignans arctiin and arctigenin act as platelet activating factor (PAF) antagonists and have exhibited anti-HIV activity, as well as cytotoxic activity in vitro and antitumor activity in vivo (Tamayo et al. 2000).

Phytochemistry: The herb contains bitter substances, cnicin (a sesquiterpene lactone), resin, mucilage, sterins, tannins, essential oils, and vitamin C. The essential oil contains n-paraffin, aromatic aldehydes (cinnamaldehyde, benzaldehyde, cuminaldehyde), and monoterpenes (citronellol, fenchone, p-cymene, etc.). The seeds contain 24–28 % semi-drying fatty oil and lignans (including arctiin and arctigenin), some of which are phytoestrogen precursors for mammalian lignans (Khalmatov 1964; Khalmatov et al. 1984; Tamayo et al. 2000).

Codonopsis clematidea (Schrenk) C.B. Clarke – Campanulaceae

Synonyms: *Glossocomia clematidea* (Schrenk) Fisch., *Wahlenbergia clematidea* Schrenk.

English name: Asian bellflower

Russian name: Кодонопсис ломоносовидный (Kodonopsis lomonosovidnyy)

Uzbek name: Qoraqurt, Dogboyut

Kyrgyz name: Конгуроодой сасык гул (Konguroodoy sasyk gul)

Description: Perennial herb, 50–80 cm tall. Root fusiform, vigorous. Stem erect or winding, densely branching from the base, pubescent or glabrous, deep-green. Leaves alternate or sub-opposite, oval, acute, petiolate except at top of plant, margins entire, short-hairy. Calyx with 5 deep lobes, glabrous or pubescent; lobes up to 2 cm long, triangular, oblong or ovate-lanceolate, during flowering becoming recurved. Corolla 2–3 cm long, widely campanulate, with 5 short lobes, whitish or bluish with darker blue veins. Fruit a compressed capsule, obconical or oval, acute. Seeds oblong, shining or dull, wingless.

Other distinguishing features: Flowering plant with strong, objectionable odor.

Phenology: Flowers and fruits in June-August.

Reproduction: By seeds.

Distribution: Toshkent, Samarqand, and Surxondaryo provinces of Uzbekistan; all of Kyrgyzstan.

Habitat: The tau zone. Wet taluses, along canyon bottoms and along mountain streams.

Population status: Fairly often, does not grow in dense populations.

Traditional use: The aboveground parts are widely used in medicine as a cholagogue for hepatitis and cholecystitis (Rakhimov et al. 2003).

Documented effects: This plant is a part of cholagogue collection by professor Khodzhimatov. Effective doses of the alkaloid codonopsin provoked general depression in mice. Codonopsin reduced blood pressure in acute experiments on cats and caused premature ventricular beats when applied intravenously to rabbits (Khanov et al. 1971). Codonopsinine and codonopsine have antibiotic properties and exhibit hypotensive activity with no observed effects on the central nervous system in animal tests (Haddad and Larchevêque 2003).

Phytochemistry: The aboveground plant parts contain the alkaloids codonopsin and codonopsinin (Yunusov 1974: Tashkhodzhaev et al. 2004).

Conium maculatum L. – Apiaceae

Synonyms: *Cicuta major* Lam., *Cicuta officinalis* Crantz, *Conium cicuta* Neck., *Conium maculosum* Pall., *Coriandrum cicutum* Crantz, *Coriandrum maculatum* (L.) Roth, *Selinum conium* (Vest) E.L. Krause, *Sium conium* Vest.

English name: Poison hemlock

Russian name: Болиголов пятнистый (Boligolov pyatnistyy)

Uzbek name: Sasik alaf

Kyrgyz name: Уу балдыркан (Uu baldyrkan)

Description: Herbaceous biennial. Stems 60–200 cm with red-brownish spots on lower portion. Basal leaves triangular in outline, petiolate, 30–60 cm long, tripinnatisect; primary and secondary segments petiolulate, tertiary segments sessile, oblong-ovate, pinnatifid. Inflorescence a compound umbel with 10–20 rays; bracts 4, lanceolate, acute; bracteoles 3–7, connate at the base. Petals 5, obcordate, white. Fruit a 2-seeded schizocarp, 3–3.5 mm long, nearly orbicular to ovate, with wavy ribs.

Other distinguishing features: Flowers have sharp smell.

Phenology: Flowers in June, fruits in July.

Reproduction: By seeds.

Distribution: All provinces of Kyrgyzstan; Toshkent, Samarqand, and Surxondaryo provinces of Uzbekistan.

Habitat: In forest glades, long-used animal corrals in the tallgrass-meadow belt of mountains, and in valleys.

Population status: Common, forming dense groups.

Traditional use: Aboveground parts of the plant and seeds are used. The plant is poisonous. It is used as a sedative, anticonvulsant, and analgesic, and to treat chorea, epilepsy, pertussis, migraine headaches, cancer, and uterine fibroids (Khalmatov 1964).

Documented effects: Extracts and plasters from this plant are rarely used externally as analgesics (Khalmatov 1964). The plant contains piperidine alkaloids that are toxic to humans and animals. These alkaloids have also been shown to cause congenital birth defects in goats and pigs (Panter et al. 1985a, b). These alkaloids have 2 modes of action. The first is similar to curare, which effects neuromuscular function and can cause respiratory failure. The second action effects the autonomic ganglia and can cause salivation, mydriasis, and tachycardia, followed by bradycardia and occasionally rhabdomyolysis and acute tubular necrosis (Frank et al. 1995; Lopez et al. 1999).

Phytochemistry: The roots contain up to 0.042 % total alkaloids, with the stems up to 0.065 % and leaves up to 0.1 %. The fruits contain up to 1 % total alkaloids, but sometimes unripe fruits contain up to 2 % (with 50 % of it being coniine). Other alkaloids include conhydrine, pseudoconhydrine, γ-coniceine, and methyl-coniine. The above parts also contain essential oils (mainly terpenes), vitamin C, carotene, and caffeic acid. Quercetin and kaempferol have been isolated from the flowers (Khalmatov 1964; Lopez et al. 1999).

▲**Codonopsis clematidea (Schrenk) C.B. Clarke**
Photos: Sasha Eisenman

►**Conium maculatum L.**
Photos: *top*: Clinton Shock; *bottom*: Alim Gaziev

▼**Cnicus benedictus L.**
Photos: Bazar Dovletov

Clematis orientalis L. Photos: Alexander Naumenko►

Convolvulus arvensis L. – Convolvulaceae

Synonyms: *Convolvulus chinensis* Ker Gawl., *Convolvulus sagittifolius* (Fisch.) T. Liou & Ling.

English name: Field bindweed

Russian name: Вьюнок полевой (V'yunok polevoy)

Uzbek name: Kuy pechak

Kyrgyz name: Чырмок (Chyrmok)

Description: Herbaceous perennial. Stems twining or prostrate. Leaves alternate, long-petiolate, ovate to oblong; apex acute, obtuse to rounded; base usually hastate. Inflorescences axillary cymes, 2–3-flowered, or flowers solitary; long-pedunculate. Corolla funnelform, up to 3 cm in diameter, pink or white. Stamens 5. Stigmas 2. Fruit a smooth, spherical capsule. Seeds 3–4 mm long, brownish or black.

Other distinguishing features: Flowers plicate when young.

Phenology: Flowers in May-September, fruits in June-September.

Reproduction: By seeds and rhizomes.

Distribution: In all provinces of Kyrgyzstan and Uzbekistan.

Habitat: In abandoned fields and waste grounds.

Population status: Common, forming dense groups.

Traditional use: The roots, leaves, stems, and flowers are used in folk medicine. The powdered leaves are used to treat wounds, cuts, and bruises. Juice from the leaves mixed with cow fat is used to treat lung and ear diseases. The root is used a laxative. A decoction of the herb is used to wash wounds and to treat skin ulcers, fungal skin diseases, and scabies. Avicenna used this species to treat asthma, lung disease, chest pains, liver and spleen diseases, and as a choleretic remedy (Khalmatov 1964).

Documented effects: In experiment on animals, this plant species had hypotensive, antispasmodic, anti-inflammatory, and styptic properties (Plant Resources of the USSR 1985). A methanol extract of the plant induced a dose-dependent relaxation of duodenal smooth muscle in rabbits (Atta and Mouneir 2004). Mice fed high doses of the plant died or had severe hepatic necrosis and gastritis after 4–7 days. Mice fed low doses of the plant had no clinical disease or large lesions, but developed mild multifocal hepatitis and gastritis (Schultheiss et al. 1995).

Phytochemistry: All parts of the plant contain alkaloids. The roots contain up to 5 % resins. The resins contain convolvine, jalapine, convolvuline, and caffeic acid. The aboveground parts contain flavonoids (quercetin and kaempferol) and caffeic acid. The leaves contain carotene and vitamin C (Khalmatov 1964). Plants from a pasture in Colorado (USA) were found to contain the tropane alkaloids tropine, pseudotropine, and tropinone, and the pyrrolidine alkaloids cuscohygrine and hygrine (Todd et al. 1995).

Convolvulus subhirsutus Regel & Schmalh. – Convolvulaceae

Synonyms: *Convolvulus chondrilloides* Boiss. var. *sericeus* Kuntze, *Convolvulus dorychium* ssp. *subhirsutus* (Regel & Schmalh.) Saad, *Convolvulus tschimganicus* Popov.

English name: Unknown

Russian name: Вьюнок жёстковолосистый, Вьюнок шерстистый (V'yunok zhyostkovolosistyy, V'yunok sherstistyy)

Uzbek name: Mingbosh

Kyrgyz name: Туктуу чырмоок (Tuktuu chyrmook)

Description: Hairy perennial herb to 40–100 cm with a thick taproot. Stems multiple, erect, with spreading branches. Leaves alternate, simple, elliptic, elongate-lanceolate or obovate, 1.5–10 cm long, narrowly acute, margins entire; petioles 0.5–2 cm long. Inflorescence a long-branched dichasium. Flowers pink, lilac, or rarely white. Corolla funnelform, 1–2.3 cm long. Fruit an ovoid capsule, 4–7 mm long, 1-seeded, glabrous. Seeds oviform or flat-elliptical, brown.

Other distinguishing features: Sepals are recurved when plant is fruiting. Seeds velutinous.

Phenology: Flowers in June-July, fruits in July-August.

Reproduction: By seeds.

Distribution: Toshkent, Andijon, Farg'ona, Samarqand, Buxoro, and Surxondaryo provinces of Uzbekistan; all of Kyrgyzstan.

Habitat: The adyr and tau zones. Shallow soil, more rarely on shallow-soiled, stony slopes.

Population status: Common, usually found in small populations.

Traditional use: In the folk medicine of Tajikistan, a decoction of the seeds is drunk to treat gastrointestinal diseases. An infusion of the herb is used as an analgesic, anticonvulsant, to heal wounds, and to treat asthma and lung tuberculosis (Khalmatov et al. 1984; Khodzhimatov 1989).

Documented effects: The alkaloids convolvine and convolamine have anesthetic properties, produce irritation of the mucous membranes of the eyes, and large doses can paralyze the central nervous system. Due to high toxicity they are not used in medical practice. Their derivatives, convocaine and tropacin, were developed as preparations for medical use. *Tropacin* is used to treat Parkinson's disease, spastic paresis and other diseases following muscle tone increase. *Tropacin* is also recommended to treat ulcers, bronchial asthma, other cases of spasms of the smooth muscular system, and poisoning with phosphorganic compounds (Mashkovskii 1953; Khalmatov et al. 1984; Khodzhimatov 1989).

Phytochemistry: The herb contains alkaloids (convolvine, convolamine, convolidine, phyllalbine, phyllalbine N-oxide, etc.) and the aminoalcohol nortropine (Razzakov and Aripova 2004; Gapparov et al. 2007). Roots collected at the end of the growing season in a Toshkent suburb (Kaplanbek) contained 4.1 % total alkaloids. The aboveground parts collected in the beginning of the growing season (March) contained 2.08 % total alkaloids (Yunusov 1974).

Cousinia lappacea Schrenk – Asteraceae

Synonyms: None

English name: Unknown

Russian name: Кузиния репейниковидная (Kuziniya repeynikovidnaya)

Uzbek name: Unknown

Kyrgyz name: Уйгактай кокуй тикен (Uygaktay kokuy tiken)

Description: Herbaceous perennial. Stems numerous, erect, 40–70 cm tall, up to 2 cm in diameter at base, usually pinkish or purple. Leaves soft, gray-hairy; basal leaves petiolate, obovate, margins spinose-toothed; cauline leaves oblanceolate, finely prickly-toothed, sessile, densely arranged. Inflorescences oviform capitula, 12–13 mm long, 5–6 mm wide, with 4–5 flowers; involucral bracts 25–30; outer bracts closely appressed, ovate, apices acuminate-hooked. Disc flowers purple; ray flowers lacking. Fruits obovoid achenes, 6 mm long, 4 mm wide, smooth.

Other distinguishing features: Inner involucral bracts purple at apices, ending with a thin hooked spine. Receptacle with smooth bristles.

Phenology: Flowers in June, fruits in July.

Reproduction: By seeds.

Distribution: Chuy, Naryn, and Jalal-Abad provinces of Kyrgyzstan; Samarqand and Farg'ona provinces of Uzbekistan.

Habitat: On stony slopes of the middle mountain belt.

Population status: Common, found as single individuals.

Traditional use: An infusion of the herb and roots is used to prevent tumor growth and to treat gastrointestinal ulcers (Plekhanova et al. 1985).

Documented effects: No data.

Phytochemistry: The aboveground parts contain monosaccharides (glucose and fructose), oligosaccharides (5.1–5.5 %), and pectic substances. The roots contain water-soluble polysaccharides (2.18–2.78 %; Plekhanova et al. 1985).

Cousinia umbrosa Bunge – Asteraceae

Synonyms: None

English name: Unknown

Russian name: Кузиния теневая (Kuziniya tenevaya)

Uzbek name: Okboshtikon

Kyrgyz name: Колоко кокуй тикен (Koloko kokuy tiken)

Description: Herbaceous perennial. Stems numerous, 60–120 cm tall, deeply grooved, branched above. Leaves green and glabrous on adaxial side, abaxial side grayish felted; basal leaves very large, petiolate, obovate, cordate, margins irregularly dentate; cauline leaves similar but becoming gradually reduced towards apex. Inflorescences ovoid capitula arranged in a panicle; involucral bracts oblong, bases appressed, spreading above and tapering to incurved hooks. Disc flowers 10–12, pink; ray flowers lacking. Fruits obovoid achenes, 6 mm long, 3 mm wide, light brown with dark spots.

Other distinguishing features: Outer involucral bracts with 1 or 2 pairs of glandular hairs; inner bracts linear, apex acuminate. Receptacles with smooth bristles.

Phenology: Flowers in June-July, fruits in July-August.

Reproduction: By seeds.

Distribution: Chuy and Jalal-Abad provinces of Kyrgyzstan; Toshkent and Qashqadaryo provinces of Uzbekistan.

Habitat: In shady places in the foothills and the lower mountain belt.

Population status: Common, found in dense groups.

Traditional use: An infusion and decoction of the aboveground and underground parts is applied to treat stomach ulcers and hypoxia in mountainous conditions, and is also used as a general tonic during recovery from a variety of diseases (Turdumambetov 1995).

Documented effects: No data.

Phytochemistry: The roots and aboveground plant parts contain oligo- and polysaccharides (fructan), pectic substances, and hemicellulose (Turdumambetov 1995; Turdumambetov et al. 2007). The fruits contain lipids, hydrocarbons, triterpene alcohols, sterols, mono- and diacylglycerides, etc. The predominant fatty acids are 16:0, 18:1, and 18:2 (Ul'chenko et al. 1999)

▲**Convolvulus arvensis L.**
Photos: *top*: Sasha Eisenman; *bottom*: Mary Backlund

▲**Convolvulus subhirsutus Regel & Schmalh.**
Photos: Evgeny Davkaev

▲**Cousinia umbrosa Bunge**
Photos: Evgeny Davkaev

◄**Cousinia lappacea Schrenk**
Photos: Georgy Lazkov

Crambe kotschyana Boiss. – Brassicaceae

Synonyms: *Crambe cordifolia* ssp. *kotschyana* (Boiss.) Jafri, *Crambe cordifolia* var. *kotschyana* (Boiss.) O.E. Schulz, *Crambe palmatifida* Regel & Schmulh., *Crambe sewerzowii* Regel.

English name: Colewort

Russian name: Катран (Katran)

Uzbek name: Katron

Kyrgyz name: Кочи катраны (Kochi katrany)

Description: Perennial herb with a thick, fleshy root. Stems 50–150(–250) cm tall, single or multiple, coarse-hairy; branches spreading. Basal leaves with long petioles up to 30 cm long, blades cordate-reniform to ovate-oblong, up to 50 cm wide, roughly lobed, coarsely toothed, coriaceous, covered with rough prominent hairs; upper leaves alternate, ca. 1 cm long. Inflorescences racemes arranged in large panicles. Petals 4, obovate, white. Stamens 6, tetradynamous, the longer 4 toothed. Fruit an elongate-spherical silique, 6–7 mm long.

Other distinguishing features: Seeds 3–4 mm in diameter, pale brown.

Phenology: Flowers and fruits from the end of March until mid-June.

Reproduction: Propagates by seeds, it can be easily cultivated.

Distribution: Karakalpakstan autonomous republic, Toshkent, Andijon, Farg'ona, Jizzax, Samarqand, Qashqadaryo, and Surxondaryo provinces of Uzbekistan; Chuy, Talas, Osh and Jalal-Abad provinces of Kyrgyzstan.

Habitat: The adyr and tau zones. On soils with a high diversity of minerals and shallow-soiled slopes with rocky debris.

Population status: Common.

Traditional use: Seeds are used to treat respiratory tract catarrh. Roots are used by veterinarians to treat gastric diseases in camels. The roots, baked or boiled, are used for food by people (Khalmatov 1964).

Documented effects: No data.

Phytochemistry: Roots contain 18–19 % sugar (monosaccharides 10.54 %, disaccharide 9.2 %), 39.62 % starch. Seeds contain up to 40 % oils (Khalmatov 1964). The aboveground parts contain a variety of lipids of which a high proportion is palmitic acid. The seeds have high erucic and linolenic acid contents (Bekker et al. 2003).

Crataegus altaica (Loudon) Lange – Rosaceae

Synonyms: *Crataegus chlorocarpa* Lenne & K. Koch, *Crataegus korolkowii* L. Henry, *Crataegus purpurea* var. *altaica* Loudon, *Crataegus sanguinea* var. *incisa* Regel, *Crataegus sanguinea* var. *inermis* Kar. & Kir., *Crataegus wattiana* var. *incisa* C.K. Schneid.

English name: Altai hawthorn, Altai mountain thorn

Russian name: Боярышник алтайский (Boyaryshnik altayskiy)

Uzbek name: Dulana zardak, Sarik dulana

Kyrgyz name: Алтай долоносу, Сары долоно (Altay dolonosu, Sary dolono)

Description: Shrubby tree with multiple trunks, up to 3–5 m tall. Bark smooth, mostly gray; 1-year old stems red-brown or green-brown, smooth, with many white lenticels; older branches gray-orange with large lenticels; some branches with short (1–1.5 cm), thick spines in the leaf axils. Leaves petiolate, broadly triangular, oval or circular, entire or 3–7-lobed, coarsely toothed. Inflorescence corymbiform with 10–30 flowers. Flowers up to 1.9 cm in diameter, with 5 white petals. Fruits spherical pomes, yellow or dark-brown with 3–5 seeds.

Other distinguishing features: Fruit 8–10 mm in diameter. Leaves glabrous or slightly pubescent.

Phenology: Flowers in May-June, fruits in August-September.

Reproduction: By seeds.

Distribution: All provinces of Kyrgyzstan; Toshkent, Farg'ona and Qashqadaryo provinces of Uzbekistan.

Habitat: Along forest edges, in the understory, and in meadows.

Population status: Common, found as single plants.

Traditional use: Flowers, fruits, leaves and bark are used in folk medicine for treatment of various illnesses. A decoction of leaves and tea from dried flowers and fruits are taken to treat hypertension, dizziness, tachycardia, insomnia, heart diseases, and common colds. Fresh fruits are recommended as a laxative (Khodzhimatov 1989).

Documented effects: Extracts from fruits or tinctures from flowers are used to treat cardiovascular diseases and hypertension, to tone cardiac muscles, and as a sedative. This hawthorn extract of is part of *Cardiovalen*, which is used to treat rheumatic heart disease, cardiosclerosis, stenocardia, and vegetative neurosis (Khodzhimatov 1989).

Phytochemistry: The bark contains tannins and the fruits contain up to 260 mg% vitamin C (Zapryagaeva 1964). Flowers contain flavonoids (hyperoside, quercetin, vitexin, and vitexin-ramnoside), triterpene saponins (ursolic and oleanolic acids) and essential oil. The fruits contain flavonoids, saponins, tannins, polysaccharides, fatty oil, and phenolcarbonic acids (chlorogenic and caffeic; Khodzhimatov 1989).

Crataegus songarica K. Koch. – Rosaceae

Synonyms: *Crataegus fischeri* C.K. Schneid.

English name: Dzhungarskei hawthorn

Russian name: Боярышник сонгорский (Boyaryshnik songorskiy)

Uzbek name: Dulana

Kyrgyz name: Сонгор долоносу (Songor dolonosu)

Description: Shrubby tree, up to 4–5 m tall with multiple trunks, each 5–9 cm in diameter. Bark reddish-gray to blackish with small cracks; young twigs green, glabrous or slightly hairy with spines up to 1.7 cm long; 1-year old twigs reddish-brown; older branches brown with smooth bark. Leaves petiolate, 6–8 cm long, 5–6 cm wide, broadly triangular to almost circular, 5–7-lobed, margins dentate. Inflorescence corymbiform with 28–35 flowers. Flowers up to 1.8 cm in diameter with 5 white petals. Fruits round pomes, dark red, with 2–3 seeds.

Other distinguishing features: Differs from other *Crataegus* species by having smooth, brown older branches and reddish-brown year-old twigs.

Phenology: Flowers in May-June, fruits in August-September.

Reproduction: By seeds.

Distribution: Chuy, Jalal-Abad, and Osh provinces of Kyrgyzstan; Toshkent, Samarqand, Qashqadaryo and Surxondaryo provinces of Uzbekistan.

Habitat: Along river valleys and on mountain slopes.

Population status: Common, found in loosely arranged groups.

Traditional use: A tea made of dried flowers and infusions of dried fruits are used to treat heart pain, dyspnea, hypertension, and gastrointestinal diseases (Khodzhimatov 1989).

Documented effects: Clinical tests of tincture of this hawthorn have yielded positive results when used against the active form of rheumatism (Kuchin 1955). An extract from the fruits and tinctures from the flowers are also used to treat cardiovascular diseases by strengthening the heart muscle, as a sedative, and to treat hypertension (Khodzhimatov 1989).

Phytochemistry: Fruits contain vitamin C, carotene, tannins (0.53–0.85 %), and the catechins epicatechin and leucoanthocyanidin (Petrova 1972).

Dactylorhiza umbrosa (Kar. & Kir.) Nevski – Orchidaceae

Synonyms: *Dactylorchis umbrosa* (Kar. & Kir.) Wendelbo, *Orchis magna* Czerniak, *Orchis orientalis* ssp. *turkestanica* Klinge, *Orchis umbrosa* Kar. & Kir.

English name: Unknown

Russian name: Дактилориза теневая, Яртышник теневой (Daktiloriza tenevaya, Yartyshnik tenevoy)

Uzbek name: Saleeb

Kyrgyz name: Колоколуу арала (Kolokoluy arala)

Description: Perennial herb, 30–50(–80) cm high, with a cluster of 1–6 finger-like tubers. Stems erect, thick, hollow. Leaves usually 6–7 in number, lanceolate to linear-lanceolate, apex acute, parallel veined; basal leaves 10–20 cm long, 2–5 cm wide; upper leaves smaller and more narrow, usually positioned up to the base of spike. Inflorescence spiciform, 5–18 cm long, from elongate-oviform to short- or long-cylindrical, dense and many-flowered. Flowers zygomorphic, with 6 tepals in 2 whorls, lilac- or violet-purple; lower inner tepal forming a large lip with a white "ω"-shaped blotch, spurred. Fruit a capsule with very small seeds.

Other distinguishing features: Leaves lack spots. Bracts narrow-lanceolate, green or violet.

Phenology: Flowers from May to the end of July. Bears fruit in July-August.

Reproduction: By seeds.

Distribution: Toshkent and Samarqand provinces of Uzbekistan; Chuy and Ferghana valleys and Talas province of Kyrgyzstan.

Habitat: The adyr and tau zones. Wet, boggy places, meadows, edges of brooks, river banks, tugai, wet slopes, and ravines from 700 to 2.800 m above sea level.

Population status: Uncommon, found as single individuals.

Traditional use: Avicenna recommended a decoction of the tubers to treat gout, paralysis, convulsions, and joint pain. In Russian folk medicine the plant is used as a diuretic, as well as to treat fevers and gynecological diseases. The crushed tubers are mixed with lard and applied to abscesses. Fresh tuber is applied to the teeth to treat toothaches and is used to stimulate hair growth. Tadjiks use a decoction of the tubers to treat hand convulsions, paralysis, stomach catarrh, kidney stones, and to stimulate blood production. The boiled roots are used to rejuvenate the elderly and people with lung tuberculosis. A tea made of the fried, crushed tubers is use to treat coughs, inflammation of the respiratory tract, to increase energy and to calm nerves. The roots boiled in milk is used to treat coughs, impotence, and gastrointestinal tract weakness. A powder of the tubers mixed with honey is used as a tonic (Khodzhimatov 1989).

Documented effects: A decoction of the tubers is used in modern medicine to coat the digestive tract as a treatment for gastritis, enterocolitis, and other gastrointestinal diseases. It is also used as an enema to treat diarrhea in children. Experiments have shown that this plant has anti-inflammatory activities (Khalmatov et al. 1984; Khodzhimatov 1989).

Phytochemistry: The main component of the tubers is a water soluble mucilage, which contains starch, sugars, mineral salts, bitters and proteins, essential oils, etc. During the fruit bearing stage, polysaccharide content in tubers reaches 68.48 % (Khalmatov et al. 1984). The aboveground parts of many species in the genus *Dactylorhiza* contain the glycoside loroglossine. This species was found to contain traces of alkaloids and saponins as well as lactone compounds in the leaves (Khodzhimatov 1989).

◄**Crambe kotschyana Boiss.**
Photos: Alexander Naumenko

▼**Dactylorhiza umbrosa (Kar. & Kir.) Nevski**
Photos: *top*: Stefano Doglio; *bottom*: Ilya Raskin

Crataegus altaica (Loudon) Lange Photos: Alexander Naumenko ►

▼**Crataegus songarica K. Koch.**
Photos: Alexander Naumenko

Datura stramonium L. – Solanaceae

Synonyms: *Datura tatula* L., *Stramonium spinosum* Lam.
English name: Jimsonweed
Russian name: Дурман обыкновенный, Дурман вонючий (Durman obyknovennyy, Durman vonyuchiy)
Uzbek name: Bangi divana
Kyrgyz name: Кадимки чочко жангак (Kadimki chochko zhangak)
Description: Herbaceous annual, up to 1 m tall. Stems erect, branching. Leaves alternate, petiolate, 8–20 cm long, up to 15 cm wide, ovate, apex acuminate, slightly lobed, margins roughly dentate. Flowers singular, in leaf and branch axils. Calyx tubular, 5-sided, up 6 cm long. Corolla white, up to 12 cm long, tubular-funnelform, 5-sided. Fruit an oviform capsule up to 5 cm long and 2.5 cm wide, densely covered with hard prickles up to 1 cm long. Seeds up to 0.5 cm long, kidney-shaped, nearly black, finely tuberculate.
Other distinguishing features: Capsule splits open into 4 valves when ripe and can contain up to 800 seeds.
Phenology: Flowers in June-September, fruits in July-September.
Reproduction: By seeds.
Distribution: Agriculture zones in all provinces of Kyrgyzstan; Karakalpakstan autonomous republic, Toshkent, Samarqand, Andijon, Sirdaryo, Jizzax and Surxondaryo provinces of Uzbekistan.
Habitat: In vegetable gardens, in orchards and waste places, near forested areas, and along rivers and canals.
Population status: Common, sometimes forming dense groups.
Traditional use: This species has been widely used since ancient times. Avicenna said that this plant makes you drunk, is too dangerous for the brain, and is the enemy of the heart. Beruni wrote that half a gram of the seeds can make you drunk and 4.2 g can kill you. A decoction of the seeds is used as a gargle for people with tooth- and headaches, as a painkiller and sedative, and to treat fevers, neuralgia, rheumatism, and radiculitis (Khodzhimatov 1989). Oil from the seeds is used to treat hemorrhoids and the leaves are laid over the eyes to treat eye aches (Khalmatov et al. 1984).
Documented effects: The plant is poisonous. Preparations of this species is used as an antispasmodic, and mainly used to treat bronchial asthma, neuralgia, and convulsions. The leaves are a component of the preparations *Asthmatin* and *Asthmatol*. A liquid extract of the leaves is used in the preparation *Solutan*, which is used to treat bronchial asthma and bronchitis (Khodzhimatov 1989). In general, preparations of this species are used as an antispasmodic to treat bronchial asthma, stomach ulcers, cholecystitis, colitis, spastic constipation, cardio-vascular diseases, and bradycardia. Preparations derived from the plant also used as a preventive treatment for sea and air sickness (Turova and Sapozhnikova 1984). The active compounds are hyoscyamine and scopolamine. The basic pharmacological action of hyoscyamine is to block N-cholinoreceptors. Hyoscyamine increases heart rate, but decreases saliva secretion, gastric and sweat production, secretions of the pancreas, and the tonus of smooth muscles in the bronchial tubes and abdominal cavity. Hyoscyamine causes prolonged mydriasis. It tones and increases the activity of the respiratory center. In experiments with frogs a tincture reduced heart beat amplitude (Turova and Sapozhnikova 1984).
Phytochemistry: The whole plant contains alkaloids, with the main alkaloids being hyoscyamine, atropine, and scopolamine (Khalmatov et al. 1984). The leaves and stems also contain coumarins (scopoletin, esculetin, and esculin), tannins, essential oils, and carotene. Seeds contain up to 25 % fatty oil, containing linoleic, oleic, palmatic, stearic, and lignoceric acids (Khodzhimatov 1989).

Daucus carota L. – Apiaceae

Synonyms: *Daucus bactrianus* Bunge, *Daucus exarmatus* Korovin, *Daucus pulcherrimus* (Willd.) Koch ex DC., *Carota sativa* Rupr., *Carota sylvestris* (Mill.) Rupr., *Caucalis carota* (L.) Crantz, *Caucalis daucus* Crantz.

English name: Wild carrot, Queen Anne's lace

Russian name: Морковь дикая (Morkov' dikaya)

Uzbek name: Yovoyi sabzi

Kyrgyz name: Жапайы сабиз (Zhapayy sabiz)

Description: Herbaceous biennial with a thin taproot. Stem up to 1 m high, branching, rough due to scattered, stiff hairs. Basal leaves petiolate, bipinnatisect with narrow, lanceolate or linear sections, sometimes glabrous on adaxial side, abaxial side hairy along veins; stem leaves alternate, becoming sessile and sheathing. Inflorescence a compound umbel with many rays, up to 10 cm wide, subtended by pinnate bracts. Petals white. Fruits schizocarps with 2 one-seeded mericarps, oval or oblong, flattened, covered with short bristles along and between the ribs.

Other distinguishing features: Umbel curving inwards in fruit and becoming spherical.

Phenology: Flowers in May-June, fruits in July.

Reproduction: Propagates by seeds. One plant bears up to 4,000 seeds.

Distribution: All regions of Uzbekistan and Kyrgyzstan.

Habitat: The chul, adyr and tau zones. Occurs fairly often as a weed in irrigated regions, predominantly in shaded places.

Population status: Common, found as single individuals.

Traditional use: Avicenna used wild carrot fruits as a diuretic. For a long time, the essential oils of wild carrot fruits have been used in medicine for making astringent and spicy extracts. The plant extract has been used as vermifuge and purgative (Khalmatov et al. 1984).

Documented effects: In the past, flavonoids isolated from the fruits were made into a preparation called *Daukarin*. This was in used in cardiology to improve coronary blood circulation as well as chronic coronary disease (Khalmatov et al. 1984). Presently the preparation is not made because more active alternatives have been developed. An extract of the fruits of wild carrot, contained in the preparation *Urolesan*, is used in medicine. This preparation has been approved for the treatment of liver and kidney diseases, for acute and chronic cholestasis, and different kinds of kidney and gallbladder stones (Gammerman et al. 1990). An extract of carrot root exhibited hepatoprotective activity in mice (Bishayee et al. 1995). Compounds isolated from the seeds showed significant inhibition of cyclooxygenase (COX) enzymes (Momin et al. 2003).

Phytochemistry: Wild carrot seeds contained up to 2.5 % essential oils, which consisted of 17 substances including 1-α-pinene, myrcene, bergamotene, β-bisabolene, caratol and asarone. Besides essential oil, the seeds contained flavonoids, coumarins, steroidal compounds and fatty oil. The roots also contain essential oils, which consisted of asarone, caratol, bisabolene and 5–9 % carotene. Large amounts of pyrrolidine and daucene were found in essential oils extracted from the herb. In the herb and flowers flavonoids, coumarins, anthocyanins, as well as large amounts of carotenoids, vitamins in the groups B and C, pantothenic acid, anthocyanidin, essential oils, umbelliferone and sugars were found (Gammerman et al. 1990).

Delphinium confusum Popov – Ranunculaceae

Synonyms: None

English name: Larkspur

Russian name: Живокость спутанная (Zhivokost' sputannaya)

Uzbek name: Isfarak

Kyrgyz name: Татыш туктуу бутоо (Tatysh tuktuu butoo)

Description: Herbaceous perennial. Stems 30–70 cm tall, tomentose; leaves aggregated on the lower half of stem. Leaves petiolate, the blade circular to kidney-shaped in outline, 5–13 cm long, 7–20 cm wide, 3-lobed with sinuses half-way into leaf blade; middle lobe elongate-obovate, with 3–5 lobules; lateral lobes of leaf with 2–3 lobules; all lobes with unequal triangular-lanceolate teeth. Inflorescence a multi-flowered raceme; bracts broadly lanceolate. Flowers with 5 dark-violet tepals, upper tepal with a spur at the base. Fruit a follicetum with 3 follicles.

Other distinguishing features: Spur straight, positioned almost horizontally, curving at the end.

Phenology: Flowers in June-August, fruits in July-September.

Reproduction: By seeds.

Distribution: Chuy, Naryn, Talas, and Ysyk-Kol provinces of Kyrgyzstan; in the western Tian-Shan (Akhangaran region) and the Alai Range in Uzbekistan.

Habitat: In alpine meadows, spruce forests, juniper stands, on stony slopes of mountains, in feather-grass steppes, and in grassy meadows.

Population status: Common, found in small groups.

Traditional use: A decoction of this plant is used to treat intestinal disorders, to increase muscle tone, and as an antiparasitic treatment for cows (Altimishev 1991).

Documented effects: In modern medicine a decoction of the stems and leaves is used in medicine to disinfect animals. Finely ground seeds mixed with butter is used to treat pediculosis (lice infestation). Tablets of condelphine are used to treat psycho-neurological diseases. The compounds delsemine and mellictine are used as anesthesia during surgical procedures (Altimishev 1991). The alkaloid condelphine has an activity similar to curare. Physicians use tablets of 0.025 g to treat conditions of excess skeletal muscle contraction, Parkinson's disease, multiple sclerosis, spastic and traumatic paralysis, etc. This preparation cannot be used by patients with conditions of reduced muscle contraction, liver and kidney diseases, or heart decompensation (Khalmatov et al. 1984).

Phytochemistry: Flowers, stems, and leaves contain condelphine, delphiline, delatine, delsine, delsoline, isobaldine, etc., as well as aconitic acid. During the bud stage buds contain up to 0.8 % alkaloids, and during the flowering stage flowers contain up to 2 % alkaloids. The roots contain up to 1 % alkaloids (Dzhakupova 1968; Vaisov and Yunusov 1987; Narsullaev et al. 1989; Altimishev 1991).

Delphinium semibarbatum Bienert. – Ranunculaceae

Synonyms: *Delphinium bitternatum* var. *leiocarpum* Freyn., *Delphinium hybridum* var. *sulphureum* Regel, *Delphinium zalil* Aitch.

English name: Unknown

Russian name: Живокость полубородатая (Zhivokost' poluborodataya)

Uzbek name: Isfarak

Kyrgyz name: Жарым сакалчалуу бутёё (Zharym sakalchaluu butyoyo)

Description: Perennial herb, 35–70 cm high. Stem unbranched or branched, glabrous or short-pubescent on lower part of stem. Basal leaves long-petiolate; leaf blades palmatifid with 5 segments with petiolules; segments tripartite with long, narrow-linear, glabrous or slightly villose lobes. Inflorescence racemiform. Flowers with 5 bright yellow, obovate tepals, the upper tepal with a spur at the base. Fruits with 3 follicles, glabrous. Seeds tiny, 3-edged, paleaceous.

Other distinguishing features: Pedicels are glabrous up to the calyx. Follicle with 3 sharply protruding longitudinal ribs and slightly-protruded ribs between them.

Phenology: Flowers in May-June, fruits in July.

Reproduction: Reproduces by seeds.

Distribution: Toshkent, Namangan, Andijon, Farg'ona, Samarqand, Qashqadaryo, Buxoro and Surxondaryo provinces of Uzbekistan; Chuy, Osh and Jalal-Abad provinces of Kyrgyzstan.

Habitat: The adyr and tau zones. On dry, mixed-grass steppes and loess slopes. It is a typical component of the plant associations in these zones.

Population status: Common.

Traditional use: A decoction of the herb mixed with barley flour is used to treat various tumors. The ashes are applied to treat eczema and scabies (Khalmatov 1964). A decoction of the herb is used to treat fever, flu, sore throat, pertussis, stomach diseases, burns, and used as an anticonvulsive. It is also used as an insecticide to kill flies and cockroaches. It is important to note that since the plant is very toxic, it should be used with extreme caution (Kulikov 1975; Khalmatov et al. 1984; Khalmatov and Kosimov 1994).

Documented effects: The preparation *Delsemine* has a muscle relaxing effect and has been used during surgery (Khodzhimatov 1989). Intravenous injection of the alkaloid delsemine at 0.5–2 mg/kg, was used to cause relaxation during surgery. At doses of 5–6 mg/kg, delsemine was used to stop autonomous breathing. Presently, delsemine is not used in medical practice. Intravenous injection of the alkaloid licoctonine caused general calming with muscle relaxation, respiratory depression, and decrease of blood pressure while pulse rate remained the same (Tulyaganov et al. 1976). Intravenous introduction of the alkaloid methyllycaconitine had a curariform effect on narcotized animals. Methyllycaconitine provoked brief hypotensive effects. The alkaloid is used for spastic paresis of pyramidal character, postencephalitic arachnoencephalitis, and spinal arachnoiditis (Dozortseva 1958, 1959). A preparation of methyllycaconitine, *Mellictin*, is used to treat Parkinson's diseases and cerebral palsy (Khalmatov et al. 1984).

Phytochemistry: All plant parts contain alkaloids. Before flowering, the top portion of the plant contained 0.25 % total alkaloids and at flowering, 0.09 % total alkaloids. Delsemine, licoctonine, delphirine, methyllycaconitine, and anthranoyllycoctonine were isolated from the total alkaloids (Yunusov 1974). The flowers contained up to 4 % pigments. From this, the flavonoids isorhamnetin, quercetin, and their glycosides were isolated (Khalmatov et al. 1984).

▲**Datura stramonium L.** Photos: *left* and *center*: Vadim Prokhorov; *right*: Alim Gaziev

▼**Daucus carota L.** Photos: *top*: Sasha Eisenman; *center*: Benjamin Zwittnig; *bottom*: Evgeny Davkaev

▲**Delphinium confusum Popov** Photo: Stefano Doglio

►**Delphinium semibarbatum Bienert.** Photos: *top right*: Evgeny Davkaev; *lower right*: John B. Taft; *lower left*: Bazar Dovlet

Descurainia sophia (L.) Webb ex Prantl – Brassicaceae

Synonyms: *Hesperis sophia* (L.) Kuntze, *Sisymbrium sophia* L., *Sisymbrium tenuissimum* Kar. & Kir., *Sophia lobelii* Rupr.

English name: Flixweed

Russian name: Дескурения Софьи (Deskureniya Sof'i)

Uzbek name: Shuvaran, sassyk kapa

Kyrgyz name: София дескурениясы (Sofiya deskureniyasy)

Description: Herbaceous annual, grayish pubescent. Stem 10–90 cm tall, erect, unbranched or with spreading branches. Stem leaves alternate, bi- or tripinnatisect with linear, acute lobes; basal and lower leaves petiolate; upper leaves sessile. Inflorescence many-flowered raceme. Flowers 3–5 mm long with 4 sepals, 4 yellow petals, and 6 stamens, pedicellate. Fruit a silique, erect, slightly arcuate. Seeds light brown, 1–1.5 mm long, 0.5–0.75 mm wide.

Other distinguishing features: The silique has a prominent vein along the septum.

Phenology: Flowers and fruits from the beginning of April until the end of May.

Reproduction: By seeds.

Distribution: All regions of Uzbekistan and Kyrgyzstan.

Habitat: The chul, adyr, and tau zones. As a weed, near roads, in fields, and in pastures.

Population status: Common.

Traditional use: A decoction of the plant is recommended as a febrifuge to treat laryngeal diseases, measles, and smallpox and is also used as a hemostatic. Fresh leaves are used to heal flesh wounds and are thought to have antibacterial action (Khalmatov 1964). The herb is often used to treat diarrhea and dysentery. Essence from the fresh flowering plant is used in homeopathy (Ogolevitz 1951). A decoction of the aboveground parts is used to treat throat diseases and as an antipyretic for smallpox and measles. In veterinary medicine, a decoction of the roots is used for diarrhea and helminthosis in cattle and horses. In Tibetan medicine, the roots are used for treatment of anthrax and ergotism. A tincture is used as a antihelmintic, diuretic, and hemostatic for internal hemorrhages (Bekker et al. 2005). A decoction of the herb is promoted as and considered a stimulant in the Russian Far East (Mamedov 2005).

Documented effects: Experiments showed that a galenic preparation of this species reduced hypotension (Khalmatov 1964). An alcoholic extract of the seeds increases the tonus of muscles responsible for intestinal contractions. This extract is also used as a laxative to treat constipation (Khodzhimatov 1989).

Phytochemistry: Leaves contain 10 mg% carotene, seeds contain 27–30 % oils and 1.5 % sinigrin glycoside, which produce 0.8–0.9 % mustard essential oil after enzymic hydrolysis. The latter consists of 60 % benzyl isothiocyanate, 15 % allylisothiocyanate, and 5 % propenyl isothiocyanate (Khalmatov 1964). The seeds contain glucosides of quercetin, kaempferol, and isorhamnetin as well as sinapic acid, ethyl ester, and trimethoxyl-cinnamic acid (Wang et al. 2004a), as well as lipids consisting of hydrocarbons, esters of fatty acids and cyclic alcohols, triacylglycerides, epoxyacylglycerides, free fatty acids, triterpenols, sterols, diacylglycerides and monoacylglycerides. Linolenic, linoleic, arachic, and erucic acids were the main components of the total lipids and triglycerides (Bekker et al. 2005).

Dianthus superbus L. – Caryophyllaceae

Synonyms: Some consider Central Asian populations to be a distinct species, *Dianthus hoeltzeri* Winkl.

English name: Unknown

Russian name: Гвоздика Гельцера (Gvozdika Gel'tsera)

Uzbek name: Unknown

Kyrgyz name: Гельцер чеге гул ("Gel'tser chege gul")

Description: Herbaceous perennial with rhizomes. Stems of 2 kinds: non-elongated vegetative (non-reproductive) and reproductive, 15–60 cm tall. Leaves opposite, linear-lanceolate, 4–6 cm long, 2–4 mm wide, opposite blades connate at base, sheathing the stem. Sheath 2–4 mm long. Flowers solitary or in groups of 2–4. Calyx cylindrical, violet-tinged. Petals 5, light pink to dark pink, deeply fringed. Fruit a cylindrical capsule.

Other distinguishing features: Capsule longer than calyx.

Phenology: Flowers in June-August, fruits in July-September.

Reproduction: By seeds.

Distribution: All provinces of Kyrgyzstan; Toshkent Province of Uzbekistan.

Habitat: In the tallgrass-meadow and forest belts, in glades and meadow slopes, and in subalpine meadows.

Population status: Common, found as single plants.

Traditional use: In folk medicine a decoction of the aboveground parts is drank to treat heart diseases, gastrointestinal diseases, and uterine bleeding. The herb is also used to treat people bitten by rabid animals (Alimbaeva and Goncharova 1971). A decoction of the aboveground parts and roots is used to treat various uterine diseases (Khalmatov 1964).

Documented effects: No data.

Phytochemistry: All plant parts contain saponins (triterpenes), alkaloids, tannins, flavonoids, and lipids (Boguslavskaya et al. 1983; Plant Resources of the USSR 1985). Phytoecdysteroids have been isolated from the plant (Saatov et al. 1990).

Dictamnus angustifolius G. Don fil. ex Sweet – Rutaceae

Synonyms: *Dictamnus albus* ssp. *turkestanicus* Wint.

English name: Burning bush

Russian name: Ясенец узколистный (Yasenets uzkolistnyy)

Uzbek name: Togturbid

Kyrgyz name: Ичке жалбырактуу диктамнус (Ichke zhalbyraktuu diktamnus)

Description: Herbaceous perennial. Stems 50–100 cm high, long- or short-hairy, but often glabrous. Leaves odd-pinnate with 5–6 pairs of large oblong or elongated-elliptic leaflets; leaflets with serrate margins and narrow-winged petiolules. Inflorescence racemose-paniculate, strongly glandular. Corollas lilac-purple, 3.5–4.5 cm long. Staminal filaments hairy, style glabrous. Ovary hairy. Seeds 4–5 mm long, black, shiny, smooth.

Other distinguishing features: During dry weather, when the flowering plant is exposed to fire, it flares up, but the plant remains intact. Therefore people had named it "burning bush".

Phenology: Flowers in May-June, fruits in June-July.

Reproduction: By seeds.

Distribution: Toshkent and Farg'ona provinces of Uzbekistan; all of Kyrgyzstan.

Habitat: Among shrubs in the tau zone.

Population status: Uncommon, found as individuals.

Traditional use: Avicenna used the plant as a purgative, but noted that it was poisonous and when taken in large doses (6–7 g), caused nausea, vomiting, and even death due to excessive vomiting. Avicenna also noted that some physicians prescribed the plant decoction to treat paralysis. An infusion of the leaves and flowers is used to treat rheumatism in Central Asian folk medicine. An aromatic water is made out of the flowers, which is used in cosmetics for facial skin care (Khalmatov 1964). *Dictamnus angustifolius*, growing in the Xin Jiang province of China, has been used as an alternative for *D. dasycarpus* in the treatment of rheumatism, bleeding, itching, jaundice, chronic hepatitis, and skin diseases, and as an anti-inflammatory agent, febrifugal, and detoxicant drug (Wu et al. 1999a).

Documented effects: The root bark has antispasmodic, vermifugal, and antihysteric actions (Khalmatov 1964). The alkaloids dubinidine, evoxin, and skimmianine caused central nervous system depression resulting in sleep and then narcosis, when introduced in active doses into animals, and also had hypothermic action and increased pain threshold (Berezhinskaya and Trutneva 1959; Polievtsev 1962a, b, Polievtsev 1965; Polievtsev et al. 1967; Sadritdinov 1968). In clinical tests at doses 0.6–0.8 g/day the alkaloid dubinidine had good sedative effect, especially on patients with severe insomnia. However, it was not recommended for clinical use (Polievtsev 1965; Evdokimova and Kurmukov 1972). The alkaloid dictamine was toxic when injected intravenously at 0.05–0.055 mg, evoked convulsions of rear extremities, decreased respiration, and eventually caused death of animals from asphyxia (Kovalenko 1946). Flavonoids from this species showed choleretic, anti-inflammatory, and capillary strengthening activity (Komissarenko and Levashova 1988, 1989). A methanolic extract of the root bark of *Dictamnus angustifolius* showed significant vascular relaxing activity (Wu et al. 1999a).

Phytochemistry: Roots contain 0.21 % and seeds contain 0.025 % total alkaloids. Skimmianine, dictamine, dubamine, dubinidine (roots), evoxin (haploperin), and other alkaloids were isolated from the total alkaloids (Yunusov 1974). Limonoids and coumarins have been isolated from the root bark (Wu et al. 1999a). The plant also contains essential oils and seeds contain 18–21 % drying oils (Khalmatov 1964).

Dipsacus dipsacoides (Kar. & Kir.) Botsch. – Dipsacaceae

Synonyms: *Cephalaria dipsacoides* Kar. & Kir., *Dipsacus azureus* Schrenk.

English name: Teasel

Russian name: Ворсянка лазоревая (Vorsyanka lazorevaya)

Uzbek name: Unknown

Kyrgyz name: Когултур топчу баш (Kogultur topchu bash)

Description: Herbaceous biennial or perennial. Stems 60–140 cm tall, branching above. Basal leaves lanceolate, up to 40 cm long, margins entire; lower leaves sessile, 15–30 cm long, 4–5 cm wide, pinnatifid towards leaf base; uppermost leaves smaller, becoming lanceolate or linear, sometimes pinnatisect. Inflorescences nearly spherical heads; involucral bracts awn-like, stiff, prickly. Corolla tubular, 4-lobed, bright blue, pubescent on the outside, each with a stiff green bract. Fruits 4-sided achenes.

Other distinguishing features: Inflorescence heads 2.5–4 cm long, with involucral bracts only slightly shorter.

Phenology: Flowers in July, fruits in August-September.

Reproduction: By seeds.

Distribution: Chuy, Jalal-Abad, and Osh provinces of Kyrgyzstan; Toshkent, Andijon, Farg'ona, Samarqand and Surxondaryo provinces of Uzbekistan.

Habitat: Among diverse grass assemblages on foothills, grassy steppes, and more often in bushy places.

Population status: Common, forming dense groups.

Traditional use: Infusions and decoctions of the herb are used to treat acute rheumatism, ulcers, and stomach cancer (Alimbaeva and Goncharova 1971).

Documented effects: In experiment on animals (mice and rats), the preparation *Dipsacozide* (total plant saponins) was non-toxic and caused short-term decrease in arterial pressure. It noticeably raised the animals' tolerance to hypoxia, as found in foothill and high mountain conditions. In lipid metabolism *Dipsacozide* caused results similar to the known preparation *Polysponin*, and it also had hepatoprotective abilities (Alimbaeva et al. 1986).

Phytochemistry: Roots contains glucose, lactose, organic acids, triterpene glycosides (18.9–31.8 %, hederagenin derivatives), alkaloids, vitamin C, phenolcarbonic acids, coumarins, and flavonoids (2.18 %). The aboveground parts contain organic acid saponins (4.51–18.3 %, hederagenin derivatives), alkaloids (gentianine), phenolcarbonic acids, coumarins, and flavonoids (0.5 %; Mukhamedziev and Alimbaeva 1969; Rakhmatullaev and Yunusov 1972a; Alimbaeva et al. 1986; Akimaliev et al. 1989; Putieva and Mukhamedziev 1998).

▲**Dictamnus angustifolius G. Don fil. ex Sweet**
Photos: Alim Gaziev

▲**Dipsacus dipsacoides (Kar. & Kir.) Botsch.**
Photos: Evgeny Davkaev

▲**Descurainia sophia (L.) Webb ex Prantl** Photos:
top: Denis Mirin;
bottom: Alim Gaziev

►**Dianthus superbus L.**
Photo: Sergey Appolonov

Dipsacus laciniatus L. – Dipsacaceae

Synonyms: None

English name: Cutleaf teasel

Russian name: Ворсянка разрезная (Vorsyanka razreznaya)

Uzbek name: Tungiztarok

Kyrgyz name: Кесиктуу ворсянка (Kesiktuu vorsyanka)

Description: Herbaceous biennial. Stems 50–150 cm tall; stems covered by prickles. Basal leaves elongate-obovate, toothed or pinnatilobate; cauline leaves opposite, up to 30 cm long, up to 15 cm wide; opposite leaves connate at base, forming a cup-shaped sheath; underside of midvein with prickles. Inflorescence a dense, elongate-oviform head; involucral bracts linear-lanceolate, coarse, prickly. Corolla tubular, 4 lobed, pale-blue to white, each with a stiff green bract. Fruits gray-brown achenes.

Other distinguishing features: Inflorescence heads 5–8 cm long. Involucral bracts usually shorter than inflorescence head.

Phenology: Flowers in July, fruits in August-September.

Reproduction: By seeds.

Distribution: Chuy and Jalal-Abad provinces of Kyrgyzstan; Toshkent, Andijon, Farg'ona, Samarqand, and Surxondaryo provinces of Uzbekistan.

Habitat: In wet meadows and near canals.

Population status: Common, found as single plants.

Traditional use: An infusion of the roots is applied to treat tuberculosis and syphilis. A decoction, ointment, or paste is used as an anesthetic for hemorrhoids, calluses, and snake bites. Preparations of the aboveground parts are used as an anti-inflammatory and diuretic, and to stimulate respiratory function, cardiovascular function, and blood circulation. A decoction is used to treat fevers, ulcers, and stomach cancer, and is applied as compresses to treat skin cancer. A decoction of the inflorescence is used to treat rheumatism (Alimbaeva et al. 1986; Plant Resources of the USSR 1990).

Documented effects: In experiment on animals, a preparation of the total saponins showed low toxicity and reduced arterial pressure for a short time (Alimbaeva and Goncharova 1971).

Phytochemistry: The roots contain triterpene glycosides (10 %), iridoids, and alkaloids (0.24 %). The aboveground parts contain triterpene glycosides (8.5 %), alkaloids (0.4 %), iridoid and phenolic glucosides, and flavonoids. Fruits contain iridoids (Alimbaeva et al. 1986; Abdallah 1991; Kocsis et al. 1993).

Dodartia orientalis L. – Phrymaceae (formerly in Scrophulariaceae)

Synonyms: *Dodartia atro-coerulea* Pavlov.
English name: Unknown
Russian name: Додартия восточная (Dodartiya vostochnaya)
Uzbek name: Takasoqol
Kyrgyz name: Чыгыш теке сакалы (Chygysh teke sakaly)
Description: Herbaceous perennial. Nearly leafless, lower leaves opposite. Stems multiple, erect, multi-branched, 25–40 cm high, younger shoots with curly hairs, becoming glabrous with age. Inflorescence a loose raceme. Flowers sessile, dark purple-violet. Calyx campanulate, 5-lobed. Corolla 1.6–2.2 cm long, 2-lipped, glabrous outside, bearded in the throat; lower lip 3-lobed, longer and broader than upper lip; upper lip short, erect. Stamens 4. Fruit a spherical capsule. Seeds multiple, oviform, deep-brown, 0.5–0.75 mm long, 0.5 mm wide.
Other distinguishing features: Multi-branched, green, almost leafless herb with dark violet flowers. Middle lobe of lower lip smaller than lateral lobes.
Phenology: Flowers in May-July, fruits in July-August.
Reproduction: By seeds and rhizomes.
Distribution: All of Uzbekistan and Kyrgyzstan.
Habitat: The adyr and tau zones. Plains and slopes of hills.
Population status: More often found as single individuals, but due to intensive vegetative propagation by rhizomes it grows as small but dense populations.
Traditional use: A decoction made of the aboveground plant parts is used as a purgative and to treat syphilis (Khalmatov 1964).
Documented effects: The plant extract has slight purgative action, which is strengthened when mixed with other drugs (Ogolevitz 1951).
Phytochemistry: The plant has barely been investigated chemically. The aboveground plant parts contain alkaloids and possibly saponins (Khalmatov 1964). Mussaneoside [mussaenoside] has been isolated from this species (Umarova et al. 1988).

Eminium regelii Vved. – Araceae

Synonyms: Some consider *E. regelii* a synonym of *Eminium lehmannii* Kuntze.

English name: Unknown

Russian name: Эминиум Регеля (Eminium Regelya)

Uzbek name: It kuchala, Korakulak

Kyrgyz name: Тамыр кучала (Tamyr kuchala)

Description: Perennial herb to 15–40 cm high, with a flat-spherical tuber, 3 cm in diameter. Leaves basal, light green, entire, oblanceolate to elliptic, the base wide-cuneate, sheathing, petiolate. Inflorescence a spadix; spathe tube 4–7 cm long, spathe blade ovate or oblong, inside velvety black-violet; spadix appendix 5–7 cm long, cylindrical, black-blue. Fruits subglobose berries, 1–2-seeded.

Other distinguishing features: Inflorescence produces the odor of rotten meat.

Phenology: Flowering and fruits in April-May.

Reproduction: By seeds and tubers.

Distribution: Toshkent, Farg'ona, and Samarqand provinces of Uzbekistan; Osh province of Kyrgyzstan.

Habitat: The adyr zone. Loess slopes of hills, on dry, shallow-soiled slopes with rocky debris.

Population status: Not common, found as single individuals.

Traditional use: The powdered tubers are used as an analgesic to treat rheumatism (Khalmatov 1964). The powdered tuber is also used internally to treat stomach aches, abdominal pain, internal diseases, and dysentery (Sezik et al. 2004; Pak 2005).

Documented effects: An extract of the tubers had strophantine-like action on the heart (Khalmatov 1964).

Phytochemistry: The tubers contain poisonous saponins, traces of alkaloids and starch. The spathe contains pigments (Khalmatov 1964). The leaves and tuber contain a number of different lipids. The leaves contain carotinoids: neoxanthine and carotene (Chernenko et al. 2005).

Ephedra equisetina Bunge – Ephedraceae

Synonyms: *E. procera* var. *typica* Regel.

English name: Ephedra, ma huang

Russian name: Хвойник хвощевой (Khvoynik khvoshchevoy)

Uzbek name: Zogoza, Kizilcha

Kyrgyz name: Кырк муундай чекенде (Kyrk muunday chekende)

Description: Large dioecious shrub, to 1.5(−2.5) m high, usually with a single, thick stem (occasionally multiple). Bark gray or brown, cracking, spongy; older branches thick, woody, erect; young branches green, opposite or whorled on older branches. Leaves opposite, scale-like, paleaceous, triangular, 2.5–3.5 mm long. Male inflorescences consist of pollen cones, solitary or in clusters of 2–4 at the nodes. Female cones usually opposite at nodes, each cone composed of overlapping bracts. Mature female cone berry-like, 6–7 mm long, spherical, fleshy, red. Seeds 4–6 mm long, elongate-ovoid, dark brown.

Other distinguishing features: Leaves reddish to deep brown in color, connate for three fourth of their length.

Phenology: Flowers in May-June, fruits in July-August.

Reproduction: Propagates by seeds and rhizomes.

Distribution: Toshkent, Farg'ona, Samarqand, Qashqadaryo, Buxoro, and Surxondaryo provinces of Uzbekistan; all of Kyrgyzstan.

Habitat: The tau zone. Stony slopes in tree-shrub belt of mountains.

Population status: Common, sometimes forming dense groups.

Traditional use: For a long time, an infusion of the green shoots has been used for acute rheumatism, scabies, malaria, ulcers and other gastric diseases, altitude sickness, fever and heart diseases (Khalmatov 1964; Khalmatov et al. 1984). The stems, inflorescences, and berries are used as a treatment for bronchial asthma (Mamedov and Craker 2001).

Documented effects: This plant is one of the main sources of ephedrine. The alkaloid d-pseudoephedrine (0.5 mg/kg intravenously) evokes pressor action in narcotized animals. Repeated injections of the alkaloid usually cause tachyphylaxis. Pseudoephedrine has some properties of sympathomimetics of indirect action (Cession-Fossion 1967; de Meyts and Cession-Fossion 1966, 1967, 1968). In dogs, the alkaloid (1–2 mg/kg) provoked heartbeat deceleration, increased the blood pressure, and it also increased oxygen content in the blood of the coronary sinus. In general, pseudoephedrine has positive effects as a vasoconstrictor for rhinitis, tracheitis, and pharyngitis. Unlike L-ephedrine, pseudoephedrine has little or no effect on hemodynamics (Rowe et al. 1965). d-pseudoephedrine has direct stimulating effects on β- adrenoreceptors, and L-ephedrine has indirect stimulating effects (Tye et al. 1967).

Phytochemistry: All plant parts contain alkaloids. Young shoots have up to 3.5 % alkaloids (ephedrine and pseudoephedrine), tannins, vitamin C, and pigments. The stems contain up to 14 % tannins. The core of the wood stems contained from 30 % to 65 % tannins. Seeds contained 4 pigments of the flavone series (Khalmatov 1964; Khalmatov et al. 1984).

▲**Dipsacus laciniatus L.** Photos: *left*: Andrei Lubchenko; *center* and *right*: Evgeny Davkaev

▲**Ephedra equisetina Bunge**
Photo: Evgeny Davkaev

▼**Dodartia orientalis L.** Photos: *left*: Evgeny Davkaev; *right*: Alim Gaziev

▼**Eminium regelii Vved.** Photos: Evgeny Davkaev

Ephedra intermedia Schrenk & C.A. Mey. – Ephedraceae

Synonyms: *Ephedra ferganensis* V. Nikitin, *Ephedra glauca* Regel, *Ephedra microsperma* V. Nikitin, *Ephedra persica* (Stapf) V. Nikitin, *Ephedra tesquorum* V. Nikitin, *Ephedra tibetica* (Stapf) V. Nikitin, *Ephedra valida* V. Nikitin.

English name: Ephedra

Russian name: Хвойник средний, Эфедра средняя, Пустынная Эфедра (Krpvoynik sredniy, Efedra srednyaya, Pustynnaya efedra)

Uzbek name: Kizilcha, Chul kizilcha

Kyrgyz name: Орточо чекенде (Ortocho chekende)

Description: Perennial, evergreen bush up to 1 m tall. Branches dense, erect, opposite or whorled, segmented, gray-green, glabrous; bark gray, fibrous. Leaves reduced, triangular to scale-shaped, opposite or in whorls, up to 3.5 mm long, leaves partially connate. Male cones usually clustered at nodes, subtended by circular or ovate bracts, connate at the base. Mature female cones berry-like, spherical, juicy, up to 6 mm long.

Other distinguishing features: Has longer micropylar tubes than other *Ephedra* species; cones with 2–3 seeds.

Phenology: Flowers in May and fruits in June-July.

Reproduction: By seeds.

Distribution: Surrounding Ysyk-Kol lake and in the Boom gorge of Kyrgyzstan; Toshkent, Farg'ona, Jizzax and Samarqand provinces of Uzbekistan.

Habitat: On stony slopes in the lower and middle belt of mountains, and on rocks.

Population status: Common, forming dense groups.

Traditional use: This is one of the three species of *Ephedra* that is officially used in the Chinese Pharmacopoeia as a source of ma huang, a stimulant and antiasthmatic that has been used for at least 2,000 years (Abourashed et al. 2003).

Documented effects: Ephedrine is obtained from the herb and is issued in the form of ephedrine chloride. It is widely used to treatment allergies (bronchial asthma, rashes, rhinitis, etc.). It acts by stimulating the central nervous system. In cases of morphine, scopolamine, and ganglioplegic poisoning, a preparation of ephedrine is used to raise arterial pressure, render positive inotropic action on the heart, increase heart rate and to tone peripheral vessels, relax smooth muscles of bronchial tubes, and to stimulate breathing. The basic mechanism of ephedrine's activity is its ability to cause the liberation of noradrenaline from its reserves in nervous fibers and inhibit the return of noradrenaline to nervous fibers. In addition, it protects noradrenaline and adrenaline from decomposition and strengthens their effects (Dobrokhotova and Chudinov 1966; Gammerman 1967). An extract of the plant exhibited antibacterial effects against *Micrococcus luteus* and *Klebsiella pneumoniae* (Shahidi Bonjar 2004).

Phytochemistry: The thin, green stems contain up to 2.2 % total alkaloids, flavonoids, pigments, and up to 8 % tannins. Of the total alkaloids, up to 75 % is pseudoephedrine, with the rest being ephedrine and others (Khalmatov et al. 1984; Kim et al. 2005). The main constituent (12.80 %) of the essential oil isolated from the dried stems was 1,4-cineole (Ji et al. 1997).

Epilobium hirsutum L. – Onagraceae

Synonyms: *Chamaenerion hirsitum* (L.) Scop., *Epilobium tomentosum* Vent., *Epilobium velutinum* Nevski, *Epilobium villosum* Thunb.

English name: Codlins and cream, great willowherb, great hairy willowherb

Russian name: Кипрей мохнатый, Кипрей волосистый (Kiprey mokhnatyy, Kiprey volosistyy)

Uzbek name: Kizilkon

Kyrgyz name: Сапсагай кипрей (Sapsagay kiprey)

Description: Herbaceous perennial. Stems up to 1.5 m tall, densely gray-hairy. Leaves sessile, oblanceolate, 4–10 cm long, 1–2 cm wide, margins serrulate, upper and lower surfaces densely pubescent. Flowers in a raceme. Calyx campanulate, lobes lanceolate, pubescent. Corolla lilac-purple, deeply lobed. Stigma deeply 4-lobed, recurved. Fruit a capsule, 4–10 cm long, pubescent. Seeds brown or light-brown, papillate.

Other distinguishing features: Leaves clasping the stem.

Phenology: Flowers in June-August, fruits in August-September.

Reproduction: By seeds and rhizomes.

Distribution: All provinces of Kyrgyzstan; the Karakalpakstan autonomous republic (delta of the Amu-Darya river) and Toshkent province of Uzbekistan.

Habitat: In wet and marshy places near rivers and canals.

Population status: Common, forming dense groups.

Traditional use: The aboveground parts are used as a hemostatic, astringent, and anti-inflammatory (Vandisheva et al. 1977).

Documented effects: In experiments on animals, a water infusion caused reduced heart rate, increased amplitude of heartbeats, and caused diuresis (Appolonova 1956). Extracts of the plant exhibited a significant inhibitory effect on the reproduction of influenza viruses (Ivancheva et al. 1992), and prolonged the lifespan of mice with 2 types of tumorous cancers (Voynova et al. 1991).

Phytochemistry: The aboveground parts contain saponins, phenols, phenolcarbonic acids, tannins, flavonoids (hyperoside, rutinoside, etc.), trace alkaloids, vitamin C, and coumarins (Plant Resources of the USSR 1987; Barakat et al. 1997).

Equisetum arvense L. – Equisetaceae

Synonyms: *Equisetum boreale* Bong., *Equisetum calderi* B.Boivin, *Equisetum saxicola* Suksd.

English name: Field horsetail

Russian name: Хвощ полевой (Khvoshch polevoy)

Uzbek name: Kirk bugim

Kyrgyz name: Талаа кырк мууну (Talaa kyrk muunu)

Description: Herbaceous, rhizomatous perennial, bearing spores. Stems dimorphic. Vegetative stems (developing later than sporebearing stems), 10–15 cm high, green, with 6–12 ribs, segmented with whorls of branches, with reduced leaves, forming a toothed sheath. Sporebearing stems (appearing in spring and die back after spores ripen), up to 40 cm tall, fleshy, reddish, brown or brownish-yellow, unbranched, topped with conical-cylindrical spore-bearing cones; sheathes longer than on vegetative stems.

Other distinguishing features: Spores green, spherical.

Phenology: Spores ripen in April-May.

Reproduction: By spores and rhizomes.

Distribution: All provinces of Kyrgyzstan and Uzbekistan.

Habitat: In wet places and sandy meadows in valleys as well as in the lower and mid mountain belt.

Population status: Common, forming dense groups.

Traditonal use: Preparations of the herb are used as a diuretic, anti-inflammatory, astringent, hemostatic, and disinfectant, and also to increase metabolism and treat skin wounds (Altimishev 1991). The herb is used to treat kidney and bladder diseases, edema, rheumatism, and stomach and intestinal growths (Kurochkin 1998).

Documented effects: Preparations of this plant have been approved as a medicinal remedy. An infusion or liquid extract as a component of a tea mixture is used as a very strong diuretic for patients with kidney or heart diseases, to treat inflammation of the bladder and urinary tract, to stop stomach, intestinal, hemorrhoidal, and uterine bleeding, and as a treatment for pleurisy and some types of tuberculosis (Khalmatov et al. 1984). The compounds onitin-9-*O*-glucoside and luteolin, isolated from the plant, exhibited hepatoprotective activity in vitro, as well as strong superoxide scavenging effects (Oh et al. 2004).

Phytochemistry: The herb contains flavonoids (equisetrine, luteolin and glycosides of luteolin, quercetin, kaempferol, etc.), up to 5 % saponins (equisetonin), alkaloids (equisetin and nicotine) resins, organic acids (malic, aconitic, and oxalic), up to 25 % silicic acid, carotene, vitamin C, tannins, etc. (Khalmatov et al. 1984; Oh et al. 2004).

Eremurus regelii Vved. – Asphodelaceae (formerly in Liliaceae)

Synonyms: *Eremurus spectabilis* ssp. *regelii* (Vved.) Wendelbo.
English name: Fox tail lily
Russian name: Шириш Регеля (Shirish Regelya)
Uzbek name: Shirach
Kyrgyz name: Регель чырашы (Regel' chyrashy)
Description: Herbaceous perennial with fusiform-incrassate roots. Stem glabrous, 80–180 cm high. Leaves wide-linear, 2.5–5 cm wide, 20–40 cm long, fluted, keeled, blue-gray, glabrous. Inflorescence a dense, multiflorous raceme, mostly erect while flowering. Flowers with 6 perianth segments, pale-pink, each with wide brownish-purple stripe. Fruits spherical capsules, latitudinally wrinkled, 6–8 mm in diameter. Seeds narrow-winged.
Other distinguishing features: When fruiting, pedicels arcuate, capsules crowded around inflorescence axis.
Phenology: Flowers in May in the foothills, in June in the mountains. Fruits accordingly in June and August.
Reproduction: Propagates by seeds and rhizomes.
Distribution: Toshkent, Samarqand, Jizzax, Qashqadaryo, and Surxondaryo provinces of Uzbekistan; Osh and Jalal-Abad provinces of Kyrgyzstan.
Habitat: The adyr and tau zones, on gentle slopes.
Population status: Common.
Traditional use: In folk medicine the leaves are used as a carminative. Roots are used to treat gastrointestinal diseases and to coat the digestive tract. The roots are a source of native mannose. The polysaccharide eremuran is used to produced glucose and mannose by acid hydrolysis. A high quality glue is produced from the roots (Khalmatov and Kosimov 1992). The powdered rhizomes are used to treat pyoderma (Mamedov et al. 2004).
Documented effects: In acute tests on narcotized animals, the alkaloid hordenine, at doses of 0.5–1 mg/kg and higher, provoked rapid breathing. These effects are due to sympathomimetic (adrenomimetic) activity; it also has moderate vasoconstrictive action (Aliev et al. 1967; Khalmatov and Kosimov 1992). A polysaccharide, isolated from the roots, was found to increase the survival rate of rabbits subjected to hemorrhagic shock and had activity similar to that of reopoliglukin (Rakhimov 1997).
Phytochemistry: The roots contain the polysaccharide eremuran. The total alkaloid content of the roots was 0.035 % out of which 0.012 % was the alkaloid eremursine. The leaves contain vitamin C and carotene (Khalmatov 1964). The leaves and roots contain polysaccharides (Yuldasheva et al. 1993; Rakhimov 1997).

►**Equisetum arvense L.** Photos: *left* and *center left*: Vladimir Epiktetov; *center right*: Dmitri Oreshkin; *right*: Vadim Prokhorov

Eremurus regelii Vved. Photos: Evgeny Davkaev▼

▲**Epilobium hirsutum L.** Photos: Dmitri Oreshkin

▼**Ephedra intermedia Schrenk & C.A. Mey.** Photos: John B. Taft

Erodium cicutarium (L.) L'Her. ex Aiton – Geraniaceae

Synonyms: *Erodium pulchellum* Karel. ex Ledeb., *Geranium cicutarium* L.

English name: Redstem stork's bill, Redstem filaree

Russian name: Аистник обыкновенный (Aistnik obyknovennyy)

Uzbek name: Lailac tumshuk, qora mashaq

Kyrgyz name: Цикутадай турна тумшук (Tsikutaday turna tumshchuk)

Description: Ephemeral annual, with shortened vegetation cycle. Stems 10–60 cm tall, prostrate or upright, loosely villous. Leaves oblong, pinnatisect, segments pinnatipartite, stipulate. Inflorescences umbelliform, axillary. Sepals 5. Petals 5, 4–6 mm long, purple-pink. Fruit a schizocarp, splitting into 5 mericarps, each attached to the stylar column by a terminal awn.

Other distinguishing features: Sepals apiculate. Fertile stamens 5, alternating with 5 staminodes. While drying, the awn twists spirally and separates from receptacle. If the soil has enough moisture in it, the terminal awn of the mericarp will penetrate into it.

Phenology: Flowers and fruits in April-June.

Reproduction: Abundantly propagates by seeds.

Distribution: It is widespread in the irrigated farming zones of Uzbekistan; all of Kyrgyzstan.

Habitat: The chul and adyr zones. Among wheat and alfalfa fields, vegetable gardens, orchards, on small hills, waste places and dry slopes.

Population status: In some areas fairly dense populations can be found; especially in wheat farming areas, vegetable gardens and *Artemisia*-rich ephemeral complexes of the adyr zone.

Traditional use: Central Asian folk medicine uses powdered leaves, as well as powder mixed with melted lamb fat to treat abscesses and as wound healing remedy. In the past the plant was widely used but its current use is limited (Khalmatov 1964). In Iraq, a decoction of the whole plant is used for treatment of anasarca and metrorrhagia (Al-douri 2000). In Turkey, a decoction of the whole plant is used externally to treat pains (Simsek et al. 2004).

Documented effects: A decoction is recommended as a hemostatic for internal uterine bleeding. An acetone-alcohol extract of the herb has been introduced as a hemostatic treatment (Aliev et al. 1972). Zavrazhanov et al. (1977) stated this species has astringent, anti-inflammatory, hemostatic, and sedative properties. The decoction of the herb is used to treat internal and uterine bleeding (Aliev et al. 1972) and also as an anticonvulsant (Fruentov 1972; Akopov 1981). A water extract, as well as a methanol extract and its fractions, were found to have antiviral effect on myxoviruses, herpes virus type 1, vesicular stomatitis and vaccinia virus (Zielinska-Jenczylik et al. 1987). In vivo, a methanol extract injected intravenously induced interferon in mice (Zielinska-Jenczylik et al. 1988). In vitro, low concentrations of a polyphenolic fraction from an extract of the plant stimulated free radical activity of human granulocytes, whereas high concentrations inhibited the activity (Fecka et al. 1997).

Phytochemistry: The herb contains bitters, 2.1 % resins, tannins, acetylcholine, 55 mg% carotene, up to 4.94 % sugar, 1.9 % general titratable organic acids, 37.5–91.85 mg% vitamin C, and 0.64 mg% vitamin K, and 12–14 % ash, which includes up to 47 % K_2O (Akopov 1981). The aboveground parts contain a variety of tannins and flavonoids (geraniin, didehydrogeraniin, corilagin, rutin, hyperin, quercetin, isoquercitrin, kaempferol, myricetin, polyphenolic acids, etc.) (Fecka and Cisowski 2005).

Eryngium biebersteinianum Nevski – Apiaceae

Synonyms: *Eryngium caucasicum* Trautv. (some consider this the correct name), *Eryngium coeruleum* M. Bieb., *Eryngium pskemense* Pavlov

English name: Bieberstein's sea-holly

Russian name: Синеголовник Биберштейна (Sinegolovnik Bibershteyna)

Uzbek name: Kok tykan, koz tykan

Kyrgyz name: Биберштейн тикен башы (Bibershteyn tiken bashy)

Description: Glabrous perennial, bluish in color with widely fusiform root. Stems up to 1 m tall, often solitary. Branches emanating from above middle of stem and branching again, forming a wide, corymbose top. Basal leaves long-petiolate, thin-coriaceous, blue-gray, the blades oblong-oval with a cordate base; upper leaves sessile, deeply divided, margins spiny-dentate. Inflorescences subglobose heads up to 10 mm in diameter. Involucral bracts stiff, spiny, 2–4 times longer than heads. Petals blue, ca. 2 mm long. Fruits composed of obovate mericarps; mericarps angular, covered with long, narrow, lanceolate scales along the edges.

Other distinguishing features: First basal leaves have smooth margins and senesce early.

Phenology: Flowers in May, fruits in July.

Reproduction: Only by seeds.

Distribution: Toshkent, Samarqand, Buxoro, Qashqadaryo, and Surxondaryo provinces of Uzbekistan; Osh and Jalal-Abad provinces of Kyrgyzstan.

Habitat: The adyr zone. As a weed in orchards, vegetable gardens, unirrigated, cultivated fields and abandoned and long-fallow fields.

Population status: Common, as single individuals.

Traditional use: Infusions of the roots of other *Eryngium* species are used as a blood cleanser and as a sedative. The infusions are also used to treat edema, scrofula, gonorrhea, headaches, heart pain, and various tumors, and are used as a treatment for pertussis, anti-convulsant for epileptics, and as cough medicine, diaphoretic and diuretic. The roots are used to treat mushroom poisoning and bites from venomous animals. The herb is recommended for anemia (Khalmatov 1964; Minayeva 1991).

Documented effects: This species has been shown to have expectorant action (Minayeva 1991).

Phytochemistry: The plant contains essential oils, saponins, and tannids (Minayeva 1991; Khalmatov and Kosimov 1992).

Erysimum diffusum Ehrh. – Brassicaceae

Synonyms: *Erysimum canescens* Roth., *Erysimum andrzejowskianum* DC.

English name: Diffuse wallflower

Russian name: Желтушник рассеянный, Желтушник серый, Желтушник раскидистый (Zheltushnik rasseyannyy, Zheltushnik seryy, Zheltushnik raskidistyy)

Uzbek name: Kulrang zhyoltushnik

Kyrgyz name: Чачырак даргын (Chachyrak dargyn)

Description: Herbaceous biennial. Stems erect, single or few, 30–80 cm tall, sometimes branched. Basal rosette leaves petiolate, linear-lanceolate, margins entire; lower cauline leaves short-petiolate; upper cauline leaves sessile, margins entire. Inflorescence a few-flowered raceme. Flowers small, perfect, pedicellate. Petals 4, yellow. Stamens 6 (tetradynamous), erect. Fruits 4-sided siliques, thin, 3–10 cm long, 1–1.5 mm wide, whitish, hairy. Seeds ellipsoid, yellow-brown, up to 1.5 mm long.

Other distinguishing features: Fruit smooth, with 4 lines of white trichomes.

Phenology: Flowers in June, fruits in July.

Reproduction: By seeds.

Distribution: All provinces of Kyrgyzstan; Toshkent, Namangan and Farg'ona provinces of Uzbekistan.

Habitat: On steppes and dry stony exposures.

Population status: Common, found as single plants.

Traditional use: In Kyrgyzstan, an infusion of the herb is used as a diuretic, sedative and anti-depressant, and to treat heart problems. It is said to be one of the best treatments for edema (Altimishev 1991). In the folk medicine of Tajikistan, the aboveground parts are used to make a tea used as a diuretic and laxative, and to treat heart weakness, tachycardia, and hypertension (Khodzhimatov 1989).

Documented effects: The preparations *Erysimine*, *Erysimoside*, *Coreside*, liquid extracts, and *Cardiovalen* (a complex preparation) are used to treat mitral failure, hypertension, and arteriosclerotic cardiosclerosis (Khalmatov et al. 1984).

Phytochemistry: All plant parts contain cardiac glycosides. The greatest quantity is found in flowers and seeds (2–6 %). More than 10 cardiac glycosides have been isolated, including erysimine, erysimoside, and others. Seeds contains up to 30–40 % fatty oil (Kurmukov 1956; Tadzhibaev et al. 1977; Khalmatov et al. 1984).

Euphorbia jaxartica Prokh. – Euphorbiaceae

Synonyms: *Euphorbia virgata* Waldst. & Kit. ssp. *jaxartica* (Prokh.) Prokh., *Euphorbia waldsteinii* (Sojak) A. Radcliffe-Smith ssp. *jaxartica* (Prokh.) Oudejans, *Tithymalus graminifolius* (Vill.) Sojak ssp. *jaxarticus* (Prokh.) Sojak.

English name: Unknown

Russian name: Молочай сырдарьинский (Molochay syrdar'inskiy)

Uzbek name: Sultama

Kyrgyz name: Сыр-Дарыя суттуу чобу (Syr-Daryya suttuu chobu)

Description: Herbaceous perennial, 30–100 cm tall, blue-gray. Root thick, vertical or obliquely descending. Stems many or few, erect, pubescent on lower portions, branching in upper parts with vegetative branches below the flowering branches. Leaves alternate, nearly sessile, oblong-linear, 4–13 cm long, 2–7 cm wide, margins entire. Inflorescences cyathia, on upper axile branches and on terminal peduncles arranged in umbels with 8–12 rays; bracts subtending inflorescences opposite, partially connate, kidney-shaped or ovate-triangular, 6–20 mm long, 8–22 mm wide; cyathia campanulate with ciliate lobes. Styles 2–3 mm long, connate nearly to the middle. Fruit an ovoid schizocarp, 3.5–4.5 mm long, 4–5 mm wide, trisulcate. Seeds oval, 2.5 mm long, whitish-gray, smooth, with a small scarious appendage.

Other distinguishing features: Leaves on vegetative branches densely arranged. Nectaries yellowish, crescent-shaped, 2-horned.

Phenology: Flowers and fruits in April-September.

Reproduction: Reproduces by rhizomes and seeds.

Distribution: Toshkent, Namangan, Andijon, and Farg'ona provinces of Uzbekistan; Naryn Ysyk-Kol, Chuy and Talas provinces of Kyrgyzstan.

Habitat: The adyr and tau zones. In river valleys, along canals, in long-cultivated fields; often found along ravines into the mountains, where it grows on stony slopes with rocky debris and in wet meadows.

Population status: Common, found in small populations.

Traditional use: The powdered root is used to treat wounds and syphilis. The latex is used to treat fungal skin diseases and scabies, and to remove corns and warts (Khalmatov 1964).

Documented effects: No data.

Phytochemistry: The plant contains a large amount of latex, which contains 1.5 % caoutchouc, resins, and euphorbin (Pavlov 1947). The plant contains triterpenes and polyphenols (quercetin-3-galactoside, kaempferol, gallic acid, etc.; Azimov and Nazirov 1969, 1970; Abdulladzhanova et al. 2003).

▲**Erodium cicutarium (L.) L'Her. ex Aiton** Photos: *left*: Alim Gaziev; *center* and *right*: Evgeny Davkaev

◄**Eryngium biebersteinianum Nevski** Photos: *left*: Komiljon Tojibaev; *right*: Vadim Prokhorov

▼**Euphorbia jaxartica Prokh.** Photos: *left* and *center*: Alim Gaziev; *right*: Evgeny Davkaev

◄**Erysimum diffusum Ehrh.** Photos: Maxim Zaitsev

Euphorbia rapulum Kar. & Kir. – Euphorbiaceae

Synonyms: *Tithymalus rapulum* (Kar. & Kir.) Klotzsch & Garcke.

English name: Unknown

Russian name: Молочай репчатый (Molochay repchatyy)

Uzbek name: Ikhrozh

Kyrgyz name: Туймоктуу суттуу чоп (Tuymoktuu suttuu chop)

Description: Herbaceous perennial up to 25(−35) cm tall with a spherical, sometimes branching tuber, 3–6 cm in diameter. Stems erect, thicker towards the base; top of plant wide-paniculiform with bifurcating, flowering branches. Basal leaves ovate, sheathed; cauline leaves alternate, 3–4 cm long, 1–2 mm wide, spatulate or lanceolate-elliptic, sometimes cordate at the base, entire, short-petiolate to sessile. Inflorescences broadly campanulate cyathia, 2–3 mm in diameter, margin of lobes densely ciliate. Styles 1–1.5 mm long, connate at the base and forked at the top. Fruit an ovoid schizocarp, 4.5–5.5 mm long, 4–5 mm wide, trisulcate, glabrous, shiny. Seeds flattened-oblong, smooth, brownish, with a short-stalked conical appendage.

Other distinguishing features: Upper flowering branches sometimes trifurcated.

Phenology: Flowers and fruits in March-June.

Reproduction: By seeds.

Distribution: Toshkent, Samarqand, and Buxoro provinces of Uzbekistan; Chuy and Jalal-Abad provinces of Kyrgyzstan.

Habitat: The adyr and tau zones. Clayey, stony slopes and slopes with red sandstone.

Population status: Common, found as single individuals or in groups with 2–3 plants.

Traditional use: Powdered root is used as a strong purgative and also for tuberculosis (Khalmatov 1964).

Documented effects: In experiments, ethanolic extracts of the aboveground parts showed high antioxidant action (Eliseeva 2005).

Phytochemistry: The tuber contains upto 5 % resins and 0.5–0.6 % caoutchouc. Caoutchouc can also be found in the stems (up to 0.24 %) and in the fruits (up to 1.4 %). The resin contains the poisonous chemical euphorbin (Khalmatov 1964).

Ferula foetida (Bunge) Regel – Apiaceae

Synonyms: *Ferula assa-foetida* L., *Peucedanum asa-foetida* (L.) Baill., *Scorodosma foetidum* Bunge.

English name: Giant fennel

Russian name: Ферула вонючая (Ferula vonyuchaya)

Uzbek name: Sassyk kavrak, kovrak, sassyk kurayi

Kyrgyz name: Жыттуу ала гул (Zhyttuu ala gul)

Description: Herbaceous, monocarpic perennial, with a large, oval, fleshy root up to 15 cm in diameter. Stem thick, up to 1–1.2 m high, upper portion branching and forming a dense globose panicle. Leaves mostly glabrous above, more or less soft-villous beneath, senescing early; basal leaves short-petiolate with broad blade, ternate with bipinnatisect lobes, lobules decurrent, 15 cm long, 5 cm wide; lower leaves alternate; upper leaves smaller and becoming reduced to sheaths. Inflorescences compound umbels; terminal umbel sessile or on a reduced peduncle, spherical, 15–20 cm wide; lateral umbels on long peduncles. Petals light-yellow, almost cream colored. Fruit a schizocarp with 2 one-seeded mericarps; mericarps flattened, pubescent, 1.6–2.2 cm long and wide.

Other distinguishing features: Ovary and fruit pubescent. Seeds have an extremely objectionable, persistent odor.

Phenology: Flowers in March-April, fruits in April-May.

Reproduction: By seeds.

Distribution: Desert areas of Karakalpakstan autonomous republic, Toshkent, Samarqand, Buxoro, Qashqadaryo, and Surxondaryo provinces of Uzbekistan; not found in Kyrgyzstan.

Habitat: The adyr zone. Plains in foothills, on stony-clay soils.

Population status: Common, found as single individuals.

Traditional use: The plant has been used in Central Asian folk medicine since ancient times, as an anticonvulsant, vermifuge, and to treat some nervous diseases. The gum-resin is used in Chinese medicine as a restorative and tonic for hysterics, neurasthenia and vegetative neurosis, and to treat some skin diseases and common colds, as an expectorant and anticonvulsant, and mixed with other drug substances to treat lung tuberculosis, exudative diathesis, lymphadenitis, and osteitis. Avicenna used this plant to treat tumors, jaundice, and other liver diseases, as well as stomach, kidney, and spleen diseases, and as a diuretic and hemostatic for uterine bleeding (Khalmatov and Khabibov 1976; Kurmukov and Akhmedkhodzhaeva 1994).

Documented effects: Extracts of various *Ferula* species and individual compounds isolated from the extract exhibit phytoestrogenic activities. Based on these compounds 2 phytoestrogenic preparations, *Tefestrol* and *Panoferol,* were developed (Prokhorova and Kurmukov 1997; Prokhorova et al. 1992b; Kurmukov and Akhmedkhodzhaeva 1994). Infusions, pills, and emulsions of the gum-resin are used as an antispasmodic asthma treatment, to treat hysteria and other nerve diseases, and as an anticonvulsant. An infusion of *Ferula*, injected intravenously, reduced blood pressure. Its hypotensive activity is due to antispasmodic action on blood vessels (Sarkisyants 1969a, 1972). Dried resin of the roots reduced platelet adhesiveness and aggregative properties, depressed blood thromboplastic activity and elongated time and intensity of bleeding (Mansurov 1967). An infusion and decoction of *Ferula foetida* stimulated stomach secretory activity, and also had an impact on activity of the gastrointestinal tract. Thirty to thirty-five days of treatment with a water infusion (10–20 % by weight) of the herb, at 0.5–1 g/kg animal mass, prevented animal death from anaphylactic shock and development of Arthus-Sakharov phenomenon, i.e. it shows anti-allergic affect (Isakov 1969; Sarkisyants 1969b; Sarkisyants and Azizova 1971; Kurmukov and Akhmedkhodzhaeva 1994). In Kuwait this species is used as an treatment for diabetes and has hypolipidemic activity (Al-Awadi and Shoukry 1988).

Phytochemistry: In the early 1930s coumarins and organic sulfides were isolated from *Ferula* spp. (Tsukervanik et al. 1935; Kurmukov and Akhmedkhodzhaeva 1994). Later, umbelliferone, ferulic and galbanic acids and coumarins were isolated from the resins (Kurmukov and Akhmedkhodzhaeva 1994). In a systematic study of 50 species of *Ferula* in Uzbekistan, more than 250 terpenoids were isolated. It was shown for the first time that plants of this genus contained complex esters of terpenoid alcohols with aliphatic and aromatic acids. The structure and stereochemical abilities of more than 150 new terpenoids were determined (Saidkhodzhaev and Nikinov 1973, 1974; Sagitdinova and Saidkhodzhaev 1977; Sagitdinova et al. 1978). The compounds isolated from species in the genus *Ferula* can be divided into 3 groups: (1) coumarins, (2) compound esters of terpenoids and sesquiterpenoid alcohols with aromatic acids, and (3) sesquiterpenoid lactones (Bagirov et al. 1978). The roots of all the species found in Central Asia have similar chemical compounds to that of *F. foetida* and contain resins, essential oil, gums, high amounts of starch, and other compounds (Kurmukov and Akhmedkhodzhaeva 1994; Khalmatov and Kosimov 1994). The gum-resin of *F. foetida* contains 4–28 % essential oils: disulfide, hexenyl-disulfides, paraoxycoumarins, 0.68 % free asaresinotannol, asaresinol and their ether with ferulic acid, umbelliferone (which is formed from ferulic acid), asaresin A, farnesferol A, B, C, and other substances (Kurmukov and Akhmedkhodzhaeva 1994).

Ferula kuhistanica Korovin – Apiaceae

Synonyms: *Ferula jaeschkeana* Vatke.

English name: Unknown

Russian name: Ферула кухистанская (Ferula kukhistanskaya)

Uzbek name: Chair

Kyrgyz name: Unknown

Description: Herbaceous perennial, monocarpic, with thick oviform root. Stem thick, stocky, up to 1 m high, upper third branching into thick, oviform panicle. Leaves quickly senescing, abaxial side glabrous, hairy beneath; leaf blades are wide-triangular in outline, ternate with bipinnatisect lobes, lobules oblanceolate. Inflorescences compound umbels; umbels of 2 kinds: the terminal umbel nearly sessile, with 20–25 rays, up to 12 cm wide; lateral umbels long-pedunculate, in clusters of 3, exceeding the terminal umbel. Petals yellow. Fruit a schizocarp with 2 one-seeded mericarps; mericarps flattened, oval, 2–3,2 cm long and 1–2,2 cm wide, reddish-brown.

Other distinguishing features: Ovary and fruit glabrous.

Phenology: Flowers in May-June, fruits in July-August.

Reproduction: By seeds.

Distribution: Andijon, Namangan, Farg'ona, Samarqand, Qashqadaryo, and Surxondaryo provinces of Uzbekistan; Naryn and Ysyk-Kol Provinces of Kyrgyzstan.

Habitat: The tau and yailau zones. Gentle mountain slopes of the tree-shrub belt.

Population status: Rare, forming dense groups.

Traditional use: The plant's resin, boiled with milk, is used to treat syphilis. It is applied externally as a treatment for persistent wounds, tumors, and other diseases (Khalmatov and Kosimov 1992; Kurmukov and Akhmedkhodzhaeva 1994).

Documented effects: Similar to *Ferula foetida*. Compounds isolated from the fruits were toxic against gram-positive bacteria, including methicillin-sensitive and methicillin-resistant *Staphylococcus aureus* (Tamemoto et al. 2001).

Phytochemistry: Refer to *Ferula foetida* for general information on the genus *Ferula*. All plant parts of *F. kuhistanica* contain resin and essential oils. The fresh roots contain 0.42–0.72 % essential oils, the fruits 0.54 % and the fresh leaves 0.08 %. From steam distillation, 11.7–14.8 % green-colored, strong smelling essential oils were extracted. Leaf oil contains 85 % d-pinene. Roots contain up to 28 % and fruits 10–11 % resins. The resin contains n-carbolic acid (12.5 %), anisic and angelic acids, and umbelliferone (Khalmatov 1964). Daucane-type sesquiterpenes and daucane esters have been isolated from the roots and stems (Chen et al. 2000).

Ferula moschata (Reinsch.) Koso-Pol. – Apiaceae

Synonyms: *Ferula pseudo-oreoselinum* (Regel & Schmalh.) Koso-Pol., Ferula sumbul (Kaufm.) Hook. f., *Ferula urceolata* Korov.

English name: Musk fennel

Russian name: Ферула сумбул (Ferula sumbul)

Uzbek name: Sumbul

Kyrgyz name: Unknown

Description: Herbaceous perennial, with multiple, thick, conjoined taproots. Stems few, up to 50 cm high, slender, pubescent becoming subglabrous, corymbiform branching above. Leaves stiff, persisting long into the growing season, abaxial side slightly hairy, long-petiolate; basal leaves oval-triangular in outline, blade tripinnatisect, leaf segments lanceolate or oblong, 20–30 mm long, 10–15 mm wide, entire or deeply dissected; cauline leaves becoming smaller, upper leaves reduced to sheaths. Inflorescences compound umbels; umbels variable; terminal with 6–10 rays, 4–6 cm wide; lateral umbels single or in pairs, distinctly below level of terminal umbel. Petals yellow. Fruit a schizocarp with 2 one-seeded mericarps; mericarps 7 mm long, twice as long as the pedicels, flattened, with filiform ribs.

Other distinguishing features: Umbellets 10–15-flowered. Fractured roots produce a specific pleasant smell.

Phenology: Flowers in June, fruits in July.

Reproduction: By seeds.

Distribution: Samarqand, Qashqadaryo, and Surxondaryo provinces of Uzbekistan; Jalal-Abad province of Kyrgyzstan.

Habitat: The tau and yailau zones. Stony open slopes among shrubs.

Population status: Rare, found as single individuals; listed in the Red Book of Rare and Endangered Species of Uzbekistan.

Traditional use: Used as a folk medicine in Eastern and European countries. The resin was often used as a tonic and as a stimulatory remedy for gastric pneumatosis, pertussis, cholera, and other diseases. Avicenna applied it to treat tumors, jaundice and other diseases of the liver, stomach, kidneys, and spleen, and he also used it as a diuretic and hemostatic for uterine bleeding (Ogolevitz 1951; Kurmukov and Akhmedkhodzhaeva 1994; Khalmatov and Kosimov 1994).

Documented effects: Similar to *Ferula foetida*. Compounds isolated from a methanol extract of the dried roots of *Ferula sumbul* showed anti-HIV activity (Zhou et al. 2000).

Phytochemistry: See *Ferula foetida* for general information on the genus *Ferula*. The roots contain up to 21.5 % resins containing phytosterin, vanillic acid, umbelliferone, fatty acids (including isovaleric), up to 4 % essential oil, consisting of linalyl acetate, citronellyl acetate, ferulene, sesquiterpenes, doremon, doremol and its acetic ester, as well as the sesquiterpene sambulene and up to 24.41 % total sugars (Tsukurvanik and Simkhaev 1948; Khalmatov 1964). The dried roots contained many different coumarins and sesquiterpene lactones (Zhou et al. 2000; El-Razek et al. 2001).

▲**Euphorbia rapulum Kar. & Kir.** Photos: Evgeny Davkaev

▲**Ferula foetida (Bunge) Regel** Photos: Alim Gaziev

▲**Ferula kuhistanica Korovin**
Photos: Alim Gaziev

◀**Ferula moschata (Reinsch.) Koso-Pol.**
Photo: Red Book of Uzbekistan

Fragaria vesca L. – Rosaceae

Synonyms: *Fragaria chinensis* Losinsk., *Fragaria concolor* Kitag., *Potentilla vesca* (L.) Scop.
English name: Woodland strawberry
Russian name: Земляника лесная (Zemlyanika lesnaya)
Uzbek name: Yavoiy klubnay
Kyrgyz name: Токой кожогаты (Tokoy kozhogaty)
Description: Herbaceous, stoloniferous perennial, 5–30 cm tall. Stems compressed, hairy. Leaves in basal rosette, petiolate, trifoliate, margins sharply toothed, lateral leaflets sessile, middle leaflet often short petiolulate. Inflorescence cymose, on an elongated stem. Sepals 5, appressed hairy, margins entire. Petals 5, white. Stamens many. Fruits small achenes, attached to surface of swollen receptacle. Receptacle berry-like, bright red, fleshy, 0.7–2 cm in diameter, elongated or nearly spherical.
Other distinguishing features: Runners develop in the axils of leaves. Pedicels appressed hairy.
Phenology: Flowers in May, fruits in June.
Reproduction: By seeds and runners.
Distribution: All provinces of Kyrgyzstan; not found in the flora of Uzbekistan.
Habitat: In the forest belt of mountains, spruce-fir forests, and glades.
Population status: Common, found in small groups.
Traditional use: In folk medicine infusions of fruits and leaves are used to treat sore throats, jaundice, hemorrhoids, fatigue, uterine bleeding, and children with diarrhea. Fresh leaves are applied to old skin ulcers (Akopov 1990). Fresh fruits are used to treat kidney stones, inflammation of the gall bladder and bile duct, gout, stomach catarrh, constipation, hypertension, and arteriosclerosis, and is used as a vermifuge. Crushed fruits are applied to the skin to treat eczema. A decoction of the dried fruits is used as a diaphoretic and of the leaves as a diaphoretic and diuretic. A decoction of the roots is used as a hemostatic (Altimishev 1991).
Documented effects: Berries of wild strawberry possess tonic, anti-inflammatory, diuretic, antiscorbutic, choleretic, and hypoglycemic properties (Maznev 2004). An alcoholic extract of the aboveground parts increased the cellular mass of the spleen and thymus, protected the mucus membrane of the stomach and decreased stress in cyclophosphane-treated mice, as well as exhibited antiulcer and stress-protective effects (Aksinenko et al. 2003; Klimentova et al. 2005).
Phytochemistry: Fruits contain pectins, sugars, citric, malic and phosphoric acids, carotene, vitamin C, essential oils, and over 10 microelements, including iron. Leaves contain of vitamin C (high amounts), carotene, tannins, flavonoids, many different organic acids, essential oils, and up to 20 micro- and macroelements. The roots are rich in tannins and iron salts (Poludenny and Zhuravlev 2000).

Fumaria vaillantii Loisel. – Fumariaceae (Papaveraceae)

Synonyms: *Fumaria vaillantii* var. *schrammii* (Asch.) Velen.

English name: Earthsmoke

Russian name: Дымянка Вайяна (Dymyanka Vayyana)

Uzbek name: Shotara

Kyrgyz name: Вайлант фумариясы (Vaylant fumariyasy)

Description: Annual herb with taproot. Stems 10–35 cm tall, erect or reclining, branching from the base. Leaves alternate, long-petiolate, tri-pinnatisect, segments linear or linear-lanceolate, margins entire. Inflorescences terminal or leaf-opposed racemes. Flowers zygomorphic with 2 small sepals. Corolla pink-violet, darker towards the apex, 5–6 mm long. Petals 4, in 2 whorls, 1 outer petal with a short spur. Fruits indehiscent capsules, subglobose, 1.5–2.5 mm in diameter, tuberculate-wrinkled, 1-seeded.

Other distinguishing features: Staminal filaments connate into 2 groups.

Phenology: Flowering and fruits in March-July.

Reproduction: By seeds.

Distribution: In all regions of Uzbekistan and Kyrgyzstan.

Habitat: The adyr and tau zones. Mainly as a weed amongst crops, in orchards and in abandoned fields.

Population status: Common, especially in abandoned fields.

Traditional use: A decoction of the herb is taken as a blood-cleanser and diuretic, to treat coughs, jaundice, headache, fever, gonorrhea, uterine bleeding, erysipelas, and for clearing the intestines. It is also used externally in a bath to treat itching, rashes, and pimples (Khalmatov et al. 1984).

Documented effects: The alkaloid protopine (fumarine) caused narcosis in amphibians and, in mammals, caused paralysis of sensory nerve endings and increased reflex excitability. The alkaloid slightly increased the effects of analeptics and induced catalepsy (Chen-Gu 1957; Cheney 1963). In acute experiments with animals under narcosis, reduced heart rate and increased heartbeat amplitude occurred and, for a short time, decreased blood pressure was observed. Protopine has antiarrhythmic action with better effects than novocainamide and quinidine (Sadritdinov and Kurmukov 1980). In a screen to determine effects on platelet aggregation, extracts of this species showed complete inhibition of aggregation. This result was found to be caused by protopine (Sener 1994). Extracts of the dried plant displayed high rates of inhibition against the enzymes acetylcholinesterase and butyrylcholinesterase, which are associated with Alzheimer's disease (Orhan et al. 2004).

Phytochemistry: The aboveground parts contain alkaloids (protopine, vaillantine, parfumine, fumaridine, fumvailline, etc.), sugars, resins, pigments, fumaric acid, traces of essential oil, and vitamins C and K_1 (Ibragimova et al. 1974; Khalmatov et al. 1984; Khodzhimatov 1989). The seeds contain phospholipids (Gazizov and Glushenkova 1997).

Galium septentrionale Roem. & Schult. – Rubiaceae

Synonyms: some consider *G. septentrionale* a subspecies of *G. boreale* [*G. boreale* ssp. *septentrionale* (Roem. & J. A. Schult.) H. Hara].

English name: Northern bedstraw

Russian name: Подмаренник северный (Podmarennik severnyy)

Uzbek name: Chakamoog

Kyrgyz name: Тундук галиум (Tunduk galium)

Description: Herbaceous perennial, with rhizomes. Stems 30–70 cm tall. Leaves in whorls of 4, 4.5–5.5 cm long, 0.7–0.8 cm wide, elongate-lanceolate, 3-nerved. Inflorescence a dense, many-flowered, terminal panicle. Flowers small, white. Corolla rotate, 4-lobed. Fruits bristly nutlets.

Other distinguishing features: Stems glabrous.

Phenology: Flowering in June-August, fruits in July-September.

Reproduction: By seeds.

Distribution: All provinces of Kyrgyzstan; not found in the flora of Uzbekistan.

Habitat: In glades, among shrubs, and on river banks.

Population status: Common, forming dense groups.

Traditional use: Infusions and decoctions of the aboveground parts are used to treat deafness, malignant tumors, and applied to eyes to treat conjunctivitis. In Tibetan medicine a decoction of the aboveground parts is used to treat heart diseases, gastritis, and gynecological diseases. The rhizomes are used to treat pneumonia and gynecological diseases (Plant Resources of the USSR 1990).

Documented effects: In experiments with frogs a tincture decreased heart beat amplitude (Turova and Nikolskaya 1954).

Phytochemistry: The roots contain steroid saponins, tannins, flavonoids, coumarins, and anthraquinones. The aboveground parts contain essential oils, triterpene acids, iridoids, steroid saponins, alkaloids, tannins, coumarins, anthraquinones, and vitamin C (Revina and Shustova 1982).

Galium verum L. – Rubiaceae

Synonyms: *Galium glabratum* Klokov.

English name: Yellow spring bedstraw, Lady's Bedstraw

Russian name: Подмаренник настоящий (Podmarennik nastoyashchiy)

Uzbek name: Tilkisoomai

Kyrgyz name: Кадимки галиум (Kadimki galium)

Description: Herbaceous perennial, with branched rhizomes. Stems 30–125 cm tall, thin. Leaves in whorls of 8–12, narrow, linear, 1–4 cm long, 0.5–3 mm wide, 1-nerved, apex acute, margins sometimes recurved. Inflorescence a long, dense-flowered panicle. Flowers bright-yellow. Corolla rotate, 4-lobed. Fruits 2-parted.

Other distinguishing features: Flowers smell like honey.

Phenology: Flowers in June-July, fruits in July-August.

Reproduction: By seeds.

Distribution: Ysyk-Kol and Chuy provinces of Kyrgyzstan; Karakalpakstan autonomous republic, Toshkent, Samarqand, Farg'ona, Andijon and Surxondaryo provinces of Uzbekistan.

Habitat: In steppes and meadow-steppes.

Population status: Common, forming dense groups.

Traditional use: An infusion of the fresh aboveground parts is drank as a hemostatic, analgesic, sedative, and diuretic for people with swelling associated with heart or kidney diseases. A bath or compresses soaked with the infusion are used to treat rheumatism, various skin diseases, scrofula, and furunculosis. The rhizomes are used as to strengthen the libido. In Tibetan medicine, the rhizomes are used to treat pneumonia and liver diseases (Shreter 1975; Akopov 1990).

Documented effects: In vitro, ethanolic extracts of the plant showed low to moderate cytotoxic activity in human lymphoblastoid Raji cells (Spiridonov et al. 2005). Rubiadin exhibited antifungal and antituberculosis activity, as well as cytotoxicity to BC and NCI-H187 cancer cell lines (Kanokmedhakul et al. 2005). Asperuloside has laxative effects (Milkowska-Leyck et al. 1999).

Phytochemistry: The herb contains anthraglycosides and anthraquinones (galiosin, rubiadin, asperuloside, etc.), flavonoids, traces of essential oils, tannins, and dyeing substances. Rhizomes contain iridoids, steroid glycosides, coumarins, and flavonoids (Akopov 1990; Muzychkina 2000; Demirezer et al. 2006; Tamas et al. 2006; Zhao et al. 2008). Cultivated callus tissue produced a variety of different anthraquinones (Banthorpe and White 1995).

◄**Fragaria vesca L.**
Photos: Sergey Appolonov

▲**Fumaria vaillantii Loisel.**
Photos: Alexander Naumenko

▼**Galium septentrionale Roem. & Schult.** Photos: Vadim Prokhorov

►**Galium verum L.** Photos: *top* and *center*: Alim Gaziev; *bottom*: Natalie Schultz

Gentiana olivieri Griseb. – Gentianaceae

Synonyms: *Gentiana regeliana* Gand., *Gentiana weschniakowii* Regel.
English name: Unknown
Russian name: Горечавка Оливье (Gorechavka Oliv'ye)
Uzbek name: Gazakut, erbahasi
Kyrgyz name: Оливье кок базини (Oliv'ye kok bazini)
Description: Herbaceous perennial with thin rhizomes. Stems several, erect, 10–40 cm high, glabrous, smooth, rounded, sheathed at the base with the older basal leaves. Basal leaves in a rosette, elongate-lanceolate to elongate-spatulate, 2–12 cm long and 0.4–1 cm wide, green on both sides, glabrous; cauline leaves opposite, 2–3 pairs, lanceolate or narrow-lanceolate. Inflorescences terminal corymbiform cymes with 1–6 flowers. Corolla conical with 5 lobes, bluish-violet, dark blue or pale blue, rarely white, 2–4 cm long. Fruit an oblong capsule, 1–2 cm long, 2-valved. Seeds many, small, wingless, seed coat thick, surface reticulately patterned.
Other distinguishing features: Stamens inserted just below middle of corolla tube. This species has multiple forms distinguished by their pedicel lengths.
Phenology: Flowers in May-June, fruits in June-July.
Reproduction: By seeds and vegetative rhizomes.
Distribution: Toshkent, Samarqand, Jizzax, Buxoro, Qashqadaryo, Farg'ona, Andijon, and Surxondaryo provinces of Uzbekistan; Naryn, Osh, Chuy and Talas provinces of Kyrgyzstan.
Habitat: The adyr and tau zones. On dry slopes with rocky debris.
Population status: Common, usually found in small populations.
Traditional use: Decoction of the flowering herb is used for gastric diseases, malaria, toothaches, bleeding gums, and as an oral rinse, as well as is applied externally to treat abscesses and tumors. Syrup, made by boiling gentian and barberry roots for a long time, is recommended for side pains, rheumatic pain and chest pains (Khalmatov 1964).
Documented effects: The alkaloid gentianine, at doses of 50 mg/kg and higher, had sedative effects. At doses of 150–200 mg/kg it had central muscle relaxant action. At 10–25 mg/kg, the alkaloid noticeably prolonged the activity of soporifics in experiments on mice, eliminated aggressive reaction in rats, provoked a hypothermic effect, depressed developed conditioned reflexes and decreased stimulant action of caffeine and benzedrine (i.e., it has sedative and tranquilizing effect; Tulyaganov and Sadritdinov 1968; Tulyaganov et al. 1971; Danilevskii et al. 1972; Sadritdinov and Kurmukov 1980). The alkaloids gentianadine, gentianamine, and oliverine had anti-inflammatory action in rabbits and rats (Sadritdinov and Tulyaganov 1967, 1972; Sadritdinov 1971a). In experiments with rats, extracts of the plant exhibited hepatoprotective effects (Orhan et al. 2003). Methanolic extracts of the plant exhibited significant hypoglycemic effects on hyperglycemic rats (Sezik et al. 2005).
Phytochemistry: The aboveground plant parts contain alkaloids and bitter glycosides (Ersoz and Calis 1991; Orhan et al. 2003; Sezik et al. 2005). The plants around Toshkent had the following bitter index: leaves 1:20,000, flowers 1:5,000, stems 1:2,500, and the total aboveground parts 1:5,000. From the aboveground parts collected in the Toshkent province (village of Kaplanbek), 0.35 % total alkaloids were isolated and these included gentianine, gentiananine, gentianaine, gentianadine, gentioflavine, gentiotibetine, oliverine, oliveridine, oliveramine, and others (Rakhmatullaev and Yunusov 1972b; Yunusov 1974).

Geranium collinum Steph. ex Willd. – Geraniaceae

Synonyms: *Geranium minutum* Ikonn., *Geranium saxatile* Kar. & Kir., *Geranium wakhanicum* (Pauls.) Ikonn.

English name: Geranium

Russian name: Герань холмовая (Geran' kholmovaya)

Uzbek name: Anzhabor

Kyrgyz name: Шалбай каз таманы (Shalbay kaz tamany)

Description: Herbaceous perennial. Stems 15–55 cm tall. Leaves hairy, palmate, circular in outline with 3–7 lobes divided more than three fourth to the base, lobes nearly rhomboid, pinnatifid with 3–5 lobules. Flowers in small groups at end of axillary branches. Sepals 5, oblong-elliptic, 4–10 mm long. Petals 5, obovate, 10–19 mm long, pink-violet, apex rounded, claw ciliate. Fruit an elongated, beaked capsule covered with short hairs.

Other distinguishing features: Beak up to 3 cm long.

Phenology: Flowers in May-July, fruits in June-August.

Reproduction: By seeds.

Distribution: All provinces of Kyrgyzstan; Toshkent, Andijon, Farg'ona, Samarqand and Surxondaryo provinces of Uzbekistan.

Habitat: In steppes, in wet meadows in the forest-meadow mountain belt, along canals, in orchards, and in boggy places.

Population status: Common, forming dense groups.

Traditional use: The underground parts are used as an astringent and hemostatic. A decoction is used to treat gastric diseases, hemoptysis, and diarrhea, and also as a hemostatic after childbirth (Khalmatov 1964). In folk medicine it used internally to treat malignant tumors, broken bones, and fever (Amirov 1974).

Documented effects: In experiments on animals preparations from the leaves inhibited malignant tumors (Amirov 1974).

Phytochemistry: Underground parts of the plant contains tannins and phenols (pyrogallol and pyrocatechin). The aboveground parts contain flavone glycosides, saponins, alkaloids, and tannins (12–27.2 %). The whole plant is rich in tannins (Chumbalov et al. 1968; Chumbalov and Bikbulatova 1970; Plant Resources of the USSR 1988).

Geum rivale L. – Rosaceae

Synonyms: None

English name: Purple avens, water avens

Russian name: Гравилат речной (Gravilat rechnoy)

Uzbek name: Shirchai

Kyrgyz name: Ийилген гулду геум (Iyilgen guldu geum)

Description: Herbaceous perennial, with thick rhizomes. Stems 20–70 cm tall. Leaves in basal rosette, petiolate, sparsely hairy, lyrate-pinnately compound, with 3–7 leaflets, terminal leaflet largest and lobed; cauline leaves smaller, simple to 3-lobed, stipulate. Inflorescence terminal, corymbiform, 2–4-flowered, often nodding. Flowers 5-merous, pedicels pubescent. Sepals reddish-purple. Petals yellow with reddish brown-purple veins. Stamens and carpels numerous, styles plumose. Fruits long-beaked achenes in a globose aggregate; achenes fusiform, 3–4 mm long, yellow villous.

Other distinguishing features: Achenes have a hooked style to aid in dispersal.

Phenology: Flowers in June, fruits in July.

Reproduction: By seeds.

Distribution: All provinces of Kyrgyzstan; Toshkent province of Uzbekistan.

Habitat: In forest glades, in long-used animal corrals in the tallgrass-meadow belt of mountains, and in valleys and along brooks.

Population status: Common, forming dense groups.

Traditional use: Infusions and decoctions of the herb are used to treat paradontosis, stomatitis, laryngitis, stomach catarrh, dysentery, vomiting, and intestinal colitis. Infusions and decoctions of the rhizomes are used to treat headaches, insomnia, eye diseases, rheumatism, and hemorrhoids, and is effective against snake venom (Krilov 1972).

Documented effects: Extracts of this plant showed anti-inflammatory activity in vitro (Tunon et al. 1995).

Phytochemistry: Rhizomes contain carbohydrates (glucose, arabinose, and ketose), pectins, organic acids (6.46 %), essential oils, saponins, alkaloids, vitamin C, and tannins. The leaves contain vitamin C, carotene, and tannins. The flowers contain tannins (7.35 %) and the fruits contain carbohydrates (Blinova 1957; Aliev et al. 1961). The roots were found to contain small amounts of proanthocyanidins and high amounts of ellagic acid (Oszmianski et al. 2007).

Glaucium fimbrilligerum Boiss. – Papaveraceae

Synonyms: *Dicranostigma iliense* C.Y. Wu & H. Chuang, *Glaucium luteum* var. *fimbrilligerum* (Boiss.) Trautv.

English name: Unknown

Russian name: Глауциум бахромчатый (Glautsium bakhromchatyy)

Uzbek name: Urmon kora

Kyrgyz name: Туктуу саргалдак (Tuktuu sargaldak)

Description: Annual or biennial herb. Stems 8–65 cm tall, branched, leafy. Lower leaves lyrate-pinnatisect, 5–30 cm long; upper leaves clasping, many-lobed. Flowers solitary, axillary. Buds 15–20 mm long, glabrous. Petals bright yellow, lacking spots, wide-obovate or round, 1.5–3.7 cm long, 2.5–3.5 cm wide, margin wavy. Fruit a silique-like capsule, 10–25 cm long, up to 0.4 cm wide, dehiscing from the top nearly to the base, straight or arching, sparsely covered with appressed bristles. Seeds kidney-shaped, 1.5–2 mm long, brown.

Other distinguishing features: Pedicels up to 20 mm long in fruit. Capsules with 2 horns at the tip.

Phenology: Flowers and fruits in May-July.

Reproduction: Seeds.

Distribution: Toshkent, Farg'ona, Samarqand, Qashqadaryo, and Surxondaryo provinces of Uzbekistan; Osh, Chuy and Talas provinces of Kyrgyzstan.

Habitat: The adyr and tau zones. On clay bluffs, stony slopes and in dry river beds with rocky debris.

Population status: Rare.

Traditional use: The crushed, roasted seeds are recommended as a hemostatic and tonic for women after childbirth. The oil has the same abilities. A decoction of the leaves and flowers, as a tea, is given as a tonic and stimulant for people recovering from diseases. Large doses have emetic and soporific effects, but can cause asphyxiation. The seeds are considered a strong laxative (Khodzhimatov 1989).

Documented effects: In acute experiments with animals under anesthesia, reduced heart rate and increased heartbeat amplitude occurred and, for a short time, decreased blood pressure was observed. Protopine has antiarrhythmic action with better effects than novocainamide and quinidine (Sadritdinov and Kurmukov 1980). The alkaloid corydine has a general activity similar to that of bulbocarpine, and like bulbocarpine can cause catalepsy (Berezhinskaya et al. 1968). In acute experiments with animals, the alkaloid chelerythrine produced 2 phases of action on arterial pressure: hypertensive effects due to the alkaloids influence on the vasomotor center and hypotensive effects due to the alkaloids influence on the muscle walls of vessels. Chelerythrine has analgesic activities, potentiates analgetic action of morphine and elongates sleep produced sleeping preparations (Chelombito and Muravyova 1971).

Phytochemistry: The entire plant contains alkaloids (protopine, corydine, sanguinarine, corytuberine, glauvine, glaunine, norcorydine, isoboldine, etc.). The seeds contain up to 30 % drying fatty oil (Yunusov et al. 1973; Yunusov and Israilov 1974; Karimova et al. 1980, 1983; Khodzhimatov 1989; Shafiee et al. 1998).

▲**Gentiana olivieri Griseb.** Photo: Stefano Doglio

▲**Glaucium fimbrilligerum Boiss.** Photos: Evgeny Davkaev

▼**Geum rivale L.** Photos: Alexander Naumenko

▼**Geranium collinum Steph. ex Willd.** Photos: *left:* Alim Gaziev; *center*: Sasha W. Eisenman; *right*: John B. Taft

Gleditsia triacanthos L. – Fabaceae

Synonyms: *Acacia americana* Cat. Long. ex Stokes, *Acacia triacanthos* (L.) Gron., *Caesalpiniodes triacanthum* (L.) Kuntze, *Gleditsia brachycarpa* (Michx.) Pursh, *Gleditsia bujotii* Neumann, *Gleditsia elegans* Salisb., *Gleditsia hebecarpa* S. McCoy, *Gleditsia heterophylla* Raf., *Gleditsia horrida* Salisb., *Gleditsia inermis* L., *Gleditsia meliloba* Walter, *Gleditsia micracantha* Loddiges ex Steudel, *Gleditsia polysperma* Stokes, *Gleditsia spinosa* Marsh, *Gleditsia triacanthus* (L.) Mill., *Melilobus heterophyla* Raf.

English name: Honey-locust

Russian name: Гледичия обыкновенная (Gledichiya obyknovennaya)

Uzbek name: Tikandarakht

Kyrgyz name: Кадимки гледичия (Kadimki gledichiya)

Description: Deciduous tree, 12–20 m tall. Trunk and branches with long, simple or branched, reddish-brown, 2–8 cm long thorns. Leaves of 2 kinds: pinnate and bipinnate, petioles pubescent; once-pinnate leaves on short lateral spurs; bipinnate leaves on long shoots; leaflets oblanceolate. Inflorescences perfect or staminate, in separate racemes arising from the short lateral spurs. Flowers yellow-green, calyx and petals pubescent. Fruit a dark brown legume, flat, often slightly twisted, up to 40 cm long. Seeds elongate-elliptic, up to 15 mm long.

Other distinguishing features: Has sweet pulp between the seeds. Thornless cultivars exist and are used as an ornamental plant.

Phenology: Flowers in May, fruits in September.

Reproduction: By seeds.

Distribution: Native to North America. Cultivated nearly everywhere in Kyrgyzstan and Uzbekistan.

Habitat: Only cultivated.

Population status: Common.

Traditional use: The plant is used to treat spastic colitis, chronic cholecystitis, stomach ulcers, and bronchial asthma (Rakhmanberdyeva et al. 2002).

Documented effects: In experiments the alkaloid triacanthine showed hypotensive and antispasmodic activity. The antispasmodic actions occurred in the bronchial smooth muscles and the intestines. The saponin triacanthocide showed anti-arrhythmic action in experiments (Khalmatov et al. 1984). A preparation of triacanthine is used to treat digestive system problems (Altimishev 1991).

Phytochemistry: Young leaves contain up to 1 % of the alkaloid triacanthine and flowers contain up to 0.3 %. Leaves contain up to 400 mg% ascorbic acid. Fruits contain olmelin, fustin, and no less than 10 triterpene glycosides. The fruit walls contain around 2.6 % anthraglycosides, 3.1 % tannins, and traces of essential oil. The pulp of the fruits contain up to 29 % sugars, and the seeds contain up to 39 % mucilage, carbohydrates, lipids, fatty acids (palmitic, oleic, linoleic, and linolenic), carotinoids, etc. (Khalmatov et al. 1984; Rakhmanberdyeva et al. 2002).

Glycyrrhiza glabra L. – Fabaceae

Synonyms: *Glycyrrhiza glandulifera* Waldst. & Kit., *Glycyrrhiza hirsuta* Pall., *Glycyrrhiza violacea* Boiss. & Noë.

English name: Common licorice

Russian name: Солодка голая (Solodka golaya)

Uzbek name: Kizilmiya, Chuchuk miya, Shirin miya

Kyrgyz name: Тукуз кызыл мыя (Tukuz kyzyl myya)

Description: Herbaceous perennial with deep root system down to 5 m. Stems erect, simple or branched, 45–120 cm high, sparsely short-hairy with scattered glands or glandular prickles. Leaves alternate, odd-pinnate, 5–20 cm long with (2–)3–9 pairs of leaflets; leaflets oblong, ovate or lanceolate, 2–4 cm long, 0.8–2 cm wide, with glands on abaxial side. Inflorescences loose racemes, 5–12 cm long. Flowers 8–12 mm long. Calyx 5-lobed, upper 2 lobes half as long as lower 3. Corolla papilionaceous, whitish-violet. Fruit a legume, 2–7-seeded, straight or slightly curved, glabrous or with dense glandular prickles. Seeds small, 3 mm in diameter, almost round, smooth, deep-brown.

Other distinguishing features: Stamens 10 (9 united). Interior of root is lemon-yellow and has a specific sweet taste.

Phenology: Flowers in April-July, fruits in May-June.

Reproduction: By seeds and rhizomes.

Distribution: All regions of Uzbekistan and Kyrgyzstan.

Habitat: The chul and adyr zones. River banks, embankments, along canals, salty-soiled areas (salanchaks), tugai, on gentle slopes of mountains and foothills, and in melon and cotton fields as a weed.

Population status: Common, sometimes found in large groups.

Traditional use: The plant root has been used to treat various diseases since ancient times. Decoctions and extracts of the root are used as a diaphoretic and purgative and also to treat cough, chest pains, and other diseases. Avicenna recommended the roots to treat renal, lung, and bladder diseases, as well as gastritis, fever, and other diseases. The root decoction is used for throat dryness and spasms and as an expectorant for coughs and respiratory tract catarrh (Khalmatov et al. 1984).

Documented effects: Modern medicine uses preparations made of the roots (syrup, thick and dry extracts), as well as in combinations with other substances such as a mixture of powders, as a slight laxative, expectorant, and to coat the stomach. The powder, thick and dry extracts, and root syrup are widely used in pharmaceutical practice to make pills, improve mixture taste, and for other purposes. It was established that the active ingredients of the roots (glycyrrhizic and glycyrrhetinic acids) have antispasmodic and antihistamine activities, similar to adrenal hormones (deoxycorticosterone and hydrocortisone) and are recommended to treat skin diseases and inflammatory processes (Mashkovskii 1984). The preparations have tonic and adaptagenic activities and are useful for recovery of general health and memory improvement (Kurmukov 1976). Licochalcone-A, an estrogenic flavonoid found in licorice root has been shown to effectively inhibit proliferation of prostate cancer cells (Fu et al. 2004). Isoliquiritigenin inhibited platelet aggregation and aldose reductase activity in vivo (Aida et al. 1990; Tawata et al. 1992), and in vitro, inhibited proliferation and induced apoptosis in prostate cancer cell lines (Kanazawa et al. 2003; Jung et al. 2006a, b).

Phytochemistry: Underground organs contain 4.6–23 % glycyrrhizin, up to 10.5 % sugars, up to 8.1 % bitters (glycyrramarin), flavonoids (liquiritin, liquirazide, liquitigenin and 2′-4,4′-trihydroxychalcone and its glycoside isoliquiritigenin), glabric acid, 0.035 % essential oil, β-sitosterol, extriol, 1–4 % asparagines, dyes, and other substances (Kurmukov 1976; Mashkovskii 1984).

Glycyrrhiza uralensis Fisch. ex DC. – Fabaceae

Synonyms: *Glycyrrhiza asperrima* var. *desertorum* Regel, *Glycyrrhiza asperrima* var. *uralensis* (Fisch. ex DC.) Regel, *Glycyrrhiza glandulifera* Ledeb.

English name: Chinese licorice

Russian Name: Солодка уральская (Solodka ural'skaya)

Uzbek name: Shirinmiya

Kyrgyz name: Урал кызыл мыясы (Ural kyzyl myyasy)

Description: Herbaceous perennial, with large rhizome. Stems simple or branched, 40–70 cm high, short pubescent with punctuate glands or raised glands. Leaves alternate, odd-pinnate, 10–25 cm long; leaflets 3–8 pairs, 2–6 cm long, 1.5–3.5 cm wide, obovate or elliptic. Inflorescences densely flowered, axillary racemes. Flowers 1.5–2.5 cm long. Calyx 8–14 mm long, toothed, pubescent. Corolla papilionaceous; petals violet, banner petal rounded (cupped) or sinuate. Fruits crescent-shaped legumes, 2–4 cm long, in dense, tangled clusters. Seeds round to kidney-shaped, brown, smooth.

Other distinguishing features: Has a more dense-flowered raceme and larger flowers than *Glycyrrhiza glabra*.

Phenology: Flowers in May-June, fruits in August-September.

Reproduction: By seeds and rhizomes.

Distribution: Southern and eastern parts of Kyrgyzstan; Surxondaryo province of Uzbekistan.

Habitat: In meadows with relatively high water tables and along canals and rivers.

Population status: Common, found in dense groups.

Traditional use: The underground parts are used as a diuretic, laxative, and carminative, and to treat pneumonia, bronchitis, asthma, and ulcers, and also as a remedy for poisoning (Khodzhimatov 1989).

Documented effects: Similar to *Glycyrrhiza glabra*. Because of the high flavonoid content in the above and below ground parts of *Glycyrrhiza uralensis*, it is used as raw material for antispasmodic and anti-ulcer preparations (Khalmatov et al. 1984). Extracts of the root exhibited apoptotic effects on human breast cancer cells (Jo et al. 2004).

Phytochemistry: The aboveground parts have up to 3.3 % total flavonoids. The below ground parts have up to 4.3 % total flavonoids (glycyrrhizic acid, glycyrrhetinic acid, fermononetin, isoliquiritigenin, etc.; Tolmachev 1976; Nakanishi et al. 1985; Wang et al. 2004b).

Haplophyllum acutifolium (DC.) G. Don. f. – Rutaceae

Synonyms: *Haplophyllum flexuosa* Boiss., *Haplophyllum perforatum* (M. Bieb.) Kar. & Kir., *Haplophyllum sieversii* Fisch., *Ruta acutifolia* DC., *Ruta flexuosa* (Boiss.) Engl., *Ruta perforata* M. Bieb., *Ruta sieversii* (Fisch.) F. Fedtsch.

English name: Unknown

Russian name: Цельнолистник остролистный (Tsel'nolistnik ostrolistnyy)

Uzbek name: Toshbakatol, Tashbakftol

Kyrgyz name: Unknown

Description: Herbaceous perennial, covered with punctate glands. Stem erect, corymbiform- branching, glabrous, 20–70 cm high. Leaves alternate, simple, broadly-oblong to narrowly-lanceolate, entire, glabrous, short-petiolate. Inflorescence paniculate-corymbiform, multiflorous. Calyx lobes 5, ovate-triangular, acute, glabrous or slightly pubescent. Petals 5, yellow, 3.5–5 mm long, tapering to a claw. Stamens 10. Style glabrous; ovary sessile. Fruit a capsule with indehiscent deciduous segments, on a very short stipe, densely tuberculate.

Other distinguishing features: Pellucid dots on leaves observable when held up to the light. Leaves produce a specific objectionable odor when bruised.

Phenology: Flowers in May-June, fruits in July-August.

Reproduction: By seeds and rhizomes.

Distribution: Toshkent, Farg'ona, Samarqand, Qashqadaryo, and Surxondaryo provinces of Uzbekistan; not found in Kyrgyzstan.

Habitat: The adyr and tau zones. Shallow soils, rarely on stony slopes.

Population status: Common.

Traditional use: Preparations of the plant are used in folk medicine as a sedative for anxiety and cardiac neurosis, as well as for hysterics, epilepsy, gastric spasms, and menstrual period disturbance (Kovaleva 1971). A leaf decoction is used to treat toothaches, chest and stomach diseases, and for bloated abdomens. A decoction and infusion of the herb, together with decoction of common wormwood (*Artemisia absinthium*), is used as a bath to treat various skin diseases (Khalmatov 1964).

Documented effects: The plant contains the alkaloids perforine, haplofolin, and haplofilidine, which in small doses produce sedative effects, in medium doses produce soporific effects, and in large doses, causes narcosis. These alkaloids are strongly pronounced antagonists against some analeptics (camphor, strychnine, and caffeine) and reinforce actions of some soporifics and narcotics. Haplofilidine eliminated fear in tested rats, but perforine did not have such an action (Akhmedhodzhaeva and Polievtsev 1963; Danilevskii et al. 1972; Akhmedhodzhaeva and Kurmukov 1975; Akhmedhodzhaeva 1978). The majority of alkaloids contained in the plant have estrogenic activity (Akhmedhodzhaeva 1978). When tested for in vitro cytotoxicity, extracts of the aboveground plant parts had strong cytotoxic activity against multiple types of cancer cell lines (Varamini et al. 2007).

Phytochemistry: Plants collected in the Qashqadaryo province of Uzbekistan contained varying amounts of total alkaloids during different phenologic periods: alkaloid content of leaves during flower bud stage was 0.1 %; alkaloid content of leaves during flowering stage was 0.4 %; alkaloid content of leaves during seed maturation was 1.2 %, alkaloid content in stems was 0.075–0.14 %; alkaloid content at the stage of full fruit maturity, in roots was 0.025 % and in seeds was 1.6 %. From different parts of plants growing in several regions of Uzbekistan, 25 alkaloids were isolated, including evoxin, skimmianine, haplofilidine, perforine, haplamine, haplopine, flindersine, glycoperine, methyl-evoxin, evodine, evoxoidine, haplofidine, anhydroperforine, perfamine, foliosidine, dubinidine, etc., and the lignan eudesmine (Akhmedzhanova et al. 1974; Razzakova et al. 1973, 1986; Yunusov 1981). Kusunokinin, β-sitosterol, oleanolic acid, cholesterol and hexadecanoic acid, as well as the alkaloids haplophytin-A and B, were isolated from the plant (Ali et al. 2001).

▲**Gleditsia triacanthos L.** Photos: *left*: Georg Slickers; *center*: Andrew Butko; *right*: Luis Fernández García

▼**Glycyrrhiza uralensis Fisch. ex DC.** Photos: Alexander Naumenko

▲**Glycyrrhiza glabra L.** Photos: Maxim Kucherov

▼**Haplophyllum acutifolium (DC.) G. Don. f.** Photo: Evgeny Davkaev

Helichrysum maracandicum Popov ex Kirp. – Asteraceae

Synonyms: None

English name: Unknown

Russian name: Цмин самаркандский, Бессмертник самаркандский (Tsmin samarkandskiy, Bessmertnik samarkandskiy)

Uzbek name: Samarkand buznoch

Kyrgyz name: Самарканд очпос гулу (Samarkand ochpos gulu)

Description: Herbaceous perennial. Stems 15–75 cm tall, striated, hairy. Leaves alternate, greenish, gray-green, to yellow-green, densely hairy; basal and cauline leaves linear to linear-lanceolate, apex very acute, margins entire, base partly sheathing stem. Inflorescences capitula, 5 mm wide, spherical, semispherical, campanulate, or elliptic; capitula in groups of 20–80 and forming dense clusters or compact corymbiform structures. Involucral bracts 40–60 in 5 rows, stiff-membranaceous, yellow. Flowers 50–80 per capitulum; corollas yellow. Fruits dark-brown achenes with pappus of 20–25 very thin, whitish-yellow bristles.

Other distinguishing features: The entire plant is densely hairy. Outer involucral bracts lanceolate to elliptical, more numerous than inner bracts. Inner bracts spatulate, glabrous, shiny.

Phenology: Flowers in June, fruits in September-October.

Reproduction: By seeds.

Distribution: Naryn, Chuy, Talas, and Jalal-Abad provinces of Kyrgyzstan; Toshkent, Andijon, Farg'ona, Samarqand and Surxondaryo provinces of Uzbekistan.

Habitat: In steppes, on stony slopes, and among bushes.

Population status: Common, forming dense groups.

Traditional use: This species is used as a replacement for *Helichrysum arenarium.* The inflorescences are collected at the beginning of flowering and are used to make a tea, which is taken to treat liver diseases, jaundice, gall and kidney stones, edema, and tuberculosis. It is also used as a hemostatic for hemorrhoidal bleeding, as a vermifuge (particularly for ascaridosis), as a common cold remedy, and as a diuretic (Khodzhimatov 1989; Sezik et al. 2004).

Documented effects: In modern medicine, water decoctions and infusions, liquid extracts, and dry concentrates of the inflorescences, as well as the preparation *Flamin,* are used as a choleretic for treating liver disease, cholecystitis and hepatocholecystitis (Khodzhimatov 1989). An ethanolic extract of the flowers and the compound naringenin chalcone (isolated from the extract) showed antiproliferative activity against mouse skin tumor cells in vitro. Application of isosalipurposide, isolated from the flowers, delayed formation of papillomas in an in vivo assay of carcinogenesis on mouse skin (Yagura et al. 2008).

Phytochemistry: The flowers contain flavonoids, glycosides, diterpenes, coumarins, sterins, vitamin K, essential oil, gum, dyeing substances, fatty acids, etc. (Khodzhimatov 1989; Baimukhamedov and Komissarenko 1990; Ul'chenko et al. 2000; Yagura et al. 2008).

Herniaria glabra L. – Caryophyllaceae

Synonyms: *Herniaria suavis* Klokov.

English name: Rupturewort, smooth rupturewort

Russian name: Грыжник голый (Gryzhnik golyy)

Uzbek name: Tuksiz saminchop

Kyrgyz name: Туксуз самын чоп (Tuksuz samyn chop)

Description: Yellowish-green perennial herb with woody taproot. Stems prostrate, sometimes ascending, 5–25 cm long, strongly branched from the base, glabrous or slightly hairy. Leaves mostly opposite, simple, elliptic to obovate, 2–7 mm long, 1–3 mm wide, short-petiolate, usually glabrous or sometimes short-ciliate. Inflorescences axillary clusters or capitate-spiciform, usually leaf opposed. Flowers sessile. Calyx 5-lobed, whitish-green, lanceolate to oblong, glabrous. Petals absent. Stamens 5. Styles 2, lower 1/3 connate. Fruit a utricle, 1–1.3 mm, usually longer than calyx.

Other distinguishing features: Differs from closely related species by having herbaceous stems (sometimes woody at base) and mostly glabrous leaves.

Phenology: Flowers and fruits in June-August.

Reproduction: By seeds.

Distribution: Toshkent, Farg'ona, Samarqand, and Surxondaryo provinces of Uzbekistan; all of Kyrgyzstan.

Habitat: The adyr and tau zones. In open, dry, sandy, stony places, along rivers, near roads, and on mountain slopes.

Population status: Not common, found as single individuals.

Traditional use: Used as diuretic to treat edema and acute catarrh of the bladder, as an astringent, to treat syphilis, pulmonary and other diseases, as well as for kidney inflammation and jaundice (Khalmatov et al. 1984).

Documented effects: Preparations have antispasmodic and diuretic activities, and are especially effective for urinary bladder inflammation. In medicine, an infusion is used for renal pain, inflammation of the renal pelvis, ureteritis, and to help excrete stones from kidneys and the urinary bladder (Khalmatov et al. 1984). A water extract from the aboveground parts increased diuresis in rats by 73 % (Khodzhimatov 1989). Treating hypertensive rats with saponins from *Herniaria glabra* resulted in a significant decrease in blood pressure (Rhiouani et al. 1999; Rhiouani et al. 2001).

Phytochemistry: The herb contains coumarins (umbelliferone and herniarine) and their derivatives, flavonoids (quercetin, rutin, quercetin triglycoside, quercetin arabinoside, quercetin galactoside, rhamnoglycoside, isorhamnetin triglycoside, etc.), triterpene saponins, essential oil, and traces of alkaloids (Khodzhimatov 1989; Akopov 1990; Schröder et al. 1993).

Hibiscus trionum L. – Malvaceae

Synonyms: *Hibiscus ternatus* Cav.

English name: Flower of an-hour

Russian name: Гибискус тройчатый (Gibiskus troychatyy)

Uzbek name: Burytaroq

Kyrgyz name: Уч айчыктуу гибиск (Uch aychyktuu gibisk)

Description: Herbaceous annual, 5–75 cm tall. Stems erect, mostly branched, lower branches elongated, stems with scattered stiff, forked and stellate hairs,. Leaves alternate, petiolate, stipulate, adaxial surface of leaf nearly glabrous, abaxial side with scattered stellate-hairs; stem leaves palmatilobate, usually with 3 oblong, pinnatilobate segments; uppermost leaves unlobed to slightly lobed. Flowers solitary in leaf axils, with long pedicels; epicalyx with 7–13 bractlets, linear, ciliate-bristly. Calyx campanulate, 5-lobed, with purple veins, bristly and stellate hairy. Corolla very deeply 5-lobed, pale-yellow with reddish-purple center, 1.5–3.5 cm wide. Staminal column 3–4 mm long. Stigmas 5, reddish-purple. Fruit a black capsule, hairy. Seeds 2.5 mm long, kidney-shaped or irregular.

Other distinguishing features: Stipules 2–7 mm long, long-ciliate. Flowers quickly fading. Calyx becomes inflated in fruit.

Phenology: Flowers and fruits in July-September.

Reproduction: Only by seeds.

Distribution: All regions of Uzbekistan and Kyrgyzstan.

Habitat: The chul zone. As a weed in cotton and melon fields, vegetable gardens, and all irrigated farming areas.

Population status: Common.

Traditional use: An infusion of the leaves is used as an expectorant to treat catarrh in the upper respiratory tract. In Romania the plant is used as a diuretic (Khalmatov 1964).

Documented effects: An infusion and extract made from different plant parts have a diuretic effect. Special diuretic properties were documented from preparations of the leaves (Khalmatov 1964). Extracts of the plant exhibit antimicrobial activity (Szabo et al. 2006).

Phytochemistry: Gossypol has been isolated from the seeds (Schmidt and Wells 1990). The main fatty acids isolated from the seed oil were linoleic acid (63.61 %), hexadecanoic acid (16.72 %), oleic acid (12.30 %), stearic acid (2.23 %), and the total content of the unsaturated fatty acids was 79.11 % (Hu et al. 2006).

Hippophae rhamnoides L. – Elaeagnaceae

Synonyms: *Elaeagnus rhamnoides* (L.) A. Nelson, *Hippophae angustifolia* Lodd. ex Dippel, *Hippophae littoralis* Salisb., *Hippophae rhamnoideum* Saint-Lager, *Hippophae sibirica* Hort. ex Steud., *Osyris rhamnoides* Scop., *Rhamnoides hippophae* Moench.

English name: Sea buckthorn, seaberry

Russian name: Облепиха крушиновая (Oblepikha krushinovaya)

Uzbek name: Chakanda

Kyrgyz name: Кадимки чычырканак (Kadimki chychyrkanak)

Description: Deciduous, dioecious shrub or small tree, 1.5–11 m tall. Branches with 2–7 cm long spines. Leaves alternate, short-petiolate, linear-lanceolate, 2–8 cm long, 2–8 mm wide, adaxial side gray-green, abaxial side brownish-silver due to scales and stellate hairs. Flowers unisexual. Staminate flowers in short spikes; flowers 5–8 mm long, 4–6 mm wide, outside covered with brown and white scales. Pistillate flowers covered with scales, very short-pedicellate, in groups of 2–5 in branch and thorn axils. Fruit a juicy, orange, red or yellow ellipsoidal drupe, 0.5–1 cm long, 3.8 mm wide. Seeds dark-brown, shiny.

Other distinguishing features: Fruits have a peculiar flavor and aroma.

Phenology: Flowers in April-May, fruits in August.

Reproduction: By seeds.

Distribution: All provinces of Kyrgyzstan; Toshkent, Farg'ona and Samarqand provinces of Uzbekistan.

Habitat: Along stream and river banks in valleys and into the mountains.

Population status: Common, found in dense groups.

Traditional use: The fruits are used as an analgesic, as a remedy for stomach pain, to improve digestion, and to treat scurvy. A decoction of the fruits is drunk to treat ulcers and is added to baths to prevent skin diseases. Fresh fruits are used to moisturize the skin, to help heal small wounds and burns, and to treat skin diseases associated with poor metabolism. An infusion of the leaves is drunk or the leaves are directly applied to the body to treat rheumatism. A decoction of the seeds is used as a laxative (Khalmatov et al. 1984; Khodzhimatov 1989).

Documented effects: Fruits are a rich source of polyvitamins. Oil from the fruits is used as an analgesic and to treat burns, frostbite, eczema, persistent wounds, as well as stomach and duodenal ulcers. The oil is used during radiation treatment for esophageal cancer (Tolmachev 1976). A study of the radioprotective action of a preparation of this species resulted in an 82 % survival rate in mice that received the treatment compared to no survival in irradiated control (Goel et al. 2002). Alcoholic extracts of leaves and fruits of sea buckthorn were found to inhibit chromium-induced free radical production, apoptosis, and DNA fragmentation, and restored the anti-oxidant status to that of control cells. These extracts also were able to arrest the chromium-induced inhibition of lymphocyte proliferation (Geetha et al. 2002). Flavonoids isolated from the plant are reported to have antioxidant, anti-ulcerogenic, and hepato-protective properties (Yue et al. 2004).

Phytochemistry: The fruits contain carotene, vitamins C, E, B_1 and B_2, folic acid, sugars, organic acids, quercetin, isorhamnetin, tannins, and semi-drying fatty oil. The leaves contain tannins, vitamin C, and flavonoids (kaempferol, quercetin, isorhamnetin and myricetin; Khalmatov et al. 1984; Yue et al. 2004).

▲**Helichrysum maracandicum Popov ex Kirp.** Photos: *left*: Evgeny Davkaev; *center* and *right*: Alexander Naumenko

▲**Herniaria glabra L.**
Photos: Maxim Kucherov

◀**Hippophae rhamnoides L.**
Photos: *top*: Vadim Prokhorov; *center*: Rostislav Lezhoev; *bottom*: Sergey Mayorov

▶**Hibiscus trionum L.**
Photos: *bottom*: Alim Gaziev; *center* and *top*: Sasha Eisenman

Hyoscyamus niger L. – Solanaceae

Synonyms: *Hyoscyamus agrestis* Kit. ex Schult., *Hyoscyamus bohemicus* F.W. Schmidt.

English name: Black henbane

Russian name: Белена чёрная (Belena chyornaya)

Uzbek name: Ming divana

Kyrgyz name: Кара мендубана (Kara mendubana)

Description: Herbaceous biennual with taproot. Stems green, 15–150 cm, villous. Leaves alternate, simple, dull green from above, gray-green below with long hairs; basal rosette leaves long-petiolate, elliptic, pinnatifid; cauline leaves sessile, elongate-lanceolate, with triangular lobes. Flowers solitary in axils or in scorpioid spikes. Calyx tubular-campanulate, with 5 broadly triangular lobes. Corolla funnelform with 5 lobes, greenish-yellow with purple reticulate veins. Fruit a bilocular capsule, circumscissile, 15–18 mm long. Seeds up to 500 per capsule, brownish-gray.

Other distinguishing features: The entire plant is densely hairy and has an unpleasant aroma.

Phenology: Flowers in May-August, fruits in July-September.

Reproduction: By seeds.

Distribution: Almost all provinces of Kyrgyzstan and Uzbekistan.

Habitat: In waste places, near houses, in vegetable gardens, and cultivated and fallow fields.

Population status: Common, forming dense groups.

Traditional use: Avicenna recommended the juice of the leaves to treat eye, ear, tooth, and uterine pain and as a hemostatic for uterine bleeding. He also suggested that a paste made with the leaves and seeds be used as an analgesic for pain associated with gout. In current folk medicine this plant is still used as an analgesic. The leaf juice is used to treat tumors and earaches. A water infusion of the seeds is used to treat convulsions and smoke from the burning seeds is used to treat toothaches. A plaster of the leaves is put on swollen abscesses to draw out pus (Khalmatov et al. 1984).

Documented effects: The plant is highly toxic. Preparations from this species are mostly used as antispasmodic and analgesic medicines. Atropine is used to treat bile ducts, stomach and duodenal ulcers, intestinal spasms, and bronchial asthma, and is used in opthalmology as a mydriatic. Scopolamine is used as a depressant of the central nervous system in surgery and psychiatry. Oil from the leaves is used as analgesic to treat rheumatism and neurological pains. The leaves are used to prepare antiasthmatic medicines (*Asthmatin*; Tolmachev 1976).

Phytochemistry: The entire plant contains alkaloids including hyoscyamine (isomer of atropine), scopolamine, and glycosides. The seeds contain essential oils. The leaves are rich in flavonoids such as rutin (Tolmachev 1976; Gammerman et al. 1990).

Hypericum perforatum L. –Hypericaceae

Synonyms: *Hypericum komorovii* Gorschk., *Hypericum nachitschevanicum* Grossh.

English name: Common St. Johnswort, St. Johnswort, Klamath weed, goat weed

Russian name: Зверобой продырявленный (Zveroboy prodyryavlennyy)

Uzbek name: Kizil-poicha

Kyrgyz name: Козонокчолуу сары чай чоп (Kozonokcholuu sary chay chop)

Description: Herbaceous perennial plant with a much-branched taproot. Stems one to many, erect, 20–100 cm tall, the upper portions branched. Leaves simple, opposite, sessile, entire, elliptic or elongate-obovate, dotted with light-colored translucent and black (along margins) glands. Inflorescences cymes or corymbiform. Flowers 1.5–2.5 cm wide. Sepals 5, lanceolate to oblong. Petals 5, yellow, twice as long as sepals with marginal black dots. Fruit a capsule, 5–9 mm long, elongate-ovoid. Seeds small, elongate, brown.

Other distinguishing features: Stamens united at base into 3–5 fascicles. Stems ridged below leaves.

Phenology: Flowers in June-July, fruits in July-August.

Reproduction: By seeds.

Distribution: All provinces of Kyrgyzstan and Uzbekistan.

Habitat: In meadow-steppes, meadows, stony slopes on foothills, along canals, and in fallow fields.

Population status: Common, not found in very large groups.

Traditional use: One of the most commonly used herbs in Central Asia. A decoction of the herb is used as an astringent, anti-inflammatory, antiseptic, tonic, and hemostatic, and is used to treat kidney diseases, heart diseases, diarrhea, and hemoptysis. The decocotion is applied externally to treat wounds (Khalmatov et al. 1984).

Documented effects: Preparations of this species are used externally as an astringent, disinfectant, and anti-inflammatory, and used internally to treat gastrointestinal diseases and acute and chronic colitis of non-bacterial origin. Oil from the plant is used to treat gingivitis and stomatitis. A tincture of the herb is used to rinse the mouth and is drank to treat colitis, gallstones, and cystitis. The antibacterial preparation *Novoimanin* is used against gram-positive bacteria, including penicillin-resistant *Staphylococcus*. Externally, it is applied to infected wounds, carbuncles, paronychia, and furuncles. *Novoimanin* is used to treat mastitis and the cracked nipples of lactating women, in stomatology, to treat stomatitis ulcers, and in otolaryngology to treat acute rhinitis, pharyngitis, laryngitis, highmoritis, chronic tonsilitis, and chronic and acute otitis (Maznev 2004). The extracts of the plant has been shown to have antidepressant, antiviral, and antibacterial effects. The flavonoid hyperforin has been identified as one of the major constituents responsible for antidepressant activity (Barnes et al. 2001).

Phytochemistry: The herb contains anthocyans (hypericin, pseudohypercin, protopseudohypercin, frangula-emodin anthronol, etc.) and flavonoids (hyperoside, rutin, quercetrin, isoquercetrin, and quercetin). The herb also contains essential oil with terpenes, sesquiterpenes, and complex esters of isovalerianic acids, tannins, carotene, ceryl alcohol, choline, and traces of alkaloids (Khodzhimatov 1989; Nahrstedt and Butterweck 1997).

Hypericum scabrum L. – Hypericaceae

Synonyms: *Drosanthe scabra* (L.) Spach, *Hypericum asperum* Ledeb.

English name: None

Russian name: Зверобой шероховатый (Zveroboy sherokhovatyy)

Uzbek name: Dalachoi, Choichoop

Kyrgyz name: Бодурлуу сары чай чоп (Bodurluu sary chay chop)

Description: Herbaceous perennial. Stems numerous, 20–70 cm tall, brown or reddish, covered with small, rigid papillae. Leaves opposite, sessile, oblong to lanceolate or elongate-linear, apex rounded or mucronate, covered with glands. Inflorescence a dense, corymbiform cyme. Sepals 5, partially connate. Petals 5, yellow with marginal black glands. Fruit a brown, ovoid to elongate-elliptical capsule. Seeds 1.5 mm long, brown.

Other distinguishing features: Stems rough, stamens in 3 fascicles.

Phenology: Flowers in May-June, fruits in July-September.

Reproduction: By seeds.

Distribution: Jalal-Abad, Naryn, Talas, and Chuy provinces of Kyrgyzstan; Toshkent, Farg'ona, Samarqand, Qashqadaryo, and Surxondaryo provinces of Uzbekistan.

Habitat: On dry, stony mountain slopes and in dry stream beds.

Population status: Common, found in small groups.

Traditional use: In folk medicine *Hypericum scabrum* is used in a similar manner as *H. perforatum* (Khalmatov 1964). The aboveground parts are collected during flowering before the appearance of unripe fruits and are used to treat coughs and liver, heart, stomach, intestinal, and bladder diseases. An infusion of the flowers is used to treat jaundice (Khodzhimatov 1989).

Documented effects: Crude extracts of *Hypericum scabrum* showed antimicrobial activity in vitro against *Bacillus cereus*, *E. coli*, *Staphylococcus aureus*, *Branhamella catarrhalis*, *Clostridium perfringens* and *Candida albicans* (Sokmen et al. 1999). Experiments demonstrated that an aqueous extract of the plant, given orally to rats, showed significant antiulcerogenic activity (Yesilada et al. 1993). Also in vitro, compounds isolated from the plant had moderate cytotoxicity against human tumor cells and mild antibacterial activity against methicillin-resistance *Staphylococcus aureus* (MRSA) and methicillin-sensitive *Staphylococcus aureus* (MSSA; Matsuhisa et al. 2002; Tanaka et al. 2004).

Phytochemistry: The total flavonoids isolated from this species are nearly identical to those of *Hypericum perforatum*. Xanthones, vitamin C, carotene, anthocyanins, essential oil, sugars, mucilage, resins, organic acids, and saponins and others have also been isolated from the plant (Plant Resources of the USSR 1986; Matsuhisa et al. 2002; Tanaka et al. 2004).

Hyssopus seravschanicus (Dub.) Pazij – Lamiaceae

Synonyms: *Hyssopus tianschanicus* Boriss., *Hyssopus ferganensis* Boriss.

English name: Tian Shan hyssop

Russian name: Иссоп тянь-шанский (Issop tyan'-shanskiy)

Uzbek name: Dorivor kukut

Kyrgyz name: Тянь-Шань иссобу (Tyan'-Shan' issobu)

Description: Subshrub. Stems 40–50 cm tall, twig-like, 4-sided, glabrous. Leaves opposite, linear, 1–3.5 cm long, 1–3 mm wide, almost glabrous, margins curled. Inflorescences 4–6-flowered verticillasters, found in narrow spikes. Calyx 5–6 mm long, blue, with sharp triangular teeth. Corolla blue-violet, about 10 mm long, 2-lipped, upper lip ovate, lower lip 3-lobed. Fruits oblong nutlets, 2 mm long, 1 mm wide, glabrous.

Other distinguishing features: Stamens 4, two equal to length of corolla and two longer. Style exserted.

Phenology: Flowers in July-August, fruits in September.

Reproduction: By seeds.

Distribution: Jalal-Abad and Talas provinces of Kyrgyzstan; Toshkent, Andijon, Samarqand, and Surxondaryo provinces of Uzbekistan.

Habitat: On stony slopes, on rocky and pebbly soils, on steppes.

Population status: Common, found in small groups.

Traditional use: An infusion is used as an expectorant, anti-inflammatory, astringent, tonic, antihelminthic, to heal wounds, and to treat bronchial asthma, gastrointestinal diseases, dyspepsia, rheumatism, anemia, stenocardia, neurosis, scrophula, meteorism and hyperhydrosis. It applied to the mouth to treat stomatitis and bad breath, and externally to heal persistent wounds. In Indian medicine it is used to treat bronchial asthma and acute respiratory infections (Zotov 1975; Dzhumaev 1980).

Documented effects: The plant has antiprotist, antibacterial, and antifungal activities, as well as lactogenic properties. The essential oil and phytoncides have antibacterial actions. In veterinary science an infusion is used to treat inflammation of the gastrointestinal tract in calves (Zotov et al. 1977).

Phytochemistry: The plant contains steroids (β-sitosterin), flavonoids (diosmine), essential oil (containing camphene, β-pinene, pinocamphone, 1,8-cineol, linalool, α-terpenyl-acetate, bornyl acetate, myrcene, limonene, etc.), triterpenoids (ursolic and oleanolic acids), vitamins B_1, B_2 and C, and phenolcarbonic acids and their derivatives. The seeds contain fatty oil including palmitic, stearic, oleinic, linoleic, and linolenic acids (Zotov 1975).

▲**Hyssopus seravschanicus (Dub.) Pazij** Photo: Dr. Petr Kocna, www1.lf1.cuni.cz~kocnaflowr_myflow_gb.htm.jpg

▲**Hyoscyamus niger L.** Photos: Alim Gaziev

▼**Hypericum perforatum L.** Photos: *left*: Mary Backlund; *right* and *center*: Sergey Appolonov

▼**Hypericum scabrum L.** Photos: *left*: John B. Taft; *right*: Vladimir Epiktetov

Impatiens parviflora DC. – Balsaminaceae

Synonyms: *Impatiens brachycentra* Kar. & Kir.

English name: Small balsam, small flower touch-me-not

Russian name: Недотрога мелкоцветная (Nedotroga melkotsvetnaya)

Uzbek name: Hinagina, Chupkhina

Kyrgyz name: Майда гулду кына (Mayda guldu kyna)

Description: Herbaceous annual with fibrous roots. Stems erect, 30–70 cm tall, succulent, glabrous. Leaves alternate, 8–17 cm long, 4–8 cm wide, elliptic or ovate, apex acuminate, margins serrate-dentate, gradually tapering to 1–2 cm long petiole. Inflorescences loose axillary racemes, with 4–12 flowers; peduncles similar in length to the leaves; pedicels thin, 1.5–2 cm long. Flowers irregular, up to 1 cm long. Sepals 3, 2 lateral sepals small, ovate; lower sepal petaloid with 4–5 mm long spur. Petals 5, lateral petals connate in pairs, 3-lobed, yellow with red spots in the throat; fifth petal sub-orbicular. Fruit an oblong capsule, 2 cm long, 3–4 mm wide, explosively dehiscing along raised longitudinal seams. Seeds oval, almost round.

Other distinguishing features: Leaf teeth glandular. Flowers directed upward or aside, not drooping. Ripe fruits burst when touched.

Phenology: Flowers in June-July, fruits in July-August.

Reproduction: By seeds.

Distribution: Toshkent, Samarqand, Qashqadaryo, and Surxondaryo provinces of Uzbekistan; Chuy and Jalal-Abad provinces of Kyrgyzstan.

Habitat: The tau zone. Grows in wet, shady places, walnut forests, in oases, and can be found in shaded areas of orchards, as a weed.

Population status: Common.

Traditional use: Used in folk medicine as a hemostatic and as a treatment for various uterine diseases (Khalmatov 1964).

Documented effects: An alcohol extract of this species had highly significant hemostatic activity. Extracts of the herb contain antibacterial substances (Khalmatov 1964). An infusion of the herb in alcohol had sedative and hypotensive effects, regulated the menstrual cycle, and accelerated childbirth delivery (Ibragimov and Ibragimova 1960). An experiment with an aqueous extract of the plant, to determine cyclooxygenase inhibition, showed negative inhibition, indicating an enzyme-stimulating effect (Tunon et al. 1995).

Phytochemistry: Aboveground parts contained flavonoid glycosides (0.43 %), alkaloids (0.016 %), resins (3.53 %), vitamin C (7.2 mg%), and traces of carotene. The compounds N-oxy-benzoic acid, vanillic, gentisinic, ferulic acid, N-coumarinic and caffeic acids, as well as 2-methoxy-1,4 naphthoquinone have been isolated from the leaves (Khalmatov 1964). Oil from the seeds contains parinaric acid (Tsevegsuren et al. 1998).

Inula britannica L. – Asteraceae

Synonyms: *Conyza britannica* (L.) Moris ex Rupr., *Inula serrata* Gilib., *Inula tymiensis* Kudô.

English name: British yellowhead, British elecampane, meadow fleabane, yellow starwort

Russian name: Девясил британский (Devyasil britanskiy)

Uzbek name: Chachalbosh

Kyrgyz name: Сары баш карындыз (Sary bash karyndyz)

Description: Herbaceous perennial, with thin creeping rhizomes. Stems mostly erect, often villous or with orange glands, 10–70 cm tall. Basal leaves elliptic, lanceolate or ovate, 3–13 cm long, 1–3.2 cm wide; cauline leaves alternate, sessile, elongate-lanceolate to lanceolate. Inflorescence a capitulum, 3–5 cm wide, single or in corymbiform groups; involucral bracts linear, 4–6 mm long, in 2 rows. Ray flowers many (ca. 40–70), 1–1.5 cm long, yellow, twice as long as bracts; disc flowers 4–6 mm long, yellow. Fruits linear-oblong achenes, ribbed, brown, with gray-white pappus.

Other distinguishing features: Pappus consists of 15–25 simple hairs.

Phenology: Flowers and fruits in May-September.

Reproduction: By seeds.

Distribution: All provinces of Kyrgyzstan; Toshkent, Farg'ona, Andijon, Jizzax and Surxondaryo provinces of Uzbekistan.

Habitat: In semi-desert areas, steppes, and meadows, along the edges of rivers and lakes, and among bushes.

Population status: Common, found in small groups.

Traditional use: This species is gathered in autumn or early spring. An infusion or decoction of the underground parts is used to treat cystitis, diabetes, jaundice, respiratory catarrh, bone tuberculosis, rheumatism, and hemorrhoids, and is used as a vermifuge, hemostatic for uterine bleeding, and to improve the appetite. An infusion of the leaves is drunk as an anti-inflammatory and astringent remedy (Khodzhimatov 1989).

Documented effects: In modern medicine *Inula britannica* is used the same way as *Inula helenium* (Khodzhimatov 1989). Flavonoids isolated from this species were shown to protect cultured rat cortical cells from cell death caused by oxidative stress (Kim et al. 2002). Results from experiments with mice suggest an aqueous extract from the flowers of *Inula britannica* ssp. *japonica* Kitam. has a preventative effect on autoimmune diabetes by regulating cytokine production (Kobayashi et al. 2002b). The sesquiterpene lactone ergolide has anti-inflammatory activity (Han et al. 2001).

Phytochemistry: The aboveground parts contain flavonoids, essential oil, tannins, ergolide, britanin, and other sesquiterpene lactones. The underground parts contain essential oil (with alantolactone and isoalantolactone), alkaloids, and inulin. The leaves contain vitamin C (Khalmatov et al. 1984; Zhou et al. 1993; Han et al. 2001; Kim et al. 2002).

Inula grandis Schrenk ex Fisch. & C.A. Mey. – Asteraceae

Synonyms: *Codonocephalum grande* (Schrenk ex Fisch. & C.A. Mey) B. Fedtsch., *Inula macrophylla* Kar. & Kir.

English name: Large-leaved elecampane

Russian name: Девясил крупнолистный (Devyasil krupnolistnyy)

Uzbek name: Sari andiz, Ok andiz

Kyrgyz name: Unknown

Description: Herbaceous perennial, 50–200 cm high with a thick, vigorous, branched root. Stem erect, branching towards the top. Leaves coriaceous, shiny, adaxial side glabrous, abaxial side glandular, margins serrate-dentate. Basal leaves widely elliptic, up to 25–85 cm long and 18–32 cm wide, petioles 10–20 cm; stem leaves elongate-elliptical, 20–37 cm long, 8–15 cm wide, sessile; upper leaves lanceolate, 3–10 cm long, 1–4.5 cm wide. Inflorescences capitula, 2–5 arranged in a corymbiform raceme; capitula 4.5–6 cm in diameter with ray and disc flowers. Ray flowers yellow, 1–3 cm long. Fruit a cylindrical achene, brown, with multiple longitudinal ribs and yellowish pappus.

Other distinguishing features: Leaves stiff, odorous, with vaguely sinuate edges and distinctly reticulate veins. Involucral bracts lanceolate, acute, coriaceous.

Phenology: Flowers in May-July, fruits in July-August.

Reproduction: By seeds.

Distribution: Toshkent, Samarqand, Andijon, Farg'ona, and Surxondaryo provinces of Uzbekistan; all of Kyrgyzstan.

Habitat: The adyr and tau zones. Shallow soils and gentle slopes.

Population status: Common, usually found in small populations.

Traditional use: A decoction made of underground organs of this plant, and related species, are used to treat brucellosis, tuberculosis, gastrointestinal diseases, and as a vermifuge. Young juicy stems, with the bark removed, are used as a restorative and to treat phthisis (Khalmatov and Kosimov 1992).

Documented effects: Preparations are used to treat ulcers and gastric catarrh, as well as duodenal ulcers (Khalmatov 1964). Compounds isolated from this species showed unique anti-oxidant activity (Kogure et al. 2004).

Phytochemistry: The roots contain essential oils (up to 3 %), the main portion of which is sesquiterpene lactones: alantolactones and isoalantolactones, proazulen and alantone. The roots also contain saponins, inulin (up to 44 %), resins, traces of alkaloids, and the sequiterpene lactones carabron and granilin (Kulikov 1975; Akopov 1981; Khalmatov and Kosimov 1992). The bark contains many mono- and sequiterpene lactones (Fu et al. 2001; Su et al. 2000, 2001).

Inula helenium L. – Asteraceae

Synonyms: *Aster helenium* (L.) Scop., *Corvisartia helenium* (L.) Mérat, *Helenium grandiflorum* Gilib.

English name: Elecampane

Russian name: Девясил высокий (Devyasil vysokiy)

Uzbek name: Kora andiz

Kyrgyz name: Бийик карындыз (Biyik karyndyz)

Description: Herbaceous perennial, with thick, fragrant rhizomes. Stem erect, 0.5–2.5 m tall, white-hairy. Lower leaves alternate, long-petiolate, elliptic, irregularly shallow-dentate, up to 50 cm long, up to 10–20 cm wide, densely hairy on abaxial side; stem leaves elongate-ovate, becoming sessile towards the top. Inflorescence a capitulum, 3–8 cm wide, in groups forming loose racemes or corymbs. Ray flowers golden-yellow, numerous (ca. 50–100), ca. 3–4 cm long, thin; disc flowers 9–11 mm long. Fruits brown achenes, quadrangular, with light colored pappus.

Other distinguishing features: Basal leaves very broad, up to 20 cm wide.

Phenology: Flowers in June-August, fruits August-September.

Reproduction: By seeds.

Distribution: In the Chuy valley (North slopes of Kyrgyz Alatau and Talas Alatau) and Ferghana Range (Arslanbob and Kara-Alma areas) in Kyrgyzstan; Toshkent, Samarqand, Jizzax, Andijon and Farg'ona provinces of Uzbekistan.

Habitat: Along rivers and streams, in the lower and mid belt of mountains.

Population status: Common, found in small groups.

Traditional use: Avicenna stated that this plant belongs to a group with the ability to act as a tonic and invigorate and strengthen the heart. It is useful to treat inflammation of the sciatic nerve and joint pain. Mixed with honey it is used as an expectorant. A decoction of the rhizomes, especially a syrup made from it, works as a diuretic and promotes menstruation. The rhizomes are still used to treat gastrointestinal diseases, malaria, cystitis, bone tuberculosis, rheumatism, radiculitis, diabetes, jaundice, edema, and respiratory catarrh. An ointment or water infusion is applied to treat eczema and scabies. A tincture of roots (in vodka) is drank to treat gastritis, stomach and duodenal ulcers, tuberculosis, nervous diseases, goiters, heart diseases, and hypertension and is used as an expectorant for treatment of chronic respiratory diseases (tracheitis, lung tuberculosis, and bronchitis). It is also used to treat gastroenteritis and diarrhea of non-infectious origins (Khodzhimatov 1989).

Documented effects: In modern medicine, a decoction of the underground parts is recommended to treat respiratory and gastrointestinal diseases. The preparation *Alanton* is used to treat ulcers (Poludenny and Zhuravlev 2000). Experimental results show that a decoction of the underground parts act as an expectorant, weak diuretic, choleretic, and weak hemostatic, and normalizes the function of the gastrointestinal tract. The essential oil acts as a very strong vermifuge, especially against *Ascaris* worms and pork and beef tapeworms. A preparation from this plant is used externally to treat skin diseases such as eczema, scabies, and neurodermatitis (Akopov 1990). Extracts of the roots, and isolated sesquiterpene lactones, showed significant inhibitory activity against a variety of cancer cell lines in vitro as well as against *Mycobacterium tuberculosis* (Cantrell et al. 1999; Konishi et al. 2002).

Phytochemistry: The underground parts contain 1–3 % essential oil (including sesquiterpene lactones such as alantolactone, isoalantolactone, dihydroalantolactone, etc.), up to 44 % inulin and other sugars, pigments, gums, mucilage, alkaloids, and acetic and benzoic acid. The aboveground parts contain alkaloids, essential oil, alantopicrine, and folic acid (Khalmatov et al. 1984; Khodzhimatov 1989; Akopov 1990; Cantrell et al. 1999; Konishi et al. 2002).

▲**Impatiens parviflora DC.** Photos: Sergey Appolonov

▲**Inula grandis Schrenk ex Fisch. & C.A. Mey.**
Photos: Alexander Naumenko

▲**Inula britannica L.** Photos: Sergey Appolonov
▼**Inula helenium L.** Photos: Sergey Mayorov

Juglans regia L. – Juglandaceae

Synonyms: *Juglans duclouxiana* Dode, *Juglans fallax* Dode, *Juglans kamaonia* (C. DC.) Dode, *Juglans orientis* Dode, *Juglans sinensis* (C. DC.) Dode.

English name: Persian walnut, English walnut

Russian name: Грецкий орех (Gretskiy orekh)

Uzbek name: Yong'oq

Kyrgyz name: Грек жангагы (Grek zhangagy)

Description: Large monoecious tree with a wide, dense crown, 15–35 m tall. Trunk diameter up to 1.5(–2.5) m wide; young trees with slightly cracked, light-gray bark; older trees have darker, strongly cracked bark. Leaves alternate, 19–54 cm long, 15–40 cm wide, dark-green, odd-pinnate with 3–5 pairs of leaflets; leaflets ovate, coriaceous, glabrous with entire margins. Male flowers arranged in catkins, each flower with 8–40 stamens. Female flowers in groups of 1–3 on ends of young branches. Fruit drupe-like, spherical, pericarp green, drying when ripe; endocarp or "shell" light brown, hard. Seed covered with thin yellow papery layer.

Other distinguishing features: The pith of young branches is chambered. Leaves produce a specific, pungent smell when crushed.

Phenology: Flowers in April-May, fruits in September.

Reproduction: By seeds.

Distribution: All regions of Uzbekistan; Osh and Jalal-Abad provinces of Kyrgyzstan.

Habitat: The tau zone. Mountain river banks, rarely on slopes, as single trees or groups.

Population status: As individual trees and groups, creates walnut forests.

Traditional use: Young, green fruits are used to prepare a concentrate of vitamins. A decoction of the nuts is drank to treat high arterial pressure, cardiac diseases, and to rinse the mouth to treat gum disease. Juice from the fruit husk is used as an ointment to treat different kinds of external ulcers, eczema, and other cases of itchy dermatosis. A tea of the leaves is drunk to treat diabetes and decrease sugar content in the urine. The leaves are used as a vermifuge and to treat skin diseases, venereal diseases, catarrh of the gastrointestinal tract, and tuberculosis. A decoction of the leaves is drank to treat scrofula and rickets. The bark from the roots is used to make a very mild laxative (Akopov 1981).

Documented effects: The leaves and the fruit husks are used to make the preparation *Juglon*. It is used externally to treat skin tuberculosis and *Staphylococcus* and *Streptococcus* lesions. It has antimicrobial and anti-inflammatory actions and is used to heal wounds. Unsaturated fatty acids from the nuts help to prevent arteriosclerosis (Nuraliev 1989). Clinical studies showed that a water extract of the leaves was effective in treating some forms of skin tuberculosis, tuberculosis lymphadenitis, and tuberculosis of the larynx (Altimishev 1991). In experiments with mice, extracts of *Juglans regia* improved glucose tolerance in hypoglycemic activity screens (Neef et al. 1995). In vitro, an extract of the nuts inhibited the oxidation of human plasma and low density lipoproteins (Anderson et al. 2001). Polyphenols and tocopherols, isolated from the nuts, exhibited antioxidative properties (Fukudu et al. 2003; Li et al. 2007).

Phytochemistry: The leaves contain α-hydrojuglone, which easily oxidizes to juglone, β-hydrojuglone, flavonoids (hyperoside, quercetin-3-arabinoside), ascorbic acid, vitamin P, B_1, tannins, carotene, pigments and essential oil. The fruit husk contains ascorbic acid, tannins, and α- and β-hydrojuglone. The seeds contain carotene, vitamins C, B_1, E, and P, and fatty oil, which contains glycerides of linoleic and oleic acids. β-sitosterol and its glycoside have been isolated from the papery layer surrounding the seed. The bark contains tannins, pigments, and gallic and ellagic acids (Khodzhimatov 1989; Colaric et al. 2005; Li et al. 2007).

Juniperus sabina L. – Cupressaceae

Synonyms: *Sabina vulgaris* Antoine
English name: Savin juniper, Savin
Russian name: Можжевельник казацкий (Mozhzhevel'nik kazatskiy)
Uzbek name: Archa
Kyrgyz name: Кара арча (Kara archa)
Description: Dioecious, evergreen, more or less prostrate shrub, or occasionally an erect small tree, to 5 m high. Bark reddish-gray. Branchlets slender, densely arranged. Leaves of 2 kinds, needle-like and scale-like; needle-like leaves present on young plants and sterile branches only, 3–7 mm long, appressed; scale-like leaves 1–3 mm long with a oval shaped gland on the back. Pollen (male) cones ellipsoid or oblong, 3–4 mm long. Seed (female) cones berry-like, brown-black, pruinose, round-oval, 2–5-seeded. Seeds brown, 4–5 mm long.
Other distinguishing features: Leaves have a characteristic smell when ground. Prostrate branches sometimes form roots.
Phenology: Flowers in April, fruits ripen in the fall or the spring of the following year.
Reproduction: By seeds.
Distribution: The Chuy and Naryn provinces of Kyrgyzstan; not found in the flora of Uzbekistan.
Habitat: In the steppe and forest belts and on stony slopes of hills and low mountains.
Population status: Common, forming dense groups.
Traditional use: An infusion of the berries is used to treat urogenital diseases. A decoction of the berries is drunk to treat kidneys and bladder illnesses, kidney stones, liver diseases, rheumatism, scurvy, jaundice, and to improve choleretic action. The berries are eaten to treat stomach ulcers and to improve appetite. A decoction of the roots is used for stomach ulcers, bronchitis, tuberculosis, kidney stones, and skin illnesses. A tincture of the bark and roots is drunk to treat arthritis. A decoction of the bark is used to raise the libido. The cones and green branches are also used in baths to treat rheumatism (Makhlayuk 1992).
Documented effects: Extracts from the fruits and branches have cytotoxic effects on cancer cell lines in vitro (Jafarian-Dehkordi et al. 2004). Results from experiments with mice indicate that the abortifacient effect of essential oil from *Juniperus sabina* is related to an implantation inhibiting effect induced by sabinyl acetate (Pages et al. 1996). Cyclolignans, isolated from the leaves, exhibits anti-cancer and anti-viral activity (San Feliciano et al. 1993).
Phytochemistry: The plant contains podophyllotoxin and other cyclolignans. Fresh branches, leaves, bark and fruits contain essential oil which contains pinene, cadinene, terpinene, terpinolene, camphene, cedrol, etc. The bark and stems contain tannins and the leaves contain vitamin C. The leaves, bark, and unripe cones contain pigments. The fruits contain sugar, juniperin, resins, pentosan, and organic acids (Makhlayuk 1992; San Feliciano et al. 1993).

Juniperus semiglobosa Regel – Cupressaceae

Synonyms: *Juniperus jarkendensis* Kom., *Juniperus sabina* var. *jarkendensis* (Kom.) Silba, *Juniperus schunganica* Kom., *Juniperus tianshanica* Sumnev., *Sabina vulgaris* var. *jarkendensis* (Kom.) Cheng-yuan Yang.

English name: Russian Juniper

Russian name: Можжевельник полушаровидный (Mozhzhevel'nik polusharovidnyy)

Uzbek name: Saur archa

Kyrgyz name: Саур-арча (Saur-archa)

Description: Dioecious or occasionally monoecious evergreen tree or shrub, up to 10 m tall. Branchlets thick, straight, loosely arranged. Leaves of 2 kinds, needle-like and scale-like; needle-like leaves usually present on young plants, rarely on adult plants, 3–7 mm long; scale-like leaves closely appressed, rhomboid-ovate, 0.9–2.5 mm long. Pollen (male) cones ellipsoid, 3–5 mm long. Seed (female) cones berry-like, 4–8 mm long, 5–10 mm wide, globose to semispherical, green-brown when unripe, black when ripe, pruinose, 2–4-seeded. Seeds up to 6 mm long, up to 3.5 mm wide, brown.

Other distinguishing features: Branchlets thick.

Phenology: Flowers in March-May, fruits ripen the following year.

Reproduction: By seeds.

Distribution: Chuy, Naryn, and Jalal-Abad provinces of Kyrgyzstan; Toshkent, Farg'ona and Samarqand provinces of Uzbekistan.

Habitat: In the vegetation of the tree-shrub belt, at elevations of 1,500–2,700 m on dry and stony slopes.

Population status: Common, found growing as single plants.

Traditional use: An infusion or decoction of the fruits is used to induce appetite, aid in better digestion, to increase urination, as a disinfectant of the urinary system, an anti-inflammatory to treat pneumonia, and as an analgesic and expectorant (Altimishev 1991). The green branches are burned in homes to provide a pleasant odor (Khodzhimatov 1989). The essential oil from the needles and fruits is used to treat skin conditions (Mamedov et al. 2004).

Documented effects: In contemporary medicine a tincture of the fruits is applied externally to treat rheumatism and gout. A decoction of the fruits is use to treat hypoacidic gastritis, cholecystitis, as a disinfectant of the bladder and to increase urination. The ground fruits are spelled to treat strong headaches (Altimishev 1991).

Phytochemistry: The wood contains 0.02 % essential oil, whereas unripe fruits and green branches contains 0.64–1.6 %. The essential oil contains up to 53 % sabinene, up to 21 % cedrol, and some sesquiterpenes and aldehydes. The fruits contain sugar and pigments. An extract from fresh branches collected in the Gissar mountains contained 0.38–0.54 % essential oil including pinene (up to 76 %), myrcene (5.4 %), cedrol (7 %), and few aldehydes (Khodzhimatov 1989).

Juniperus seravschanica Kom. – Cupressaceae

Synonyms: *Juniperus excelsa* var. *macrocarpa* Regel, *Juniperus kulsaica* Dmitr., *Juniperus polycarpos* var. *seravschanica* (Kom.) Kitam., *Juniperus polysperma* Dmitr., *Juniperus pseudosabina* var. *typica* Regel, *Juniperus sabina* var. *globosa* Regel, *Juniperus sabina* var. *macrocarpa* Regel, *Juniperus taurica* Lipsky, *Juniperus zaaminica* Dmitr., *Sabina seravschanica* (Kom.) Nevski.

English name: Unknown

Russian name: Можжевельник зеравшанский (Mozhzhevel'nik zeravshanskiy)

Uzbek name: Qora archa

Kyrgyz name: Кызыл арча, Кара арча (Kyzyl archa, Kara archa)

Description: Dioecious, evergreen tree up to 5–25 m tall, or sometimes a stocky bush with a dense oval or conical crown. Bark reddish or brick-brown in color. Branches spreading, relatively short with many smaller branches. Leaves scale-like, elongate-oval, apex acute, with long vein and gland on the lower surface. Female cone berry- or drupe-like, 9–12 mm long, globular; young cones green, mature cones deep-brown, heavily covered with a gray coating, contains 2–3(–4) seeds. Seeds 5–7.5 mm long, vaguely triquetrous-oval, curved, with longitudinal furrows on the sides.

Other distinguishing features: Seedlings with needle-like leaves in whorls of 3. Seeds white when immature, brown when ripe.

Phenology: Pollen released in March-April. Cones ripen in September-October of the following year.

Reproduction: By seeds.

Distribution: Toshkent, Farg'ona, Samarqand, and Buxoro provinces of Uzbekistan; Naryn, Osh, Jalal-Abad and Batken provinces of Kyrgyzstan.

Habitat: The upper adyr and tau zones. Stony, shallow-soiled mountain slopes with rocky debris.

Population status: Common, found as individual trees and also in small groves.

Traditional use: Smoke from burning branches is used in Central Asian folk medicine to treat rheumatism. The fruits, mixed with sesame oil, are applied to treat deafness. The powdered plant is sniffed to treat headaches. The essential oil is used to treat wounds and skin diseases (Khalmatov et al. 1984; Mamedov et al. 2004). An infusion of the dry fruits is used to treat the urogenital path, and a decoction is drunk to treat scurvy, liver disease and rheumatism. The fruits are also used to treat edema and nervous disorders. Fresh fruits are eaten to treat stomach ulcers and to increase the appetite, and a decoction is drank as a choleretic and to treat jaundice. A decoction of the roots is recommended to treat stomach ulcers, bronchitis, lung tuberculosis, kidney stones, and skin diseases. An infusion of the roots and bark is used to treat arthritis. A decoction of the bark is drank to treat impotency. A decoction of the fruits and green branches is used in a bath to treat rheumatism (Khodzhimatov 1989).

Documented effects: A 5 % solution of cedrol (isolated from the essential oil) in castor oil is used as a treatment for festering and slowly healing wounds and chilblain ulcers, and is applied as a salve on bandages or dressings. For osteomyelitis, this solution is poured into bone cavities (Gammerman 1960; Khalmatov et al. 1984). Essential oil from the leaves is used to treat trichomoniasis (Khodzhimatov 1989). Some terpenoids isolated from the fruits showed moderate antimalarial activity (Okasaka et al. 2006).

Phytochemistry: Fresh branches contain 0.45–0.75 % essential oils with d-pinene, d-camphene, myrcene, cedrene, and other sesqueterpenes. The bark, young branches, and unripe fruits contain 7–8 % tannins. Ripe fruits contain yellow pigments and up to 18.6 % sugar. Leaf samples from Tajikistan contained 120–140 mg% vitamin C (Khalmatov et al. 1984). A variety of compounds, including diterpenes and sesquiterpenes, has been isolated from the dried fruits (Okasaka et al. 2006).

Juniperus turkestanica Kom. – Cupressaceae

Synonyms: *Juniperus intermedia* V.P. Drobow, *Juniperus pseudosabina* Fisch. & C.A. Mey., *Juniperus pseudosabina* var. *turkestanica* (Kom.) Silba, *Juniperus pseudosabina* var. *typica* Regel.

English name: Unknown

Russian name: Можжевельник туркестанский (Mozhzhevel'nik turkestanskiy)

Uzbek name: Urik archa, Balik archa

Kyrgyz name: Орук арча, Жапалак арча (Oruk archa, Zhapalak archa)

Description: Dioecious evergreen tree to 18(–25) m tall or a stocky shrub up to 2 m high, with a dense crown. Bark brownish-gray. Branches spreading, ascending or horizontal. Leaves scale-like, about 2 mm long, ovate or rhobic, bright green, slightly pointed with a prominent gland on the back or with a prominent keel. Fruit a berry-like or drupe-like cone, 10–15 mm long and 8–10 mm wide, juicy, globular, sometimes oblong, with a single seed. Young cones green, mature cones black, shiny with a light gray coating. Seeds 6–10 mm long, oblong, pointed on the base, striated on the edges, with dark stripe on upper half; seed-coats thick, woody.

Other distinguishing features: Seedlings with needle-like leaves in whorls of 3. Cones taste sweet.

Phenology: Pollen released in April-June, fruits the next year in September-November.

Reproduction: By seeds.

Distribution: Toshkent, Samarqand, and Buxoro provinces of Uzbekistan; Naryn, Osh, Jalal-Abad and Talas provinces of Kyrgyzstan.

Habitat: The tau and yailau zones. Stony and shallow-soiled slopes with rocky debris.

Population status: Common, found as single individuals and in groups.

Traditional use: Fruit decoction is recommended by folk medicine as mouth-wash to treat gingivitis (Khalmatov et al. 1984). A decoction and ointment made with the fruits are used to treat eczema, tuberculosis, skin diseases and as a diuretic (Khodzhimatov 1989).

Documented effects: The fruits are used as a diuretic, for swelling due to kidney ailments, to treat kidney stones and are combined with other preparations to treat chronic respiratory disease and as an expectorant. The cedrol fraction, from the essential oil of young branches, together with castor oil is used as a remedy for persistent wounds and ulcers (Minayeva 1991).

Phytochemistry: The fruits contain essential oil which has up to 100 components including pinene, camphene, borneol, caphor, and other terpenes, the fruits also contain 40 % sugars, resins, flavonoids, pectic substances, etc. (Minayeva 1991).

▲**Juglans regia L.** Photos: *left*: Thomas Molnar; *right*: Alim Gaziev *center*: Dimitri Oreshkin

▼**Juniperus sabina L.** Photos: Vladimir Epiktetov

▼**Juniperus seravschanica Kom.** Photos: Evgeny Davkaev

Korolkowia sewerzowii (Regel) Regel – Liliaceae

Synonyms: *Fritillaria sewerzowii* Regel.

English name: Unknown

Russian name: Корольковия Северцова (Korol'koviya Severtsova)

Uzbek name: Olgi

Kyrgyz name: Северцов алгысы (Severtsov algysy)

Description: Herbaceous perennial with a spherical bulb, 3–7 cm wide. Stem thick, glabrous, erect, 30–60 cm high. Leaves simple; lower leaves opposite, wide-lanceolate to ovate, up to 20 cm long; upper leaves alternate. Inflorescence a loose terminal raceme. Flowers funnelform-campanulate, with 6 lobes, nodding; lobes greenish-brown or reddish-brown. Stamens 6, slightly shorter than the perianth. Fruit a capsule, 3–5 cm high and wide. Seeds flat, light-brown.

Other distinguishing features: Capsules erect, winged.

Phenology: Flowers in April-July, fruits in May-August, depending on altitude of location.

Reproduction: By seeds.

Distribution: Toshkent, Farg'ona, and Samarqand provinces of Uzbekistan; Osh, Chuy and Jalal-Abad provinces of Kyrgyzstan.

Habitat: The adyr and tau zones. Clay-soiled slopes.

Population status: Common, found as single individuals.

Traditional use: In folk medicine the bulb is used as a strong diaphoretic. The bulbs contain high amounts of starch and are used as food (Khalmatov 1964).

Documented effects: The alkaloid alginine has local anesthetic effects, and is 4 times less toxic than cocaine. The hydrochloric salt of alginine and the total alkaloids of the plant in the form of hydrochloric salt are recommended as a local anesthetic to be used in medical practice (Khalmatov 1964). Alginine, like novocaine, acts as a conduction anesthesia and a 3–4 % solution causes widening of the pupils (Sadritdinov and Kurmukov 1980). In experiments with narcotized animals, the alkaloid alginidine (3–20 mg/kg) decreased arterial pressure and ganglioblocking effects were observed (Ishmukhamedov and Sultanov 1965).

Phytochemistry: Bulbs collected in the Chatkal Valley (Uzbekistan), when aboveground parts had nearly senesced, contained 0.8–0.92 % total alkaloids . Plants collected near Toshkent, during the flowering stage contained 2.3 % alkaloids in the aboveground parts and 1.4 % in the bulbs. More than 20 alkaloids have been isolated from the total alkaloids, including alginine, korseverinine, alginidine, korseveramine and korseveridine, etc. (Yunusov 1981; Samikov et al. 1989; Harrison 1990; Abdullaeva and Shakirov 2006).

Lachnophyllum gossypinum Bunge – Asteraceae

Synonyms: Unknown
English name: Unknown
Russian name: Шерстистолистник хлопковидный (Sherstistolistnik khlopkovidnyy)
Uzbek name: Momuq, Oq momuq
Kyrgyz name: Пахтадай лахнофиллум (Pakhtaday lakhnofillum)
Description: Annual, 10–50 cm high, densely covered with soft, grayish, felted hairs, with abundant glands. Stems erect, often heavily branched. Lowest leaves obovate, apex obtuse or rounded, 1.5–4.5 cm long, 0.5–1.7 cm wide, leaf narrowing to petiole; middle leaves sessile, slightly amplexicaul with auricles on the base; upper leaves acute, narrow. Flowers in thick pubescent capitulum, heterogamous; marginal flowers ligulate, female, lilac-bluish; disc flowers bisexual, yellow. Fruits oblanceolate achenes, 2–3 mm long, flat, villous.
Other distinguishing features: The whole plant smells nice, like ripe melon.
Phenology: Flowers and fruits in July-October.
Reproduction: By seeds.
Distribution: All regions of Uzbekistan; Osh, Chuy and Jalal-Abad provinces of Kyrgyzstan.
Habitat: The chul (bordering adyr), adyr and tau zones. Stony slopes with rocky debris.
Population status: Uncommon, found as single individuals.
Traditional use: Fresh leaf juice is used to heal wounds. The juice is heated and brought to a thicker consistency and this is applied over the surface of old, slow-healing wounds and furuncles (Khalmatov 1964).
Documented effects: Small doses of the crystalline material from the essential oil has strong effects on the sympathetic nervous system (Khalmatov 1964).
Phytochemistry: A crystalline material, $C_{11}H_{12}O_2$, has been isolated from the essential oil (Khalmatov 1964). The principal components of the essential oil, from plants collected in the Moiynkumy desert of southern Kazakhstan, were methyl lachnophyllate (80.1 %), β-pinene (4.8 %) and caryophyllene (1 %). Other compounds isolated from the essential oil were a-pinene, β-myrcene, limonene, camphor, caryophyllene oxide, etc. (Sadyrbekov et al. 2006b).

Lagochilus gypsaceus Vved. – Lamiaceae

Synonyms: None
English name: Unknown
Russian name: Заячья губа гипсовая (Zayach'ya guba gipsovaya)
Uzbek name: Bozulbang
Kyrgyz name: Unknown
Description: Subshrub. Stems 30–40 cm high, woody at the very base, erect, often branched, with white, shiny bark, villous, subsequently becoming glabrous. Leaves petiolate, villous, rhomboid or wide-ovate in outline, 3–5-lobed, base cuneate; lobes oval or large-dentate. Inflorescences verticillasters with 4–6 flowers. Bracteoles awl-shaped, 3-sided, stiff. Flowers sessile. Calyx campanulate with spinescent lobes. Corolla 2-lipped, white or pink, with brown veins, 20–25 mm long. Fruits glabrous nutlets, 4–5 mm long.
Other distinguishing features: Corolla 1–1.5 times longer than the calyx. Differs from related species by having villous stems.
Phenology: Flowers in May-August, fruits in June-September.
Reproduction: Only by seeds.
Distribution: Endemic plant of Qashqadaryo and Surxondaryo provinces of Uzbekistan; not found in Kyrgyzstan.
Habitat: The adyr zone. Slopes of foothills with rocky debris and areas with a high diversity of soil minerals including gypsum.
Population status: Uncommon, sometimes in small populations.
Traditional use: Infusions and decoctions of the plant are used to stop bleeding and as a sedative tea (Akopov 1981).
Documented effects: In experiments with animals, an intravenous injection of a 10 % infusion of the plant extract accelerated coagulation of the blood by 30 % in 30 min, decreased the time of recalcification by 38 %, increased toleration of plasma to hepatitis by 35 %, and decreased blood pressure by 7 %. Preparations (infusion and tincture) made of the aboveground parts have hemostatic and sedative effects and decrease blood pressure. This plant is used in modern medicine as a preventive and therapeutic agents for various kinds of hemorrhage (traumatic, uterine, hemorrhoidal, pulmonary, lung, and nasal), and also to treat hemophilia and Werlhof's disease (Akopov 1981).
Phytochemistry: Leaves contain the diterpene alcohol lagochilin, 0.03 % essential oil, 11–14 % tannins, 7–10 mg% carotene, vitamin C, organic acids, calcium, and iron salts, and 0.6–0.7 % flavonoid glycosides. Lagochilin (1.98 %), tannins (2–2.7 %), ascorbic acid (106.29 mg%), carotene (4.39 mg%), and essential oils (0.083 %) were isolated from air dried plants (Akopov 1981).

Lagochilus platyacanthus Rupr. – Lamiaceae

Synonyms: *Lagochilus iliensis* C.Y. Wu & S.J. Hsuan, *Lagochilus keminensis* Isakov, *Lagochilus macrodontus* Knorr.

English name: Unknown

Russian name: Зайцегуб плоскоколючий (Zaytsegub ploskokolyuchiy)

Uzbek name: Unknown

Kyrgyz name: Жалпак тикендуу ак тикен (Zhalpak tikenduu ak tiken)

Description: Herbaceous perennial. Stems 15–45 cm, branching from the base, covered with bristly hairs. Leaves pinnatisect with linera or ovate lobes, ciliate-margined. Lower leaves rhomboid, winged-petiolate; upper leaves more rounded. Inflorescence a verticillaster with 4–8 flowers; bracteoles lanceolate to linear-lanceolate, stiff-spinescent, densely covered with glandular hairs. Calyx narrowly campanulate, tomentose, with ovate or triangular teeth. Corolla pale pink, 2-lipped, twice as long as calyx, upper lip 2 or 3 lobed. Fruits brown nutlets.

Other distinguishing features: Bracteoles 7–12 mm long.

Phenology: Flowers in June-July, fruits in August.

Reproduction: By seeds.

Distribution: Ysyk-Kol, Chuy, and Naryn provinces of Kyrgyzstan; not found in the flora of Uzbekistan.

Habitat: In steppes and tree-shrub belts, and on pebbly to stony slopes.

Population status: Common, found in small groups.

Traditional use: No data.

Documented effects: In experiments, an infusion of the aboveground parts showed low toxicity and hemostatic and sedative effects equal to, and hypotensive effects surpassing, those of *Lagochilus inebrians* Bunge. An infusion promoted blood coagulation and possessed antibacterial activity (Rakhimova and Pulatova 1972).

Phytochemistry: The aboveground parts contain flavonoids, organic acids, essential oil, diterpenoids (lagochilin), alkaloids (stachydrine), vitamin C, tannins, coumarins, lipids, etc. (Plant Resources of the USSR 1991).

Lagochilus platycalyx Schrenk ex Fisch. & Mey. – Lamiaceae

Synonyms: *Chlainanthus platycalyx* (Schrenk ex Fisch. & Mey.) Briq.
English name: Unknown
Russian name: Зайцегуб плоскочашечный, Заячья губа широкочашечная (Zaytsegub ploskochashechnyy, Zayach'ya guba shirokochashechnaya)
Uzbek name: Unknown
Kyrgyz name: Жазы чейчокчолуу ак тикен (Zhazy cheychokcholuu ak tiken)
Description: Perennial subshrub with woody roots. Stems herbaceous, erect, simple or branched, 20–50 cm tall, densely covered with fine hairs. Leaves opposite, with winged petioles, rhomboid in outline, pinnatisect nearly to mid-vein, scattered-hairy; lobes ovate, elongate or linear; upper leaves spiny or awl-like. Inflorescences verticillasters with 4–6 flowers. Calyx narrow-campanulate, with short triangular (sometimes merged) lobes, appressed-hairy. Corolla pale pink, with dark veins, 2-lipped, upper lip with 2 short lobes, lower lip with 3 wide lobes, lateral lobes elongate, oblong. Fruits nutlets, glabrous.
Other distinguishing features: Bracts 3–7 mm long, hairy. Corolla as long or 1.5 times as long as the calyx.
Phenology: Flowers in May-June, fruits in July.
Reproduction: By seeds.
Distribution: Jalal-Abad, Chuy, and Talas provinces of Kyrgyzstan; Toshkent province of Uzbekistan.
Habitat: On pebbly to stony slopes of foothills, in dry steppes and on exposures.
Population status: Common, found in small groups.
Traditional use: No data.
Documented effects: Effects on the cardio-vascular system and blood coagulability is equal to that of *Lagochilus inebrians* Bunge (Abdurakhmanov 1962). A tincture of the aboveground parts had low toxicity, hemostatic and sedative properties similar to those of *L. inebrians*, and hypotensive effects which surpassed those of *L. inebrians*. Clinical tests established the efficacy of the tincture for the treatment of hypertonic illness and as a hemostatic. A tincture of leaves and flowers exhibited hypotensive and sedative effects, and increased the speed of blood coagulation without increase of the prothrombin time (Alimbaeva 1961).
Phytochemistry: The plant contains essential oil, alkaloids (stachydrine, etc.), organic acids (chlorogenic, caffeic, hydroxycinnamic, and citric), flavonoids, diterpenoids (lagochilin), vitamin C, and tannins (Plant Resources of the USSR 1991; Nasrullaev and Makhsudova 1991; Zainutdinova et al. 1994; Kotenko et al. 1994).

▲**Lachnophyllum gossypinum Bunge** Photo: Alim Gaziev

◀**Lagochilus platycalyx Schrenk ex Fisch. & Mey.** Photos: *top*: Evgeny Davkaev; *bottom left* and *bottom right*: Alexander Naumenko

◀**Korolkowia sewerzowii (Regel) Regel** Photos: *top* and *bottom*: Evgeny Davkaev; *center*: Alexander Naumenko

▼**Lagochilus platyacanthus Rupr.** Photos: Vladimir Epiktetov

Lallemantia royleana (Benth.) Benth. – Lamiaceae

Synonyms: *Dracocephalum inderiense* Less. ex Kar. & Kir., *Dracocephalum royleanum* Benth., *Nepeta erodiifolia* Boiss.

English name: Unknown

Russian name: Лаллеманция Ройла (Lallemantsiya Royla)

Uzbek name: Mallachoi

Kyrgyz name: Ройл лаллеманциясы (Royl lallemantsiyasy)

Description: Annual herb covered with dense, short pubescence. Stems simple or branching, 5–30 cm tall. Lower leaves petiolate, ovate, 1.5–4 cm long, 0.8–2.5 cm wide, margins crenate; upper leaves smaller, subsessile. Flowers in whorls of 4–6, arranged in erect, interrupted, terminal, spiciform inflorescences. Bracteoles up to 1.5 cm long, with 2–4 awned teeth. Calyx tubular, prominently nerved with short obtuse lobes. Corolla 2-lipped, 6.5–9 mm long, azure, outside pubescent and glandular. Fruits oblinear nutlets, 2.5–3 mm long, trigonous, glabrous, smooth, dark-brown.

Other distinguishing features: Corolla slightly exceeding calyx in length. The leaves produce a distinct smell when crushed.

Phenology: Flowers in April-July, fruits in May-July.

Reproduction: Only by seeds.

Distribution: All regions of Uzbekistan; Osh, Chuy, Jalal-Abad and Talas provinces of Kyrgyzstan.

Habitat: The chul, adyr, and tau zones.

Population status: Common, as part of the ephemeral vegetation of foothills in the adyr zone.

Traditional use: A decoction of the fruits is used in folk medicine as a diuretic and expectorant and to treat gastric diseases and asthenia. An infusion of the herb is recommended for coughs and gastric pains (Khalmatov 1964).

Documented effects: Pharmacological studies have proved the diuretic action of the aboveground plant parts. A water extract of the herb increased diuresis in rats to 52 %, but the plant was toxic, and caused the death of 20 % of the tested animals (Khalmatov 1964).

Phytochemistry: Forty-six compounds were detected in the essential oil from the aboveground parts of *Lallemantia royleana*. Among them, verbenone and *trans*-carveol were found to be the major components of the oil (Ghannadi and Zolfaghari 2003). Plants collected in the Toshkent region contained traces of essential oils. The seeds of another related species, *Lallemantia iberica*, contain 27–35 % semi-drying oil. This oil is used for industrial purposes, as a food, and to produce soap (Ogolevitz 1951).

Leontice ewersmanni Bunge – Berberidaceae

Synonyms: *Leontice leontopetalum* ssp. *ewersmannii* (Bunge) Coode.

English name: Unknown

Russian name: Леонтица Эверсмана (Leontitsa Eversmana)

Uzbek name: Yersovun

Kyrgyz name: Эверсман леонтицасы (Eversman leontitsasy)

Description: Herbaceous perennial with a large, ovate tuber, 5–15 cm wide. Stem 20–60 cm tall, with subterranean part nearly as long. Basal leaves 1 or 2, with 3 petioluled lobes; each lobe trisected, middle lobe tripartite with sessile, bisected lateral lobes; upper leaves 3–5, lower 2 similar to basal leaves, the most upper leaves smaller and less divided or entire. Inflorescence apical, paniculiform, formed of racemes with 20–40 flowers. Flowers on long, horizontally spreading pedicels. Sepals yellow, petaloid. Petals 6, reduced, yellow. Stamens 6. Fruit an inflated capsule, ca. 15 mm in diameter. Seeds 1–2 per fruit, 5 mm wide, spherical, smooth.

Other distinguishing features: Petioles of basal leaves originating below ground. Tubers can grow 15–40 cm under the soil surface and weigh more than 1 kg (2.2 lbs).

Phenology: Flowers in March, fruits in April.

Reproduction: By seeds.

Distribution: Toshkent, Jizzax, Samarqand, Namangan, Qashqadaryo, and Surxondaryo provinces of Uzbekistan; Chuy province of Kyrgyzstan.

Habitat: The chul and adyr zones. Sandy and clay deserts, loess hills in the foothills.

Population status: Uncommon.

Traditional use: The powdered tuber is used in folk medicine to treat wounds and is smoked to treat syphilis. An infusion of the tuber is drunk for treating delayed menstruation and bladder stones (Khalmatov 1964).

Documented effects: In acute tests on animals with the alkaloids pachycarpine and d-lupanine, doses of 2–10 mg/kg decreased arterial blood pressure. Starting from the dose of 1–3 mg/kg it depressed and at the dose of 5 mg/kg it completely blocked parasympathetic cardiac ganglions. At the same dose, it potentiated hypotensive effect of acetylcholine and hypertensive effect of adrenaline and decreased reaction of cat's third eyelid, arterial pressure and respiration caused by cytizine introduction. Seventy to eighty percent of d-lupanine is excreted from the body: 50–70 % with urination, 10–14 % through defecation and 30–40 % turned into oxylupinine (Wittenburg and Nehring 1965). The alkaloid had minor tonic action on uterine muscles and had anticholinesterase action (Trutneva and Berezhinskaya 1960).

Phytochemistry: Roots contain tannins, up to 1.5 % total alkaloids (leontidine, leontine, leontamine, pachycarpine, and d-lupanine) and up to 30 % starch. The plant contains saponins with a hemolytic index of 1:240 in the aboveground portion of the plant and 1:6,000 in the tubers. Taspine, methylcytidine, and isoleontine were isolated from aboveground portion of the plant (Yunusov 1981).

Leonurus turkestanicus V. Krecz & Kuprian. – Lamiaceae

Synonyms: *Leonurus cardiaca* ssp. *turkestanicus* (V. Krecz. & Kuprian.) Rech.

English name: Turkestan motherwort

Russian name: Пустырник туркестанский (Pustyrnik turkestanskiy)

Uzbek name: Arslon kuirug

Kyrgyz name: Туркстан дулой чалканы (Turkstan duloy chalkany)

Description: Perennial herb with woody rhizome. Stems 50–150 cm tall, purple-red, branched, pubescent or glabrous. Leaves opposite, petiolate, wide-ovate to nearly circular in outline, palmatipartite; lobes pinnately divided into broadly lanceolate lobules. Flowers sessile, 15–20 per vertillaster, forming spiciform inflorescences. Bracts awl-like, pubescent. Calyx 8–9 mm long, funnelform, short-pubescent, with triangular spinescent lobes. Corolla pink-lilac, ca. 1 cm long, 2-lipped, villous outside. Fruits triquetrous nutlets, light brown.

Other distinguishing features: Upper corolla lip obovate, lower lip with 3 lobes, middle lobe larger than lateral lobes.

Phenology: Flowers in June-July, fruits in July-August.

Reproduction: By seeds.

Distribution: Toshkent, Samarqand, and Surxondaryo provinces of Uzbekistan; all of Kyrgyzstan.

Habitat: The tau zone. On stony, shallow-soiled slopes, among trees and shrubs.

Population status: Uncommon.

Traditional use: A decoction of the aboveground parts is used to treat heart, stomach and nervous system diseases (Khalmatov 1964). A tea and an infusion of the aboveground parts are used to treat nervous disorders, hypertension, hysteria, epilepsy, tachycardia, gastrointestinal, and female diseases, and are used as soporific, anti-inflammatory, diaphoretic, and laxative remedies (Khodzhimatov 1989).

Documented effects: Studies show that a tincture of the herb has a sedative effect, which is twice as strong as the effect of a valerian tincture. The tincture also causes decreased arterial pressure and strengthens the contraction of uterus muscles (Khalmatov 1964). Stachydrine exhibited protective effects when given to rats with experimental myocardial ischemia-reperfusion injury (Ma and Yang 2006).

Phytochemistry: The aboveground parts contain alkaloids (stachydrine), flavonoids, essential oils, tannins, saponins, resins, bitter substances and other compounds (Khalmatov 1964; Pulatova 1969; Khalmatov and Kosimov 1994).

Lepidium perfoliatum L. – Brassicaceae

Synonyms: *Crucifera diversifolia* E.H.L. Krause, *Nasturtium perfoliatum* (L.) Besser, *Nasturtium perfoliatum* (L.) Kuntze.

English name: Clasping pepper-grass

Russian name: Клоповник пронзенный (Klopovnik pronzennyy)

Uzbek name: Unknown

Kyrgyz name: Кучакталган сасык кычы (Kuchaktalgan sasyk kychy)

Description: Herbaceous biennial. Stems up to 20–25 cm tall, erect, branched, hairy at the bottom, glabrous towards the top. Leaves alternate, dimorphic; basal leaves (in rosette) and lower cauline leaves lanceolate, bi- or tripinnatisect with acute, simple or trilobate segments, hairy; upper cauline leaves sessile, ovate, cordate or nearly round, acute, amplexicaul, glabrous. Inflorescence racemose. Sepals 4. Petals 4, ca. 1.5 mm long, pale-yellow. Stamens 6. Fruits glabrous siliques, orbicular or rhombic, thin, 4–5 mm long. Seeds 1.5–2.5 mm long, 0.75–1.5 mm wide, dark-brown.

Other distinguishing features: Seeds have a narrow wing around their entire edge.

Phenology: Flowers in March-May, fruits in April-June.

Reproduction: Only by seeds.

Distribution: All regions of Uzbekistan and Kyrgyzstan.

Habitat: The chul and adyr zones.

Population status: Common, found in small populations as a part of ephemeral associations.

Traditional use: A decoction of the herb is taken to treat headaches, and ground seeds mixed with other pharmaceuticals are recommended for treatment of general weakness and to reinforce the nervous system. Avicenna applied the plant as a dressing or ointment with honey to treat "hard" and malignant tumors, as well as podagra, and used as an expectorant mixed with other drugs (Khalmatov 1964).

Documented effects: No data.

Phytochemistry: The herb contains glycosides, which are produced after the enzymic hydrolysis of mustard essential oil. Seeds contain 12–19 % drying oil. There is a possibility for the presence of prussic acid in young plants in the spring (Khalmatov 1964). The plant contains the flavonoid lepidoside (Fursa and Litvinenko 1970). The seeds were found to contain quercetin derivatives, as well as 18.71 % oil, in which alpha-linolenic, oleic, erucic and eicosenoic acid were the most abundant (Dolya et al. 1973a, b).

Lepidolopsis turkestanica (Regel & Schmalh.) Poljakov – Asteraceae

Synonyms: *Chrysanthemum turkestanicum* (Regel & Schmalh.) Gilli, *Crossostephium turkestanicum* Regel & Schmalh.
English name: Unknown
Russian name: Лепидолопсис туркестанский (Lepidolopsis turkestanskiy)
Uzbek name: Zarkuloq
Kyrgyz name: Туркстан лепидолопсиси (Turkstan lepidolopsisi)
Description: Perennial herb with thick rhizome. Stems solitary or few, 40–100 cm high, erect, leafy, with long and short hairs, later becoming glabrous. Basal leaves and lower stem leaves up to 10–15 cm long and 3.5 cm wide, petiolate; blades blue-gray-green, sparsely hairy, oblanceolate in outline, bi- or tripinnatisect with narrow-linear segments, terminal segments with short, cartilaginous tips; upper leaves reduced, sessile. Inflorescences composed of many small capitula arranged in compressed spicate-panicles, 15–30 cm long; involucres 4–6 mm in diameter, often golden-tinged. Flowers all tubular disc florets, yellow. Fruits achenes, 1.5–1.75 mm long, angled on top, with a paleaceous corona that is deeply divided into 8–12 narrow teeth.
Other distinguishing features: Basal leaves senesce early. Receptacle glabrous.
Phenology: Flowers in June-August, fruits in July-September.
Reproduction: Only by seeds.
Distribution: Toshkent, Samarqand, Farg'ona, Andijon, and Surxondaryo provinces of Uzbekistan; Osh, Jalal-Abad and Batken provinces of Kyrgyzstan.
Habitat: Loess foothills, most often in the adyr zone.
Population status: Uncommon. Found in small populations but more often as single individuals.
Traditional use: A decoction of flower heads is used in folk medicine to treat chest pains, heavy breathing, malaria, and delayed menstruation, and is also used as a vermifuge and diuretic remedy (Khalmatov 1964).
Documented effects: Pharmacological investigations of the plant extract showed effects on uterine activity (Khalmatov 1964).
Phytochemistry: This species contained traces of alkaloids (Khalmatov 1964), and at flowering it contained 0.12–0.13 % essential oil (Kudryashev 1932).

▲**Lallemantia royleana (Benth.) Benth.** Photo: Alexander Naumenko

▼**Lepidium perfoliatum L.** Photos: *top*: Maxim Kucherov; *bottom*: Clinton Shock

▲**Leonurus turkestanicus V. Krecz & Kuprian.** Photos: Evgeny Davkaev

▼**Leontice ewersmanni Bunge** Photos: Alexander Naumenko

Lithospermum officinale L. – Boraginaceae

Synonyms: None

English name: European stoneseed, common gromwell

Russian name: Воробейник лекарственный (Vorobeynik lekarstvennyy)

Uzbek name: Ilonchoop

Kyrgyz name: Дары таранчы чоп (Dary taranchy chop)

Description: Herbaceous perennial, with stout rhizome. Stems single to many, 30–100 cm tall, branched above. Leaves opposite, nearly sessile, lanceolate to oblanceolate, 3–8 cm long, 5–15 mm wide, leaves on upper portion of stems crowded. Inflorescences dense cymes, in upper leaf axils. Calyx lobes 5, oblong-linear. Corolla 3–6 mm long, tubular with 5 lobes, yellowish or greenish-white. Stamens inserted at middle of corolla tube. Stigma capitate. Fruits ovoid nutlets, about 4 mm long, white or light brown, shiny.

Other distinguishing features: Stems and leaves are scabrid-hairy.

Phenology: Flowers in May-June, fruits in June-July.

Reproduction: By seeds.

Distribution: Ysyk-Kol, Chuy, and Osh provinces of Kyrgyzstan; Toshkent, Farg'ona, Andijon, Samarqand, Buxoro and Surxondaryo provinces of Uzbekistan.

Habitat: In the tallgrass-meadow belt, in meadows, river floodplains, and among bushes.

Population status: Common, found growing as single plants.

Traditional use: The freshly ground plant is applied to heal bruises and cuts (Khalmatov 1964).

Documented effects: In preclinical tests an extract of the plant showed satisfactory results for treatment of hyperpituitarism and displayed antihormonal properties (Vyazovskaya 1963). When administered together with thyroid stimulating hormone (TSH), an extract of the plant blocked the TSH-induced increase in endocytotic activity of the thyroid glands followed by a strong decline of thyroid hormone levels. When the extract was injected alone it caused a decline in endogenous TSH-levels as well as in thyroidal secretion and thyroid hormone levels (Winterhoff et al. 1983). A decoction of the aboveground parts is used to treat the gastrointestinal tract (Utkin 1931). A water extract possesses antigonadotropic, contraceptive, and spermatocidic properties. A decoction of the ground fruits is parturifacient, and is used to treat dysmenorrhea, kidney diseases, kidney stones, and dyspepsia. Roots show protisticide activity (Dilman et al. 1968).

Phytochemistry: All plant parts contain cyclitols, organic acids (citric, malic, maleic, succinic, and fumaric), steroids, phenyalcarbonic acids and their derivatives, tannins, and flavonoids. The underground parts contain carbohydrates (glucose, saccharose, glucofructose, and fructose), cyanogenic compounds, phenylcarbonic acids, naphthoquinones, and fatty acids. The aboveground parts contain organic acids, flavonoids, and phenylcarbonic acids. Fruits contain cyclitols, aliphatic alcohols, steroids, vitamin E, phenylcarbonic acids and their derivatives, tannins, fatty oil, fatty acids, and pyrrolizidine alkaloids. Seeds contain carbohydrates, aliphatic alcohols, steroids, fatty oil and fatty acids (Dilman et al. 1968; Krenn et al. 1994).

Lycopus europaeus L. – Lamiaceae

Synonyms: None

English name: Gypsywort

Russian name: Зюзник европейский (Zyuznik evropeyskiy)

Uzbek name: Khorok, Tadzh

Kyrgyz name: Европа ликопусу (Evropa likopusu)

Description: Herbaceous perennial, with rhizomes and stolons. Stems single or many, 20–90 cm tall, erect. Leaves opposite, 3–9 cm long, 1–4 cm wide, oblong-elliptic to lanceolate-elliptic, coarsely dentate, base attenuate, apex acuminate. Inflorescences dense verticillasters, 18–20-flowered; bracteoles linear-subulate. Calyx with 4 or 5 triangular-lanceolate lobes. Corolla 2-lipped, 3 mm long, white with reddish-purple spots. Fruits oblong nutlets, glabrous.

Other distinguishing features: Two exserted and two reduced stamens. Upper leaves coarsely dentate.

Phenology: Flowers in June-July, fruits in August-September.

Reproduction: By seeds and rhizomes.

Distribution: Jalal-Abad, Chuy, and Osh provinces of Kyrgyzstan. Karakalpakstan autonomous republic, Toshkent, Andijon, Farg'ona, Samarqand, Buxoro, Surxondaryo and Xorazm provinces of Uzbekistan.

Habitat: On along rivers and in wet meadows, sometimes in water.

Population status: Common, sometimes found in small groups.

Traditional use: The herb is used to reduce swelling and as a hemostatic for uterine bleeding (Akopov 1990). A decoction and infusion of the aboveground parts is used to normalize increased heart rate due to stress, as a sedative and an antipyretic, and to treat uterine bleeding (Khalmatov 1964). Extracts from the plant are traditionally used to treat mild forms of hyperthyroidism (Vonhoff et al. 2006).

Documented effects: Clinical studies showed that the herb normalizes the function of the thyroid gland, has sedative and hypotensive effects, and dilates the coronary arteries. An infusion or tincture is recommended for the above effects (Akopov 1990). In Azerbaijan an infusion and decoction is used to treat heart diseases and lung tuberculosis. In Bulgaria a decoction is used to treat rheumatis. Based on preclinical tests, an alcoholic solution (of the polyphenols) and ointment accelerated healing of wounds and were effective in treating purulent otitis. Preparations of this species are proposed for treatment of atherosclerosis, hypertension, and coronary insufficiency. In experiments, a water infusion normalized production of thyroxine, slowed development of goiters, lowered metabolism in cases of exophthalmic goiters, showed low toxicity, and was recommended for clinical studies as a treatment for thyroidtoxicosis. A liquid extract possessed antithyroid activity, normalized the gas content of blood, reduced the ability of the thyroid gland to accumulate iodine, and positively influenced lactation. An ether extract also showed antibacterial and antifungal activity (Plant Resources of the USSR 1991). High doses of an extract of the plant caused a reduction of thyroid hormone levels in animal experiments, whereas in hyperthyroid patients treated with low doses, an improvement of cardiac symptoms was reported without major changes in thyroid hormone concentrations. Extracts diminished thyroidal secretion, reduced the plasma concentration of thyroxine (T_4), triiodothyronine (T_3), and inhibited the conversion of thyroxine to triiodothyronine. *Lycopus* extract also reduced heart rate and blood pressure and alleviated cardiac hypertrophy (Vonhoff et al. 2006). Two diterpenes isolated from the plant caused twofold potentiation of the activities of tetracycline and erythromycin against two strains of multi-drug resistant *Staphylococcus aureus* (Gibbons et al. 2003).

Phytochemistry: The plant contains organic acids (tartaric, citric, and malic), essential oil, diterpenoids (phytol), triterpenoids, steroids, saponins, alkaloids, choline, vitamin C, carotene, phenylcarbonic acids (rosmarinic acid), tannins, coumarins, flavonoids, cardiac glycosides, carbohydrates and anthocyanins (cyanin and pelargonin; Akopov 1990; Plant Resources of the USSR 1991).

Marrubium anisodon K. Koch. – Lamiaceae

Synonyms: *Marrubium alternidens* Rech. f., *Marrubium kusnezowii* N.P. Popov.

English name: Horehound

Russian name: Шандра очерёднозубая, Шандра очерёднозубчатая (Shandra ocheryodnozubaya, Shandra ocheryodnozubchataya)

Uzbek name: Devoltegiuit

Kyrgyz name: Ар тишчелуу маррубиум (Ar tishcheluu marrubium)

Description: Herbaceous perennial. Stems simple or branched, woolly-hairy, 30–100 cm high. Leaves opposite; lower leaves elliptic, orbicular or almost oblong, 5 cm long, 4 cm wide, base wide-cuneate, serrate-crenate, petiolate; upper leaves similar to lower leaves but smaller, wrinkled, densely pubescent. Inflorescences axillary verticillasters, multiflorous. Calyx hairy, with 10 teeth, 5 long alternating with 5 short. Corolla 2-lipped, 9–11 mm long, pale pink, pale yellow or white, stellate-hairy on the outside. Fruits obovoid nutlets, triquetrous, 1.5 mm long, glabrous, dark-brown or black.

Other distinguishing features: Corolla 1.5 times as long as calyx.

Phenology: Flowers and fruits in May-September.

Reproduction: By seeds.

Distribution: All regions of Uzbekistan and Kyrgyzstan.

Habitat: The chul, adyr, and tau zones. A weed, mainly a ruderal.

Population status: Common.

Traditional use: A decoction of the herb is used in folk medicine to treat chronic catarrh of the respiratory tract and throat diseases; also used as a mouthwash to treat toothaches (Khalmatov and Kosimov 1994).

Documented effects: Used as a sedative, it exceeds twice the action of valerian tincture; decreased blood pressure and increased tone of uterine muscles. The alkaloid stachydrine is slightly toxic. At doses of 5–100 mg/kg there was almost no effect on blood pressure. At the dose of 5 mg/kg there was a positive chronotropic action on the heart, where as at doses 10–100 mg/kg, there was a negative chronotropic action on the heart. Injected intravenously, starting from the dose of 5 mg/kg, stachydrine stimulated blood coagulation in dogs (Akopov et al. 1958). Research showed that marrubinic acid has choleretic effects (Khalmatov 1964). An extract of the plant showed significant antibacterial activity against *Escherichia coli* (Fazly Bazzaz and Haririzadeh 2003).

Phytochemistry: The aboveground plant parts contained 0.4 % essential oils, 1.12 % flavonoids, the alkaloid stachydrine, resins, 116.57 mg% vitamin C, and other compounds (Khalmatov and Kosimov 1994). Two diterpenoids, vulgarol and marubiin, have been isolated from the plant. (Sagitdinova et al. 1996). The total amino acids, polysaccharides, tanning agents, acids, flavonoids, phenols, essential oils, coumarins, saponins and alkaloids, from plants at the budding, flowering, and fruiting stages, were quantified by Kurbatova et al. (2003).

Mediasia macrophylla (Regel & Schmalh.) Pimenov – Apiaceae

Synonyms: *Athamanta macrophylla* (Regel & Schmalh.) Korovin, *Seseli macrophyllum* Regel & Schmalh.

English name: Unknown

Russian name: Медиазия крупнолистная (Mediaziya krupnolistnaya)

Uzbek name: Hunich, Alkor

Kyrgyz name: Unknown

Description: Herbaceous perennial with a 1.5–3 cm thick root. Stems several, 0.5–1.5 m tall, hollow, round, striated, covered with thick, short hairs. Leaves alternate, broadly ovate, 20–60 cm long and 20–40 cm wide, bipinnatipartite, coated with short hairs, long-petiolate; leaflets 4.5–12 cm long, 3.5–10 cm wide, bases cordate. Inflorescences compound umbels, apical, 5–10 cm wide, with 13–23 unequal rays; umbellets 5–6 mm wide, ca. 20-flowered. Flowers white or greenish-yellow, very hairy outside. Fruit a schizocarp with 2 mericarps; mericarps flattened, oval in outline, 5–6 mm long, hairy.

Other distinguishing features: Leaflets of upper leaves often trilobate. Leaves produce a pleasant odor when crushed.

Phenology: Flowers in June-July, fruits in August.

Reproduction: By seeds.

Distribution: Toshkent, Samarqand, Farg'ona, Qashqadaryo, and Surxondaryo provinces of Uzbekistan; Osh, Chuy and Jalal-Abad provinces of Kyrgyzstan.

Habitat: Tau zone. Stony slopes and large-fragmental taluses in the tree-shrub belt.

Population status: Common, found as individual plants.

Traditional use: A decoction of the roots is taken as a hemostatic. Local people use the fruit as a spice (Khalmatov 1964).

Documented effects: No data.

Phytochemistry: This species contained 1.67 % essential oils, the composition of which included aldehydes (Khalmatov 1964). According to Baser et al. (1997b), a total of 33 compounds were found in the essential oil, of which the principal components were *p*-cymene (27.2 %), thymol (15.1 %), carvacrol (12.5 %), and palmitic acid (2.9 %). A variety of neutral-, glyco- and phospho-lipids were isolated and identified from the leaves, and various free and bound fatty acids and carotenoids were quantified (Chernenko et al. 2002).

▲**Lithospermum officinale L.** Photos: *left*: Sergey Mayorov; *right*: Petr Filippov

▼**Lycopus europaeus L.** Photos: *top left* and *lower right*: Rostislav Lezhoyev; *top right*: Sergey Appolonov

▲**Marrubium anisodon K. Koch.** Photos: Vladimir Epiktetov

▼**Mediasia macrophylla (Regel & Schmalh.) Pimenov** Photos: Alim Gaziev

Melilotus officinalis (L.) Pall. – Fabaceae

Synonyms: *Melilotus graveolens* Bunge, *Melilotus suaveolens* Ledeb., *Trifolium officinale* L.

English name: Yellow sweetclover

Russian name: Донник лекарственный (Donnik lekarstvennyy)

Uzbek name: Kashkar beda

Kyrgyz name: Дары кашка беде (Dary kashka bede)

Description: Herbaceous biennial, with branching taproot. Stems one to many, up to 2 m tall. Leaves alternate, trifoliate, petiolate, stipulate; leaflets oblanceolate to obovate, serrulate, terminal leaflet stalked. Inflorescences axillary racemes, 5–15 cm long. Flowers 5–7 mm long. Calyx 2–2.5 mm long, toothed. Corolla papilionaceous, yellow. Fruit an oval legume, 3–5 mm long, with a beaked tip, 1–2-seeded. Seeds greenish-yellow.

Other distinguishing features: Stipules lanceolate, entire.

Phenology: Flowers in May-June, fruits in July-August.

Reproduction: By seeds.

Distribution: Almost all provinces of Kyrgyzstan and Uzbekistan.

Habitat: In meadows, fallow fields, along rivers and roads, and in cultivated fields.

Population status: Common, sometimes found in dense groups.

Traditional use: An infusion of the herb is drank to treat chronic catarrh of the bronchial tubes, migraines, and hypertension, for bladder and kidney pain, and is used during menopause. It is applied externally in the form of compresses, plasters, and as a wash, which are used as an emollient and analgesic to treat furuncles, carbuncles, purulent wounds and inflamed, pus-producing infections of the middle ear (Maznev 2004).

Documented effects: Coumarins from this species suppress the central nervous system and possess anticonvulsive and narcotic properties. After radiation treatments leucopenia patients treated with coumarins had increased leucocytes. The preparation *Dicumalin* (*Dicumarol*) has the ability to inhibit blood coagulation and is widely used as an anticoagulant and anti-vitamin K_1 to treat thrombophlebitis and heart attacks. This preparation must be used very carefully, not only because of inhibition of blood coagulation, but also because of increased permeability of capillaries and prolonged bleeding (Khalmatov et al. 1984). An extract of the plant had anti-inflammatory effects due to the activation of circulating phagocytes and lowered citrulline production (Plesca-Manea et al. 2002). A coumarinic extract from the plant was effective in reducing lymphedema in 79 % of patients with chronic lymphedema of the upper arm, caused by lymphadenectomy for breast cancer (Pastura et al. 1999).

Phytochemistry: The herb contains coumarins, melilotin, melilotic acid, melilotocide, purine derivatives, essential oil, vitamins C and E, carotene, protein, and lipids. The seeds contain protein, fatty oils, starch, and alkaloids (Tolmachev 1976; Khalmatov et al. 1984; Chikov 1989; Martino et al. 2006).

Melissa officinalis L. – Lamiaceae

Synonyms: *Melissa bicornis* Klokov.
English name: Lemon balm
Russian name: Мелисса лекарственная (Melissa lekarstvennaya)
Uzbek name: Limonuit
Kyrgyz name: Дары мелиссасы (Dary melissasy)
Description: Herbaceous perennial. Stems up to 120 cm tall, 4-sided, hairy. Leaves opposite, ovate, up to 7.5 cm long and 2–4 cm wide, serrate-crenate, hairy above, nearly glabrous below, long-petiolate. Inflorescences axillary verticillasters, 2–14-flowered. Calyx 2-lipped, angular, lobes about 2/3 as long as tube. Corolla white, yellowish or pinkish, 2-lipped, upper lip 2-lobed, lower lip 3-lobed. Fruits obovoid nutlets, brown.
Other distinguishing features: Crushed leaves have a lemon scent.
Phenology: Flowers in June-July, fruits August-September.
Reproduction: By seeds.
Distribution: Jalal-Abad, Osh, and Chuy provinces of Kyrgyzstan; Toshkent and Surxondaryo provinces of Uzbekistan.
Habitat: In Persian walnut forests, deciduous forests, on shady slopes, and among shrubs.
Population status: Common, forming dense groups.
Traditional use: This herb is widely used to treat migraines, insomnia, gynecological diseases, gout, dizziness, and anemia (Poludenny and Zhuravlev 2000). It is used as an antispasmodic for cardiovascular disease, as an analgesic, sedative, hypotensive, diuretic, and to improve digestion and to treat tympanites and pregnancy toxicosis (Kurochkin 1998).
Documented effects: Preparations of this species are used as a sedative, anticonvulsive, analgesic, and anti-flu medicine. It is used as a cardiac remedy and acts by slowing down the rate of breaths and heartbeats and by reducing tachycardia, palpitations, shortness of breath, and chest pain (Maznev 2004). An extract of the plant given orally produced a significantly better outcome in cognitive function in patients with mild to moderate Alzheimer's disease, than a placebo given to the control group (Akhondzadeh et al. 2003). Healthy people who received an extract of the herb orally exhibited a reduction in negative effects of laboratory induced stress and, at a higher dose, significantly increased the speed of mathematical processing with no reduction in accuracy (Kennedy et al. 2004). The essential oil has anti-tumor and anti-oxidant activities (de Sousa et al. 2004).
Phytochemistry: The leaves contain tannins, caffeic, oleanolic, ursolic acids, and essential oil (including citral, citronellol, myrcene, and geraniol). The aboveground parts contain ascorbic acid, potassium, calcium, magnesium, iron, manganese, copper, zinc, molybdenum, chromium, aluminium, barium, tungsten, silicon, nickel, sulfur, lead, and selenium. The seeds contain fatty oil (Volinsky et al. 1983; Carnat et al. 1998; de Sousa et al. 2004; Maznev 2004).

Melo agrestis (Naudin) Pang. – Cucurbitaceae

Synonyms: *Cucumis agrestis* (Naudin) Grebensc., *Cucumis melo* var. *agrestis* Naudin.

English name: Unknown

Russian name: Дыня полевая (Dynya polevaya)

Uzbek name: It qavun

Kyrgyz name: Жапайы коон (Zhapayy koon)

Description: Herbaceous taprooted annual with stiff, rough hairs. Stems prostrate, multiple, spreading, branched, slightly edged, 30–100 cm long. Leaves alternate, oblong or oblong-oval, 4–6 cm wide, slightly 3–5-lobed, seldom deeply notched, bristly-hairy, petiolate. Flowers uni- or bisexual. Staminate flowers in umbelliform inflorescences; pistillate flowers solitary. Calyx campanulate, 5-lobed, densely hairy. Corolla yellow, broadly funnelform, usually with 5 deep lobes. Fruit an oval-oblong berry (pepo), 2–5 cm long, usually green, yellowing at maturity, almost no aroma; rind rough, with a pattern in the form of deep-green, solid or interrupted, longitudinal stripes. Seeds small, white-yellowish, oval.

Other distinguishing features: Male flowers have 5 stamens, 4 in pairs and the fifth free. The pulp of the fruit tastes sour or bitter and is greenish-white with a large amount of cucumber-like placenta.

Phenology: Flowers in June-September, fruits in August-September.

Reproduction: Only by seeds.

Distribution: All regions of Uzbekistan; Osh, Chuy and Jalal-Abad provinces of Kyrgyzstan.

Habitat: The chul zone. As a weed in cotton and melon fields, rarely along canals and river banks.

Population status: Not common, found as single individuals.

Traditional use: A root decoction is used to treat edema and jaundice, and is used as mouthwash to treat bumps in the mouth. A fruit decoction is prescribed externally to treat eczema (Khalmatov 1964).

Documented effects: Cultivated forms of this species are used as a food with medicinal value to treat asthenia, constipation, and hepatitis and are used as a diuretic and prophylaxis to prevent arteriosclerosis and anemia (Karimov and Shomakhmudov 1993).

Phytochemistry: Roots collected in the Toshkent region contained 1.16 % tannins and up to 2 % sugars. Stems contained 0.87 % tannins, up to 4 % sugars, 0.4 % titratable organic acids, and alkaloid traces. Leaves contain 1.74 % tannins, 0.53 % titratable organic acids, and alkaloid traces; fruits contain up to 2 % sugars, 1.07 % titratable organic acids, and alkaloid traces (Khalmatov 1964).

Mentha asiatica Boriss. – Lamiaceae

Synonyms: *Mentha kopetdaghensis* Boriss., *Mentha longifolia* (L.) Huds. var. *asiatica* (Boriss.) Rech. f., *Mentha vagans* Boriss.

English name: Asian mint

Russian name: Мята лесная (Myata lesnaya)

Uzbek name: Yalpeez

Kyrgyz name: Жалбыз (Zhalbyz)

Description: Herbaceous perennial, with rhizomes. Stems erect, 40–100 cm tall, 4-sided, branched, finely hairy. Leaves short-pertiolate, ovate, elliptic or oblanceolate, margins serrate-dentate, both sides finely hairy, very glandular on underside; upper leaves sessile. Inflorescences verticillasters in terminal, cylindrical spikes; bracts awl-shaped, equal in length to the calyx. Calyx campanulate with linear teeth, densely hairy. Corolla 4–5 mm long, lilac, funnelform. Fruits ovoid nutlets.

Other distinguishing features: Leaves in inflorescence tapering to a point and extending past the verticillasters.

Phenology: Flowers in July-September, fruits September-October.

Reproduction: By seeds and rhizomes.

Distribution: Jalal-Abad, Osh, and Chuy provinces of Kyrgyzstan; Karakalpakstan autonomous republic, Andijon, Farg'ona, Samarqand, Buxoro and Surxondaryo provinces of Uzbekistan.

Habitat: In wet places and near canals, springs and streams.

Population status: Common, forming dense groups.

Traditional use: In Tibetan, Chinese, Mongolian, Indian and Central Asian folk medicine an infusion and decoction of this plant is used as an anti-inflammatory, hemostatic, and is used to treat wounds, gastritis, dysenteria, diarrhea, colitis, gastralgia, tuberculosis, respiratory infections, pertussis, and toothaches. An infusion of the leaves and inflorescences is used as a choleretic and to treat gall bladder diseases (Minayeva 1991).

Documented effects: In an evaluation for antimicrobial activity, essential oils from the related species *Mentha longifolia* ssp. *longifolia* and *Mentha sylvestris* L. exhibited activity against 30 different microorganisms including *Bacillus subtilis*, *Micrococcus luteus*, *Escherichia coli*, *Serratia marcescens* and *Aspergilus oryzae* (Carvalho et al. 1999; Gulluce et al. 2007).

Phytochemistry: Aboveground parts contain essential oil with menthol, menthone, carvacrol, and pulegone (Khalmatov 1964; Gulluce et al. 2007). Thirty-seven compounds were characterized representing 97 % of the total components detected. The major constituents of the oil were trans-piperitone oxide (64.51 %) and piperitenone oxide (12.34 %; Baser et al. 1997a). The seeds contain a variety of different fatty acids (Gusakova et al. 1976).

▲**Melilotus officinalis (L.) Pall.** Photos: *left* and *right*: Evgeny Davkaev; *center*: Rostislav Lezhoyev

▼**Melo agrestis (Naudin) Pang.** Photos: Sergey Zelentsov

▼**Melissa officinalis L.**
Photos: *top*: Andrei Lubchenko; *bottom*: Sasha Eisenman

▼**Mentha asiatica Boriss.**
Photos: *left*: Alim Gaziev; *top right* and *bottom right*: Evgeny Davkaev

Morus alba L. – Moraceae

Synonyms: None
English name: White mulberry
Russian name: Шелковица белая (Shelkovitsa belaya)
Uzbek name: Oq toot
Kyrgyz name: Ак тыт (Ak tyt)

Description: Monoecious or dioecious tree, up to 15(−18) m tall. Bark light-brown with shallow furrows; branches gray or gray-brown, young branches pubescent. Leaves alternate, ovate or rarely oblong-oval, 8–14 cm long, 5–10 cm wide, often lobed with 2–5 lobes, base rounded or slightly cordate, sometimes uneven, apex acute, margins entire or serrate, long-petiolate. Inflorescences catkins; male catkins cylindrical; female catkins oviform, densely-flowered. Flowers unisexual, sessile, glabrous. Fruits small, drupelet-like, arranged a syncarp; syncarp 0.5–2.5 cm long, white, pink, or red.

Other distinguishing features: The leaves are often glossy and the sap is milky. The stigmas are covered with papillae.

Phenology: Flowers in April, fruits in May-June.

Reproduction: By seeds, cuttings, and grafts.

Distribution: Cultivated throughout all of Uzbekistan, especially in the plains and lower mountain zones, often becoming naturalized; in agricultural zones of all provinces of Kyrgyzstan.

Habitat: The chul, adyr, and tau zones. Cultivated lands near canals and backyards.

Population status: Common.

Traditional use: In the folk medicine of Central Asia, mulberry leaves are used to treat angina. Fresh leaf juice is used to treat toothaches, and fruits and fruit juice are used to treat oral and throat bumps, dysentery, anemia, and as a diuretic and hemostatic for uterine bleeding, rashes, and scarlet fever. Fresh fruits are used to treat ulcers and the duodenum, and as a blood purifier, as well as an antipyretic and diuretic to improve heart function for cases of myodystrophy (Khalmatov 1964; Gammerman et al. 1990).

Documented effects: Resins from the leaves decrease blood pressure. An infusion of the leaves was shown to slightly reduce blood sugar levels (Gammerman et al. 1990). The flavonoid leachianone G was isolated from the root bark and showed potent antiviral activity against herpes simplex type 1 virus (Du et al. 2003). Two flavonoids isolated from the leaves significantly inhibited the growth of a human leukemia cell line (Kim et al. 2000). Flavonol glycosides, isolated from an extract of the leaves, showed some inhibition of low-density lipoprotein (LDL) oxidation (Katsube et al. 2004, 2006).

Phytochemistry: The leaves contained tannins (3.2–3.7 %), flavonoids (up to 1 %), coumarins, organic acids, resins, and small amounts of essential oils (0.03–0.04 %). Rutin, hyperoside, and quercetin were isolated from the total flavonoids and ostchol was isolated from the coumarins. The fruits contained up to 12 % sugars (occasionally up to 23 %), flavonoids, carotene, pectin, organic acids, small amounts of vitamin C, and tannins (Gammerman et al. 1990).

Nepeta pannonica L. – Lamiaceae

Synonyms: *Nepeta nuda* L.
English name: Unknown
Russian name: Котовник венгерский (Kotovnik vengerskiy)
Uzbek name: Zoofo
Kyrgyz name: Венгер непетасы (Venger nepetasy)
Description: Herbaceous perennial. Stems erect, up to 120 cm tall. Leaves 3.5–6.5 cm long, 1.5–2.5 cm wide, oblong-ovate to lanceolate, above green, nearly glabrous, pale beneath, pubescent, margin crenate or serrate. Inflorescences terminal paniculiform cymes, bracts narrow-linear. Calyx tubular, pubescent. Corolla pale-violet, pink or white, 2-lipped, upper lip 2-lobed, lower lip 3-lobed. Fruits oblong, brown nutlets.
Other distinguishing features: Plants branching above middle of stem, inflorescences long and narrow.
Phenology: Flowers in June-August, fruits in August-September.
Reproduction: By seeds.
Distribution: All provinces of Kyrgyzstan; Toshkent, Farg'ona and Surxondaryo provinces of Uzbekistan.
Habitat: In tallgrass meadows, meadow-steppes and steppes, and forest belts.
Population status: Common, found growing as single plants.
Traditional use: An infusion of the herb is used to treat asthenia and syphilis. The essential oil is used in perfumery (Plant Resources of the USSR 1991).
Documented effects: The entire plant shows antibacterial activity (Plant Resources of the USSR 1991).
Phytochemistry: Aboveground plant parts contains essential oil with 60 components with cineole and nepetalactone as the major constituents (Kobaisy et al. 2005), iridoids, steroidal saponins, flavonoids, alkaloids, and tannins. Seeds contain fatty oil, steroids, sterols, and sterol esters (Stepanenko et al. 1980; Plant Resources of the USSR 1991).

Nigella sativa L. – Ranunculaceae

Synonyms: *Nigella indica* Roxb. ex Flem., *Nigella truncata* Viv.
English name: Black cumin, fennel-flower, love-in-a mist
Russian name: Чернушка посевная (Chernushka posevnaya)
Uzbek name: Sedana
Kyrgyz name: Сейдана ундоосу (Seydana undoosu)
Description: Herbaceous annual. Stem 20–75 cm high, striated, oval, simple, slightly glandular-hairy. Leaves 1.5–3 cm long, bi- or tripinnatisect with linear, acute lobules; lower leaves petiolate, early-senescing; upper leaves sessile, similar to lower leaves. Flowers solitary, terminal or in leaf axes, 10–15 mm long, 15 mm wide, short-pubescent. Sepals 5, 1–1.5 cm long, petaloid with a short stalk, blue. Petals developed into 2-lipped nectaries. Stamens many. Fruits composed of 5 inflated follicles, ~1.5 cm long, connate nearly to apices, with erect, ribbed beaks. Seeds triquetrous, wrinkly-tuberculate, light brown.
Other distinguishing features: Follicles granular-tuberculate. Seeds have a specific bitter taste.
Phenology: Flowers and fruits in May-June.
Reproduction: Only by seeds.
Distribution: Toshkent and Samarqand provinces of Uzbekistan; all of Kyrgyzstan.
Habitat: The chul and adyr zones. Among crops in cultivated areas.
Population status: Rare.
Traditional use: Avicenna used this plant to treat headaches, facial paralysis, and eye cataracts, and when mixed together with honey in hot water to remove bladder and kidney stones. An infusion of the seeds is used to treat toothaches, gastric and intestinal diseases and chest pains, and is used as a, diuretic, soporific, and vermifuge for children (seeds in vinegar), as well as to treat angina and stimulate milk production in women (Karimov and Shomakhmudov 1993).
Documented effects: An infusion of the seeds had positive inotropic and negative chronotropic action, and reduced heart function due to increased cardiac output (Ogolevitz 1951). In a variety of experiments, extracts of the plant exhibited antibacterial activity, and in animals, increased bile and uric acid secretion, protected against histamine induced bronchospasm, shortened bleeding time, and inhibited fibrinolytic activity. Volatile oil from the seeds caused a dose-dependent increase in respiratory rate and intracranial pressure in anesthetized guinea pigs and reduced heart rate and blood pressure in anesthetized rats. Ingestion of the seeds caused reduction in cholesterol and blood glucose levels in humans. The seeds were also found to enhance immunity and had anti-cancer activity against malignant cells in mice and in humans. An aqueous extract of the seeds had anti-inflammatory and analgesic activity (Al-Ghamdi 2001).
Phytochemistry: Seeds contain 0.4–1.5 % essential oil (with a pleasant aroma), up to 40 % fatty oil, vitamin C, flavonoids (quercetin and camphorol), steroid alkaloids, coumarins, quinones, saponins, mineral salts, etc. (Karimov and Shomakhmudov 1993; Ali and Blunden 2003). The major components of the essential oil were thymoquinone, ρ-cymene, carvacrol, trans-anethole, 4-terpineol and longifolin (Ali and Blunden 2003).

Onopordum acanthium L. – Asteraceae

Synonyms: *Acanos spina* Scop.

English name: Scotch thistle

Russian name: Татарник обыкновенный (Tatarnik obyknovennyy)

Uzbek name: Okkarrak

Kyrgyz name: Кадимки коко тикен (Kadimki koko tiken)

Description: Herbaceous biennial, white-gray tomentose. Stems erect, 35–300 cm tall, spiny-winged. Leaves sinuate-lobed or toothed, teeth and lobes tipped with sharp spines; basal leaves up to 30 cm long, petiole winged; cauline leaves sessile. Inflorescences ovoid-spherical capitula, single or in corymbiform groups; involucral bracts linear, arranged in many rows, ending in sharp spines. Disc flowers purple, many; ray flowers absent. Fruits achenes, elongate-obovate, dark-gray with brown spots, pappi brownish.

Other distinguishing features: Leaves and stems white-gray tomentose. Involucral bracts are linear.

Phenology: Flowers in June-July, fruits in July-August.

Reproduction: By seeds.

Distribution: All provices of Kyrgyzstan; Karakalpakstan autonomous republic, Toshkent, Samarqand, Jizzax, Sirdaryo and Surxondaryo provinces of Uzbekistan.

Habitat: As weed in waste grounds, fallow fields, and pastures.

Population status: Common, often forming dense groups.

Traditional use: The inflorescences, roots, seeds, and late developing leaves (with spines removed), are harvested for use in folk medicine. They are used internally to treat inflammation of the bladder and urinary system, bronchial asthma, pertussis, scrofula, hypostasis of various origins, common colds, hemorrhoids, as a blood cleanser, and for treating skin diseases. The plant is used externally in the form of compresses, lotions, and fresh juice, which is especially effective, to treat skin diseases, purulent wounds, ulcers, and furuncles. An infusion of the top of the stem collected during flowering is drunk to treat nervous breakdowns, common colds, and inflammation of the respiratory system, and is put in baths for frightened children (Turova and Sapozhnikova 1984; Maznev 2004).

Documented effects: Experiments have shown that preparations of this species has very low toxicity and even after long periods of use show no side effects. It possesses cardiotonic, hemostatic, styptic, diuretic, and bacteriocidic properties and raises arterial pressure and causes narrowing of the blood vessels. In small doses preparations of this plants work as a tonic, and in larger doses, depress the central nervous system (Khalmatov et al. 1984). In some countries the herb is used to treat skin cancer and as a prophylactic after removal of a tumor (Akopov 1990). An aqueous extract of the plant exhibited anti-tumor activity in vitro (Abuharfeil et al. 2000).

Phytochemistry: Leaves contain alkaloids, glycosides, bitter substances, sesquiterpene lactones (arctiopicrin and onopordopicrin), vitamin C and K_1, resins, titratable acids, sugars, tannins, terpenoids (taraxasteryl acetate), etc. Seeds contain alkaloids, acetates of lupeol and amyrin, and drying fatty oil (Khalmatov et al. 1984; Ul'chenko et al. 1993; Khalilova et al. 2004).

▲**Nepeta pannonica** L. Photos: Andrei Lubchenko

▲**Morus alba L.** Photo: Bruce Hamilton

▲**Nigella sativa L.** Photo: Pamela J. Eisenberg

▼**Onopordum acanthium L.** Photos: *lower left* and *center*: Rostislav Lezhoyev; *top left* and *right*: Sasha Eisenman

Origanum tyttanthum Gontsch. – Lamiaceae

Synonyms: *Origanum vulgare* var. *genuinum* O. Fedtsch., *Origanum vulgare* var. *prismaticum* Gaudin, *Origanum vulgare* var. *viride* (Boiss.) Hayek.

English name: Kyrgyz oregano

Russian name: Душица мелкоцветная (Dushitsa melkotsvetnaya)

Uzbek name: Тог райхон, жамбил

Kyrgyz name: Майда гулдуу кок чай чоп (Mayda gulduu kok chay chop)

Description: Herbaceous perennial, rhizomatous. Stems many, 20–90 cm tall, erect, branched, square, villous. Leaves opposite, oval or oblong, 0.5–3 cm long, adaxial side nearly glabrous, abaxial side villous along veins, covered with punctuate glands, short-petiolate. Inflorescence a complex panicle, 10–30 cm long. Flowers nearly sessile. Calyx campanulate, 3 mm long, short pubescent. Corolla pale-pink, 5 mm long. Fruits nutlets, deep brown, less than 1 mm long.

Other distinguishing features: It differs from *Origanum vulgare* by having a more narrow inflorescence and smaller flowers.

Phenology: Flowers in June-August, fruits in July-September.

Reproduction: By seeds and rhizomes.

Distribution: Toshkent, Samarqand, Andijon, Farg'ona, and Surxondaryo provinces of Uzbekistan; Osh, Jalal-Abad, and Batken provinces of Kyrgyzstan.

Habitat: The adyr and tau zones. Grows on rocky and pebbly slopes.

Population status: Common, often makes large populations.

Traditional use: A decoction of the herb is used in folk medicine to stimulate the appetite and to improve digestion, to treat inflammation of mucous membranes in the upper respiratory tract, and decrease nervous excitability. Infusions and decoctions are applied externally as compresses for abscesses, and are also used in a bath to treat children who have rickets or scrofula. Water extractions of the aboveground plant parts are used to treat acute and chronic gastritis, bronchitis, cholecystitis, pneumonia, and urolithiasis and are also used as a cholagogue. A tea is used to treat tympanites, laryngitis, stomatitis, and angina, and as an oral and throat rinse (Khalmatov et al. 1984; Khodzhimatov 1989).

Documented effects: It is an effective remedy to treat hypertension, atherosclerosis, kidney, liver, and epilepsy (Kovaleva 1971). It is a sedative for excitement of the central nervous system (Turova 1974). A decoction of the dried leaves and flowers is used to treat intestinal atonia and as an expectorant. The plant is a component of a diaphoretic tea and is added to baths. The leaves are used as a spice and in liquor production (Tsitsina 1962). The essential oil has shown antimicrobial, hypocholesteremic, and hypolipidemic activity (Nuraliev and Zubaidova 1994; Takeda et al. 2008).

Phytochemistry: The flowering plant contain 0.17–0.6 % essential oil, which includes 35–66 % phenols (mostly thymol and carvacrol). The seeds contain up to 25 % fatty oils (Khalmatov 1964; Khalmatov et al. 1984). The plant contains phenolic glycosides, lipids and coumarins (Takeda et al. 2008).

Origanum vulgare L. – Lamiaceae

Synonyms: *Micromeria formosana* C. Marquand, *Origanum creticum* Lour., *Origanum dilatatum* Klokov, *Origanum normale* D. Don, *Origanum puberulum* (G. Beck) Klokov.

English name: Oregano, wild marjoram, Greek oregano

Russian name: Душица обыкновенная (Dushitsa obyknovennaya)

Uzbek name: Jambil

Kyrgyz name: Кадемки кок чай чоп (Kademki kok chay chop)

Description: Herbaceous perennial, with rhizomes. Stems purplish, erect or prostrate, pubescent, 20–60 cm tall. Leaves opposite, petiolate, broadly ovate to oblong, 1–4 cm long and 0.5–1.5 cm wide, densely hairy below. Inflorescences spikes in corymbiform or paniculiform clusters; bracts ovate-elliptic, green or purple. Calyx with triangular-lanceolate teeth, dark-purple. Corolla 5–7 mm long, light-purple, lilac-pink or white, 2-lipped; upper lip erect, apex 2-lobed; lower lip 3-lobed. Fruits nutlets, orbicular, bluntly 3-sided, brown.

Other distinguishing features: Inflorescence wider, and flowers bigger, than the closely related species, *Oreganum tyttanthum* Gontcsh.

Phenology: Flowers in June-July, fruits in August-September.

Reproduction: By seeds.

Distribution: Ysyk-Kol and Chuy Provinces of Kyrgyzstan; not found in the flora of Uzbekistan.

Habitat: On northern slopes in tallgrass-meadow belts and forests, among bushes, and along forest edges.

Population status: Common, found in dense groups.

Traditional use: An infusion is used to treat stomach ailments, common colds, and gynecological problems. It is used externally as a lotion, compress, and in bathes for the treatment of eczema, infected skin diseases, and to wash wounds (Gammerman et al. 1990).

Documented effects: An infusion of the herb is used to treat insomnia, hypo- and anacidic gastritis, and atonia of the intestines. It is also used as an expectorant for bronchitis and bronchiectasis, as well as to increase appetite (Turova and Sapozhnikova 1984). An extract of the herb is used as a component in the preparation *Urolesan*, which is used as an antispasmodic, antiseptic, and anti-inflammatory for the urinary tract, as well as to help eliminate ureter stones, and to increase bile production. The preparation increases diuresis and improves blood circulation through the liver (Gammerman et al. 1990). The essential oil of *Origanum vulgare* ssp. *vulgare* exhibited significant antimicrobial activity against 10 species of bacteria and 15 fungal species (Sahin et al. 2004). Essential oils from *Origanum vulgare* ssp. *hirtum* exhibited high levels of antimicrobial activity against 8 strains of gram-positive and gram-negative bacteria. The essential oil also exhibited high levels of cytotoxicity against 4 permanent animal cell lines including 2 derived from human cancers (Sivropoulou et al. 1996). Aqueous and methanolic extracts of oregano have been shown to have effective antioxidant properties (Cervato et al. 2000).

Phytochemistry: The herb contains essential oil (with phenols such as thymol and isomers of carvacrol), bi and tricyclic sesquiterpenes, free alcohols, tannins, ascorbic acids, and flavonoids (Akopov 1990; Sivropoulou et al. 1996). A total 62 constituents were identified from the essential oil of *Origanum vulgare* ssp. *vulgare,* with the main constituents being caryophyllene, spathulenol, germacrene-D, and terpineol (Sahin et al. 2004).

Orthurus kokanicus (Regel & Schmalh.) Juz. – Rosaceae

Synonyms: *Geum kokanicum* Regel & Schmalh.

English name: Unknown

Russian name: Прямохвостник кокандский (Pryamokhvostnik kokandskiy)

Uzbek name: Yerchoy, Shirchoy

Kyrgyz name: Unknown

Description: Herbaceous perennial. Stems erect, 15–50 cm high, hairy. Basal leaves in dense rosette, lyrate-pinnatisect, lateral segments many, large and small, usually in opposite or sub-opposite pairs, ovate-rhomboid, single- or bi-dentate, terminal segment larger, slightly lobed, bidentate; cauline leaves few, small, short-petiolate, oval to nearly round, tri-lobed, large-dentate. Inflorescences cyme-like with 2–7 flowers, crowded, but becoming loose with age, stiff-hairy. Hypanthium widely campanulate. Outer and inner sepals 5. Petals 5, yellow, about as long as sepals. Style erect with 2 parts; upper glabrous and deciduous; lower part persistent, longer than fruitlet, covered with retrorse bristles. Fruitlets 4–10, stiff-hairy.

Other distinguishing features: When fractured, the roots produce pleasant a eugenol smell.

Sepals wider and gynophore shorter than *Orthurus heterocarpus*.

Phenology: Flowers in June-July, fruits in July-August.

Reproduction: By seeds.

Distribution: Toshkent, Samarqand, Qashqadaryo, and Surxondaryo provinces of Uzbekistan; not found in Kyrgyzstan.

Habitat: The tau zone. Stony mountain slopes in juniper stands.

Population status: Uncommon.

Traditional use: A decoction of the roots is used in folk medicine internally for chest pains and as an astringent. A decoction of roots and leaves is used to rinse the mouth and throat. The roots are also used a tea substitute (Khalmatov 1964).

Documented effects: The essential oil, which is rich with eugenol, is used in dentistry instead of imported clove oil. A decoction or infusion of the roots are recommended as an astringent for gastrointestinal diseases (Khalmatov 1964). Essential oil, isolated from the plant, had strong antibiotic effects against *Shigella dysenteriae, Bacillus subtilis*, and *Aspergillus flavus* (Faramarzi et al. 2008).

Phytochemistry: Roots contain 22–25 % tannins, 10–13 % sugars, essential oils (up to 0.45 % eugenol), resins, and organic acids (Khalmatov 1964). The major compounds in the essential oil, distilled from underground parts, were eugenol (80.9 %) and myrtenol (5.2 %) (Faramarzi et al. 2008).

Padus avium Mill. – Rosaceae

Synonyms: *Padus racemosa* (Lam.) Gilib., *Prunus padus* L., *Prunus racemosa* Lam.

English name: Bird cherry

Russian name: Черёмуха обыкновенная (Cheryomukha obyknovennaya)

Uzbek name: Unknown

Kyrgyz name: Кадемки моюл (Kademki moyul)

Description: Deciduous tree, 2–10 m tall. Bark light tan to black-gray, cracked; young branches brown with white-yellow lenticels. Leaves simple, alternate, short-petiolate, glabrous, elliptic, margins serrate. Inflorescences hanging racemes, 8–12 cm long. Hypanthium cup-shaped, glabrous outside, hairy inside, with 5 short sepals. Petals 5, obovate, white. Fruits black, spherical drupes, 8–10 mm in diameter.

Other distinguishing features: Differs from *Padus asiatica* Kom. by having glabrous young branches, shorter racemes, and smaller corollas.

Phenology: Flowers in May to the beginning of June, fruits in July-August.

Reproduction: By seeds and rhizomes.

Distribution: Osh and Chuy provinces of Kyrgyzstan; cultivated in Uzbekistan.

Habitat: On floodplains.

Population status: Rare, found in small groups.

Traditional use: Bark, leaves, flowers, and mature fruits are used in folk medicine. Fruits are used as an astringent remedy to treat diarrhea of non-infectious origins and other intestinal disorders, as well as a secondary treatment for infectious colitis and diarrhea (Kurochkin 1998).

Documented effects: Mature fruits are used as a bactericide, anti-inflammatory, to normalize intestine and stomach function, as a source of vitamins, and as a tonic. The bark is used as a diaphoretic, antipyretic, and diuretic. Leaves are used to treat diarrhea and as a source of vitamins. Flowers are used as an anti-inflammatory. Preparations from this species are counter-indicated during pregnancy (Maznev 2004). An extract of the seeds had antibacterial activity against 5 different species, including methicillin-resistant *Staphylococcus aureus*, and had strong anti-oxidant activity (Kumarasamy et al. 2002, 2007).

Phytochemistry: Leaves, flowers, bark and seeds contain glycosides (amygdalin, prulaurasin, and prunasin). Prussic acid is found in the bark and leaves. Fruits contain malic and citric acids, sugar, astringent substances, ascorbic acid, and flavonoids (Maznev 2004; Deineka et al. 2004).

▲**Origanum vulgare L.**
Photo: Sasha Eisenman

◄**Origanum tyttanthum Gontsch.**
Photos: *top left*: Evgeny Davkaev; *top right*: John B. Taft; *bottom*: Alim Gaziev

▼**Padus avium Mill.** Photos: Sergey Appolonov

Papaver pavoninum Schrenk – Papaveraceae

Synonyms: *Papaver ocellatum* Woronow.

English name: Peacock poppy

Russian name: Мак павлиний (Mak pavliniy)

Uzbek name: Lola qizg'aldak

Kyrgyz name: Кызгалдак апийими (Kyzgaldak apiyimi)

Description: Herbaceous annual. Stem simple or branched from the base, 10–50 cm high, densely coated with bristles. Leaves multiple, bipinnatisect; segments oval-oblong, sessile, with bristles on adaxial side. Flowers often in groups of 3; buds rounded or oval, 8–15 mm long. Calyx coated with long whitish or reddish bristles, with 2 long, hollow, prominent apical horns. Petals ca. 2.5 cm long and 4 cm wide, bright red with an arching black spot at the base. Fruit an ovoid capsule, roundish, 5–10 mm long, 8 mm wide, ribbed, coated with stiff bristles. Seeds 1 mm long, light brown-gray, oblong.

Other distinguishing features: Cauline leaves often have an elongated terminal lobe.

Phenology: Flowers and fruits in March-June.

Reproduction: By seeds.

Distribution: Toshkent, Buxoro, Samarqand, Andijon, Farg'ona, and Surxondaryo provinces of Uzbekistan; Chuy, Ysyk-Kol, Talas, Osh, Jalal-Abad and Batken provinces of Kyrgyzstan.

Habitat: The chul and adyr zones. Clay deserts, on loess or stony slopes and in unirrigated winter wheat fields.

Population status: Common.

Traditional use: Juice from the petals is used as a drink to treat heatstroke (in children) and as a rinse for eye ailments. The dried petals of related species, *Papaver rhoeas, Papaver orientale*, and *Papaver bracteatum*, are used to prepare a tea to treat coughs (Khalmatov 1964; Seredin and Sokolov 1969).

Documented effects: Protopine has been shown to strongly inhibit induced platelet aggregation in vitro. In vivo, pretreatment with protopine protected rabbits from the lethal effects of specific platelet aggregation agonists. Protopine also inhibited carrageenan-induced rat paw edema and had 3 times the potency of aspirin (Saeed et al. 1997).

Phytochemistry: The herb contains 0.1 % total alkaloids, from which a-allocryptopine, protopine, and roemeridine have been isolated (Khalmatov 1964). In another study less than 0.05 % total alkaloids were found in the plant and a β-carboline was the dominant alkaloid (Taborska et al. 1988).

Patrinia intermedia (Hornem.) Roem. & Schult. – Valerianaceae

Synonyms: *Fedia intermedia* Hornem., *Fedia rupestris* var. *intermedia* (Hornem.) Vahl, *Patrinia nudiuscula* Fisch.

English name: Unknown

Russian name: Патриния средняя (Patriniya srednyaya)

Uzbek name: Unknown

Kyrgyz name: Орто патриния (Orto patriniya)

Description: Herbaceous perennial, with taproot that branches towards the top. Stems single or few, 20–75 cm tall, up to 3 cm in diameter, short-haired. Leaves opposite, 4–18 cm long, 2–5 cm wide, gray-green, glabrous; basal leaves long-petiolate, elongate-oblong, strongly dentate, pinnatilobate or pinnatisect; lower stem leaves sessile, in 2–5 pairs, pinnatisect; upper stem leaves lanceolate-ovate, 3-nerved. Inflorescence corymbiform-paniculate. Corolla bright yellow, campanulate, 5-lobed. Fruits slightly hairy achenes.

Other distinguishing features: Flowers have 4 stamens.

Phenology: Flowers in June-July, fruits in July-September.

Reproduction: By seeds.

Distribution: All provinces of Kyrgyzstan; not found in the flora of Uzbekistan.

Habitat: On stony floodplains of mountain rivers, on stony slopes and in steppe and forest-meadow belts of mountains.

Population status: Common, found in loosely arranged groups.

Traditional use: An infusion or decoction of the roots is used like valerian to treat nervous excitement and cardiac neurosis (Turova and Sapozhnikova 1984).

Documented effects: The biological activity is due to the presence of saponins, and removal of the saponins from the tincture leads to complete loss of the pharmacological properties (Ivanova 1963). The sedative effect of this species is nearly twice as strong as that of *Valeriana* (Tolmachev 1976). The roots of this species reduce excitability of the nervous system. Clinical tests showed that application of an alcohol infusion stopped or noticeably reduced chest pain as well as nervous and cardiovascular excitation caused by hypodermic introduction of caffeine (Akopov 1990).

Phytochemistry: Roots contain triterpene saponins (patrinoside A, B, C and interoside B), inulin, organic acids, tannins, and essential oils. The seeds contain alkaloids (Khodzhimatov 1989).

Peganum harmala L. – Zygophyllaceae

Synonyms: None

English name: Syrian rue

Russian name: Гармала обыкновенная (Garmala obyknovennaya)

Uzbek name: Isiriq

Kyrgyz name: Адыршаман (Adyrshaman)

Description: Herbaceous perennial, with thick woody taproot. Stems few, 20–80 cm tall, heavily branched, glabrous, slightly grooved. Leaves 3–8 cm long, sessile to short-petiolate, irregularly pinnatisect with linear-lanceolate segments, stipulate. Flowers in groups of 1–3, pedicillate, terminal on branches. Calyx deeply divided into 5 linear lobes, 1.5–2 cm long. Petals 5, white or pale yellow, elliptic, 1.5–2 cm long. Fruit a globular capsule, ca. 1 cm in diameter, 3-valved, splitting when ripe. Seeds many, triquetrous, dark brown.

Other distinguishing features: Calyx persistent in fruit. Dry leaves and plants with fruits have a specific smell when burned.

Phenology: Flowers and fruits in May-September.

Reproduction: By seeds.

Distribution: All regions of Uzbekistan and Kyrgyzstan.

Habitat: The chul and adyr zones. Waste places, abandoned fields, around the periphery of wells and in villages. In clayey and sandy soils, rich with nitrates.

Population status: Common, found in small populations.

Traditional use: This is a well-known herb to all Central Asian people. Avicenna used the plant as an analgesic for patients with sciatic nerve inflammation. In folk medicine the herb is used in baths to treat rheumatism, scabies, and other skin diseases. A decoction or infusion of the plant is drunk to treat common colds, malaria, fever, syphilis, neurasthenia, and epilepsy, and is also used as a mouth wash to treat gum disease. The smoke of the burning herb is good for headaches; for epileptic diseases the patient's room is filled with the smoke. A decoction of the seeds mixed with flax seeds is recommended for asthma and breathlessness, it is mixed with chili pepper to treat syphilis, and it is used as a diuretic and diaphoretic (Khalmatov et al. 1984; Gammerman et al. 1990). In Tajikistan smoke from the plant is used to treat paralytics. The leaves are used as a poultice to treat swelling (Khodzhimatov 1989).

Documented effects: An infusion of the roots and a preparation, *Salyanokisli garmine*, are used for Parkinson's disease after lethargic encephalitis (von Economo's disease), for epilepsy, and as a soporific. The alkaloid harmine is a reversible inhibitor of monoamine oxidase (Gorkin 1964; Coates and Cox 1972). The strong impact of harmine on the central nervous system often causes major mental disorders. Due to these effects, it is classified as a psychomimetic substance of adrenergic action (Sadritdinov and Kurmukov 1980). The alkaloid peganine depresses acetyl-cholinesterase as well as butyryl-cholinesterase (Sharapov 1959). In acute tests on cats and chronic tests on dogs the alkaloid increases bile flow up to 40–100 %, at the dose of 5 mg/kg. At the same time, secretion of bilirubin also increases (Rabinovich et al. 1966). Deoxypeganine exhibited strong anticholinesterase activity in vitro and in vivo (Tulyaganov 1976; Tulyaganov et al. 1986). In in vitro tests, the alkaloids peganol and peganidine have inhibitory action on activity of acetyl- and butyryl-cholinesterase of the blood and brain (Rustamov et al. 1974).

Phytochemistry: The aboveground parts of plants collected during the early the vegetative stage in the Buxoro and Surxondaryo provinces contained 2.1 % total alkaloids, with the young roots containing 3.32 % and older roots 1.68 % total alkaloids. At the bud stage, the aboveground parts contained 2–2.3 % total alkaloids, which decreased at the stage of flowering to 1.86–1.95 %. More than 15 alkaloids were isolated from the total alkaloids including harmine, harmaline harmalol, peganine, vasicinone, deoxypeganine, pegamine, peganidine, peganol, and dipegene, etc. (Yunusov 1981).

Perovskia abrotanoides Kar. – Lamiaceae

Synonyms: *Perovskia artemisioides* Boiss.

English name: Russian sage, Caspian Russian sage, Caspian Perovskia

Russian name: Перовския полынная (Perovskiya polynnaya)

Uzbek name: Khapri, Abrik

Kyrgyz name: Шыбактай коен томук (Shybaktay koyen tomuk)

Description: Perennial subshrub. Stems up to 100 cm tall, bases woody, white-hairy. Leaves petiolate, oblong-ovate, 2–7 cm long, 1–3 cm wide, bi-pinnatipartite. Inflorescences verticillasters, found in loose panicles; bracts lanceolate-linear. Calyx around 4.5 mm long, tubular-campanulate, 2-lipped, violet, often densely hairy. Corolla violet, funnelform, 2-lipped; upper lip 4-lobed, middle 2 lobes smaller; lower lip entire. Fruits smooth, brown nutlets.

Other distinguishing features: Differs from *Perovskia atriplicifolia* Bentham by having bi- pinnatipartite leaves.

Phenology: Flowers in June-August, fruits in August.

Reproduction: By seeds and rhizomes.

Distribution: Ysyk-Kol, Naryn, Osh, and Chuy Provinces of Kyrgyzstan; not found in the flora of Uzbekistan.

Habitat: In pebbly, dry stream beds and on dry, stony places in the mountains.

Population status: Common.

Traditional use: The aboveground parts are used to heal wounds. A decoction is used to treat scabies (Massagetov 1932). The plant is used externally to treat leishmaniasis in Iran (Moallem and Niapour 2008).

Documented effects: Compounds isolated from the roots exhibited leishmanicidal activity in vitro and inhibited growth of cultured malaria parasites, human lymphocytes, and human carcinoma cell lines (Sairafianpour et al. 2001). Compounds isolated from the aboveground parts exhibited cytotoxic activity against leukemia cells (Aoyagi et al. 2006).

Phytochemistry: The plant contains tanshinones (Sairafianpour et al. 2001). Water-distilled essential oils from leaves collected in Arslonbob (Kyrgyzystan) contained cineole, pinene, epi-13-manool, bornyl acetate, camphene, camphor, caryophyllene, caryophyllene oxide, humulene, caryophylladienol, borneol, and other compounds (Basher et al. 1997).

▲**Papaver pavoninum Schrenk** Photos: *left* and *right*: Alim Gaziev; *center*: Alexander Naumenko

▲**Patrinia intermedia (Hornem.) Roem. & Schult.** Photos: Lina Valdshmit

►**Peganum harmala L.** Photos: *top*: John B. Taft; *center*: Alim Gaziev; *bottom*: Evgeny Davkaev

▼**Perovskia abrotanoides Kar.** Photos: *left*: John B. Taft; *right*: Bazar Dovletov

Picea schrenkiana Fisch. & C.A. Mey – Pinaceae

Synonyms: *Abies schrenkiana* (Fisch. & C.A. Mey) Lindl. & Gordon, *Picea morinda* ssp. *tianschanica* (Rupr.) Berezin, *Picea obovata* Ledeb. var. *schrenkiana* (Fisch. & C.A. Mey) Carrière, *Picea prostrata* Isakov, *Picea robertii* P. Vipper, *Picea tianschanica* Rupr.

English name: Schrenk's spruce

Russian name: Ель Шренка (El' Shrenka)

Uzbek name: Неизвестно

Kyrgyz name: Archa

Description: Evergreen tree, up to 40 m tall, with narrow conical crown. Bark grayish-brown with thick plates. Leaves (needles) arranged radially, 20–25 mm long, linear, 4-sided, apex acute. Seed (female) cones ellipsoid-cylindric, 6–15 cm long, up to 3.5 cm wide. Seed scales triangular-ovate, apex rounded, brown. Seeds up to 4 mm long, flat-ovoid to fusiform, winged, brown.

Other distinguishing features: Branchlets pendulous, pale yellow.

Phenology: Seeds ripen in September.

Reproduction: By seeds.

Distribution: Ysyk-Kol, Naryn, Talas, and Chuy provinces of Kyrgyzstan; not found in the flora of Uzbekistan.

Habitat: On mountain slopes from 1,000 to 3,500 m.

Population status: Common, forming forests.

Traditional use: An infusion of needles from spring branchlets and cones are drunk to treat persistent common colds and is added to baths to treat rheumatism. An infusion of young branchlets in vodka is used to treat lung tuberculosis. The ground bark, mixed with wax and butter or lard, is applied in the form of a plaster to treat furuncles. The needles are used to prevent and treat scurvy and as a source of vitamins (Bykov 1950; Gan 1970).

Documented effects: None.

Phytochemistry: Needles and young branches contain vitamin C, essential oil (with up to 40 components such as camphene, myrcene, bornyl acetate, and others), flavonoids, and microelements (iron, manganese, chromium, aluminium and copper; Bykov 1950). Thirty-eight diterpenoids have been identified in the oleoresin (Raldugin et al. 1991). Sesquiterpenoids, diterpenoids, triterpenoids, steroids, and tocopherol were isolated from needles and twigs. Dehydroabietol, patchouli alcohol, guaiol, β-sitosterol, and campesterol were the main components of the unsaponifiable matter (Zhou 2001).

Plantago lanceolata L. – Plantaginaceae

Synonyms: None
English name: Narrowleaf plantain
Russian name: Подорожник ланцетолистный (Podorozhnik lantsetolistnyy)
Uzbek name: Nishtarsimon bargizub, Zabturum
Kyrgyz name: Бака жалбырак (Baka zhalbyrak)
Description: Herbaceous perennial. Leaves in basal rosettes, narrow-elliptic to lanceolate-elliptic, 7.5–35 cm long, 0.5–3.5 cm wide, with 3–5 parallel veins, apex acute, narrow petiolate. Inflorescence a dense, erect, cylindrical spike, 1.5–8 cm tall; peduncles 15–60 cm tall, with 5 ribs; bracts ovate, acute. Corolla 4-lobed. Fruit a 2-seeded, circumscissile capsule. Seeds elongate-oval.
Other distinguishing features: Leaves narrow-elliptic to lanceolate-elliptic. Stamens exserted.
Phenology: Flowers in June-August, fruits in August-September.
Reproduction: By seeds.
Distribution: All provinces of Kyrgyzstan; Karakalpakstan autonomous republic, Toshkent, Samarqand, Farg'ona, Buxoro, Andijon, Namangan, Surxondaryo and Xorazm provinces of Uzbekistan.
Habitat: Along canals and roads and in fallow fields.
Population status: Common, found in small groups.
Traditional use: A decoction of the leaves is used as a diuretic and as a treatment for cystitis, gastric diseases, lung tuberculosis, headaches, and to detoxify snake bites. Decoctions, infusions, extracts, and juice are used as a bacteriostatic, anti-inflammatory, antispasmodic, and expectorant, and also to treat enterocolitis, stomach ulcers, liver diseases, malaria, bronchitis, pertussis, bronchial asthma, allergic conjunctivitis, as well as to heal wounds, furuncles, skin ulcers, and purulent wounds (Plant Resources of the USSR 1990).
Documented effects: Preparation from this species are used as a hemostatic (Zemlinsky 1958) and to treat chronic bronchitis (Nosal and Nosal 1959). Compounds in the herb showed inhibitory effects on mouse ear edema (Murai et al. 1995). Results of experimental research confirmed anti-inflammatory, antispasmodic, and immunostimulatory actions (Wegener and Kraft 1999).
Phytochemistry: Roots contain steroids (sitosterin, stigmasterin, cholesterol, and campesterin) and higher fatty acids. The aboveground parts contain iridoids, phenolcarbonic acids, flavonoids, carbohydrates, organic acids, and protocatechins. Seeds contain iridoids, carbohydrates, muscilage, and fatty oil (Plant Resources of the USSR 1990; Murai et al. 1995).

Plantago major L. – Plantaginaceae

Synonyms: *Plantago borysthenica* (Rogow.) Wissjul., *Plantago dregeana* Decne., *Plantago latifolia* Salisb., *Plantago officinarum* Crantz.

English name: Common plantain, broadleaf plantain

Russian name: Подорожник большой (Podorozhnik bol'shoy)

Uzbek name: Zupturoom, Buzchi, Bakayaprok

Kyrgyz name: Чон бака жалбырак (Chon baka zhalbyrak)

Description: Herbaceous perennial. Leaves in basal rosettes, broad-elliptic to broad-ovate, 4–21 cm long, 3–14 cm wide, 3–9 parallel veins, sheathing petiolate. Inflorescences dense, erect, narrow-cylindric spikes, 5–15 cm tall; peduncles 15–70 cm tall; bracts ovate, acute. Corolla greenish or yellowish white with 4 reflexed lobes. Fruit a 2-seeded, circumscissile capsule. Seeds 1–1.5 mm long, densely reticulate.

Other distinguishing features: Differs from *Plantago rugelii* Decne. by having fruits dehisce near the middle rather than far below the middle.

Phenology: Flowers in June-August, fruits in August-September.

Reproduction: By seeds.

Distribution: Almost all provinces of Kyrgyzstan and Uzbekistan.

Habitat: In meadows, along streams and canals, and in orchards and wet places.

Population status: Common, forming dense groups.

Traditional use: This species has been used for centuries. Avicenna used the leaves as a hemostatic, to heal wounds, tumors, eye inflammation, chronic skin ulcers, and elephantitis as well as for liver and kidney diseases. In more recent times the plant has been used to treat lung tuberculosis, pertussis, stomach catarrh with low acidity, acute gastritis, enterocolitis, stomach and duodenum ulcers, and as a hemostatic (Khalmatov et al. 1984). A tea made from the dried leaves is used to treat coughing, diarrhea, dysentery (with tea from seeds is most effective), inflammation of the bladder, and malaria, and as an expectorant (Altimishev 1991).

Documented effects: Experiments with animals showed that a 20 % extract of leaves healed wounds, decreased pus volume, stimulated epithelialization of the wound surface, had sedative and soporific effects, and reduced blood pressure (Aliev 1945). The triterpenoid, ursolic acid, and isolated from the plant showed significant COX-2 inhibitory activity (Ringbom et al. 1998). Five compounds, including caffeic and chlorogenic acids, isolated from extracts of the plant exhibited potent antiviral activity (Chiang et al. 2002). The preparation *Plantaglucid*, made from a water extract of the plant, is used as an anti-ulcer treatment and to heal wounds (Poludenny and Zhuravlev 2000). A preparation of the plant is used to treat respiratory tract diseases, pertussis, lung tuberculosis, and chronic nephritis (Khodzhimatov 1989).

Phytochemistry: Leaves contain the glycoside aucubin, phenolic compounds (caffeic acid, chlorogenic acid, ferulic acid, and p-coumaric acid), triterpenes (oleanolic acid and ursolic acid), bitter substances, tannins, carotene, vitamin C and K, high amounts of potassium, mucilage, organic acids, saponins, essential oil, flavonoids (baicalein, scutellarin, apigenin, etc.), and small amounts of alkaloids. The seeds contain mucilage, fatty oil, carbohydrates, saponins, etc. (Khalmatov et al. 1984; Poludenny and Zhuravlev 2000; Chiang et al. 2002).

Polemonium caucasicum N. Busch – Polemoniaceae

Synonyms: *Polemonium caeruleum* ssp. *caucasicum* (N. Busch) V.E. Avet.
English name: Unknown
Russian name: Синюха кавказская (Sinyukha kavkazskaya)
Uzbek name: Unknown
Kyrgyz name: Кавказ полемону (Kavkaz polemonu)
Description: Herbaceous perennial, with rhizomes. Stems 50–100 cm tall. Leaves alternate, 7–20 cm long, odd-pinnatisect with 5–21 pairs of segments; segments lanceolate, sessile. Inflorescence many-flowered, corymbiform. Calyx 6–8 mm long, glandular-hairy. Corolla rotate, 8–15 mm long, blue or seldom white, 5-lobed. Stamens 5. Fruit a capsule, almost spherical, 5–7 mm long. Seeds brown, 3–3.5 mm long.
Other distinguishing features: Stamens exserted. Seeds angular and rugose.
Phenology: Flowers in June-July, fruits in July-August.
Reproduction: By seeds.
Distribution: Ysyk-Kol, Naryn, and Chuy Provinces of Kyrgyzstan; not found in the flora of Uzbekistan.
Habitat: In tallgrass meadows, subalpine meadows, and meadow-steppes.
Population status: Common, found in small groups.
Traditional use: An infusion of the underground parts is used to treat fevers. A decoction is used as a bath to treat spazmophilia. An infusion of the leaves is used as a sedative and to treat syphilis. An infusion of the flowers is used internally to treat leucorrhoea (Plant Resources of the USSR 1990).
Documented effects: The closely related species *Polemonium caeruleum* L. contains saponins that act as an expectorant. The plant has also been shown to have hemostatic effects and acts as a highly effective sedative (8–10 times that of *Valeriana*), but can be fatal to experimental animals at high doses. Preparations are used as expectorants, sedatives, treatments for stomach and duodenum ulcers, epilepsy, and chronic and acute bronchitis (Akopov 1990).
Phytochemistry: The entire plant of the closely related species, *Polemonium caeruleum*, contains triterpene saponins, triterpene glycosides, resins, organic acids, essential oils, fatty oils, and many macro- and micro-elements (Akopov 1990; Kurochkin 1998).

▲**Picea schrenkiana Fisch. & C.A. Mey** Photos: *left* and *right*: Vladimir Epiktetov; *center*: Evgeny Davkaev

▼**Plantago lanceolata L.** Photos: *left*: Dmitri Oreshkin; *right*: Sergey Appolonov

▼**Polemonium caucasicum N. Busch** Photo: Vladimir Epiktetov

►**Plantago major L.** Photos: *left*: Andrei Lubchenko; *right*: Vadim Prokhorov

Polygala hybrida DC.– Polygalaceae

Synonyms: *Polygala comosa* var. *altaica* Chodat, *Polygala comosa* Schkuhr var. *hybrida* (DC.) Petelin, (some consider *P. hybrida* a synonym of *P. comosa* Schkuhr).

English name: Milkwort

Russian name: Истод гибридный (Istod gibridnyy)

Uzbek name: Unknown

Kyrgyz name: Аргын истод (Argyn istod)

Description: Herbaceous perennial. Stems 15–40 cm tall, short-hairy. Leaves sessile, 1.5–4.5 cm long, 2–4 cm wide, elliptic or lanceolate, margins entire. Inflorescences densely flowered, terminal racemes. Calyx with 3 outer, elliptic-lanceolate sepals, and 2 inner, large petaloid, elliptic sepals. Corolla with 3 petals, purple or pink, keel shorter than lateral petals. Fruits winged capsules, 6 mm long. Seeds densely covered with appressed hairs.

Other distinguishing features: Filaments connate for the entire length. Capsules oblong. Seeds arillate.

Phenology: Flowers in June-July, fruits in July-August.

Reproduction: By seeds.

Distribution: All provinces of Kyrgyzstan; Toshkent, Farg'ona and Samarqand provinces of Uzbekistan.

Habitat: In subalpine and alpine meadows and meadow-steppes.

Population status: Common, found growing as individual plants.

Traditional use: An infusion of the herb is used to treat heart and gastrointestinal illnesses and rabid dog bites. The herb is also used to treat croupous pneumonia, coughs, asphyxia, fainting, sore throats, and oral diseases. In Mongolian medicine it is used as an expectorant to treat tuberculosis, purulent pleuritis, and as a hemostatic to treat uterine bleeding. In the Tibetan and Mongolian medicine the seeds are used to treat myopathy, obesity, tumors and wounds, and as a hemostatic (Plant Resources of the USSR 1988).

Documented effects: Other species of *Polygala* have been shown to contain biologically active saponins that exhibit significant immunological properties in vitro (Desbène et al. 1999; Estrada et al. 2000).

Phytochemistry: The roots contain carbohydrates, saponins, tannins, and fatty oil. The aboveground parts contain alkaloids (Turova and Sapozhnikova 1984; Lugmanova et al. 2007).

Polygonum aviculare L. – Polygonaceae

Synonyms: *Polygonum aequale* Lindm., *Polygonum agreste* Sumner, *Polygonum aphyllum* Krock., *Polygonum araraticum* Kom., *Polygonum arenastrum* Boreau, *Polygonum berteroi* Phil., *Polygonum heterophyllum* Lindm., *Polygonum retinerve* Vorosch., *Polygonum striatum* K. Koch, *Polygonum uruguense* H. Gross.

English name: Prostrate knotweed, Yard knotweed

Russian name: Горец птичий (Gorets ptichiy)

Uzbek name: Kiziltasma

Kyrgyz name: Тошолгон кымыздык (Tosholgon kymyzdyk)

Description: Herbaceous annual, with a slightly-branched taproot. Stems prostrate or suberect, 7–60 cm long. Leaves alternate, of 2 sizes; early leaves lanceolate, 2.5–6 cm long, 4–15 mm wide; later leaves much reduced; ocreae 4–8 mm long, membranaceous, lacerate. Flowers many, very small, in groups of 2–5 at nodes. Tepals 5, partially connate, white, greenish or pink-red. Fruits triquetrous, dark-brown achenes.

Other distinguishing features: The fruits are equal to or slightly exserted past the tepals.

Phenology: Flowers and fruits in May-October.

Reproduction: By seeds.

Distribution: Almost all provinces of Kyrgyzstan and Uzbekistan.

Habitat: In meadows and fallow or cultivated fields, from foothills up to the alpine belt of the mountains.

Population status: Common, found in dense groups.

Traditional use: A decoction and infusion of the herb is used to treat stomach spasms, intestinal infections, diarrhea and as a tonic, hemostatic and diuretic. The plant is used in a bath to treat bacterial and fungal skin diseases and rashes. The fresh herb is put on tumors, wounds, and skin ulcers (Khalmatov et al. 1984). An infusion of the herb is used to wash the head to increase the health of hair and encourage hair growth. A decoction of the herb in milk is taken to treat convulsions (Poludenny and Zhuravlev 2000).

Documented effects: Water and alcohol extracts of the plant have been shown to increase the rate of blood coagulation, decrease blood pressure, increase inhalation volume, improve lung function, tone uterine muscles, and increase diuresis. The preparation *Avicularen* is used in gynecological practice as a hemostatic (Khalmatov et al. 1984). The infusion of the herb is recommended to improve metabolism and treat diabetes (Kurochkin 1998). Experiments indicate that a methanol extract of the plant has anti-fibrotic effects on rats with induced liver fibrosis (Nan et al. 2000).

Phytochemistry: The plant contains essential oils, vitamin K_1, sugars, saponins, coumarins, mucilage, anthraglycosides, etc. (Khalmatov et al. 1984). Leaves contains tannins, flavonoids (avicularin), vitamin C, carotene, and silicic acid compounds (Tolmachev 1976).

Polygonum coriarium Grig. – Polygonaceae

Synonyms: *Aconogonon bucharicum* (Grig.) Holub, *Aconogonon coriarium* (Grig.) Soják, *Aconogonon coriarium* ssp. *bucharicum* (Grig.) Soják, *Pleuropteropyrum bucharicum* (Grig.) Nevski, *Polygonum bucharicum* Grig.

English name: Unknown

Russian name: Горлец дубильный (Gorlets dubil'nyy)

Uzbek name: Taran dubil'nyy

Kyrgyz name: Ашаткыч кымыздык (Ashatkych kymyzdyk)

Description: Herbaceous perennial, with a large rhizome (up to 5–8 kg). Stems up to 1–1.5 m tall, abundantly branched, glabrous. Leaves alternate, short-petiolate, ovate or ovate-lanceolate, 6–10 cm long, 2.5–5 cm wide, base wide-cuneate or rounded, abaxial side (sometimes both sides) densely hairy, rarely glabrous, margin bristly-ciliate. Ocreae membranous, tubular, 1.5–2.5 cm long, brown, not persisting. Inflorescence large panicle, branched, dense, up to 35 cm long and 25 cm wide. Perianth usually 2.5–3.5 cm long, with 5 white tepals. Fruit a triquetrous nutlet with sharp edges, 3–4.5 mm long, shiny, slightly exserted from perianth.

Other distinguishing features: Stamens 8, styles 3. Branches of inflorescence nodding in fruit.

Phenology: Flowers and fruits in June-August.

Reproduction: By seeds and rhizomes.

Distribution: Toshkent, Samarqand, and Surxondaryo provinces of Uzbekistan; Talas, Osh and Jalal-Abad provinces of Kyrgyzstan.

Habitat: The tau zone. Stony, shallow soil on wet slopes of mountains.

Population status: Common.

Traditional use: A decoction of the underground plant parts is used in folk medicine as an astringent for treatment of diarrhea with and without blood (Khalmatov 1964).

Documented effects: Astringent preparations effective for gastrointestinal tract diseases of alimentary origin were prepared from root powder in combination with protein (called *Taranalbin*) or formaldehyde (called *Taranform*). An extract obtained from the plant roots in the dose of 20 mg/kg increases stability for exercise stress (physical activity) and elongates swimming time of mice up to 61%. Proanthocyanidin and catacin (katacine) have distinct antihypoxic action and decrease the oxygen-need of tissues (Kurmukov et al. 1991b), which is connected to its influence on energy metabolism (Nazrullaev et al. 1990).

Phytochemistry: Underground organs contain up to 28–35 % tannins, mainly of the pyrocatechin group (proanthocyanidin; Ogolevitz 1951). Many proanthocyanidins have been isolated from the roots (Makhmatkulov et al. 1992, 1994; Keneshov et al. 1997a, b). The leaves contain flavonoids (Chumbalov and Omurkamzinova 1968).

Portulaca oleracea L. – Portulacaceae

Synonyms: *Portulaca consanguinea* Schltdl., *Portulaca intermedia* Link ex Schltdl., *Portulaca marginata* Kunth, *Portulaca mundula* I.M. Johnst., *Portulaca neglecta* Mack. & Bush, *Portulaca pilosa* L., *Portulaca pusilla* Kunth, *Portulaca retusa* Engelm.

English name: Purslane, Little hogweed

Russian name: Портулак огородный (Portulak ogorodnyy)

Uzbek name: Semiz ut

Kyrgyz name: Огород портулагы (Ogorod portulagy)

Description: Herbaceous annual. Stem 10–35 cm long, glabrous, fleshy, prostrate, spreading, branched from the base. Leaves alternate or sub-opposite, obovate or spatulate, 4–28 × 2–13 mm, apex rounded to obtuse, fleshy, sessile. Flowers 3–10 mm wide, solitary or in small clusters of 2–3 in branch and leaf axils. Sepals 2, deciduous. Petals usually 5, yellow, obovate. Fruit a circumscissile capsule, ovoid, 5–8 mm long, many-seeded. Seeds orbiculate or elongate, flattened, surface covered with tubercules, black to dark brown, slightly shining, 0.7–1 mm long, 0.25 mm wide.

Other distinguishing features: Stamens usually 8–15. Stigmas 3–6.

Phenology: Flowers and fruits in June-October.

Reproduction: By seeds.

Distribution: All regions of Uzbekistan and Kyrgyzstan.

Habitat: The chul, adyr and tau zones. A weed of irrigated agricultural areas.

Population status: Common.

Traditional use: A decoction of the herb is used in Chinese medicine to alleviate pain and swelling, as an anti-inflammatory and diuretic, and for gonorrhea, kidney and liver diseases, bacterial dysentery, syphilitic arthritis, and palsies of infectious origin. In Central Asia this herb is used as a choleretic, an antipyretic for fevers associated with hepatitis, nephritis, and cystitis, and as a treatment for intestinal ulcers and bloody diarrhea. The plant is also used to treat intestinal infections (Khalmatov 1964; Chen et al. 2003).

Documented effects: An extract of this herb sharply increases blood pressure, due to its high noradrenaline content. Hemostatic action for internal hemorrhaging has been documented (Khalmatov 1964). In experiments with mice and rats, an ethanolic extract of the dried aboveground parts showed significant antiinflammatory and analgesic effects after intraperitoneal and topical, but not oral, administration (Chan et al. 2000). Studies indicated that the consumption of the plant may help to reduce the occurrence of cancer and heart diseases. Catecholamines (noradrenaline and dopamine) contained in the plant are generally considered to be the effective component in the treatment of shock. Studies have also shown that noradrenaline is a modulator of the immune system and may have anti-cancer properties (Chen et al. 2003).

Phytochemistry: The herb contains alkaloids, glycosides, traces of saponins, and bitter substances (Khalmatov 1964). 250 mg% noradrenaline has been obtained from the fresh herb (Khalmatov 1964). The plant contains an abundance of the catecholamines noradrenaline and dopamine, free oxalic acids, alkaloids, coumarins, flavonoids, cardiac and anthraquinone glycosides, proteins, high amounts of beta-carotenes and has a higher content of omega-3 fatty acids (especially α-linolenic acid) than many other vegetables (Guil-Guerrero and Rodríguez-García 1999; Chen et al. 2003; Fontana et al. 2006).

▲**Polygala hybrida DC.** Photos: Lina Valdshmit

▼**Polygonum coriarium Grig.** Photos: Vladimir Epiktetov

▲**Portulaca oleracea L.**
Photos: *top*: Andrei Lubchenko; *bottom*: Mary Backlund

▼**Polygonum aviculare L.** Photos: *top right* and *bottom right*: Vadim Prokhorov; *bottom left*: Denis Mirin

Potentilla canescens Bess. – Rosaceae

Synonyms: *Potentilla adscendens* Waldst. & Kit. ex Willd., *Potentilla inclinata* Vill.

English name: Hoary cinquefoil, ashy cinquefoil

Russian name: Лапчатка седоватая (Lapchatka sedovataya)

Uzbek name: Unknown

Kyrgyz name: Агыш туктуу казтаман (Agysh tuktuu kaztaman)

Description: Herbaceous perennial. Stems few, erect, 10–50 cm tall, pubescent. Leaves palmately compound with 5–7 leaflets, stipulate, petioles pubescent; leaflets obovate or obovate-lanceolate, pubescent, margins coarse serrate. Inflorescence many-flowered, corymbiform or cymose-paniculiform. Flowers pedicellate, ca. 10 mm in diameter. Sepals 5, epicalyx segments 5, alternating with sepals. Petals 5, yellow, ovate, slightly longer than sepals. Fruits wrinkled achenes.

Other distinguishing features: Lower side of leaflets tomentose. Base of style thickened.

Phenology: Flowers in June, fruits in August.

Reproduction: By seeds.

Distribution: Naryn, Talas, and Chuy provinces of Kyrgyzstan; Toshkent, Andijon, Namangan, and Farg'ona provinces of Uzbekistan.

Habitat: In the foothills and steppe belt of mountains, along roads, fallow fields and in lowland steppes.

Population status: Common, found in small groups.

Traditional use: An infusion of the underground parts is used to treat menorrhagia, diarrhea, and hematuria. An infusion of the aboveground parts is used to treat laryngitis (Plant Resources of the USSR 1987).

Documented effects: No data.

Phytochemistry: The rhizomes and flowers contain traces of alkaloids. Leaves and flowers contain vitamin C (Plant Resources of the USSR 1987).

Prangos pabularia Lindl. – Apiaceae

Synonyms: *Hippomarathrum sarawschanicum* Regel & Schmalh., *Hyalolaena sewerzowii* Regel & Herd., *Koelzella pabularia* (Lindl.) Hiroe, *Prangos cylindrocarpa* Korovin, *Prangos hissarica* Korovin, *Prangos lamellata* Korovin, *Prangos seravschanica* (Regel & Schmalh.) Korovin.

English name: Unknown

Russian name: Прангос кормовой (Prangos kormovoy)

Uzbek name: Tulky kuyruq

Kyrgyz name: Тоют аюу чачы (Toyut ayuu chachy)

Description: Herbaceous perennial with thick taproot. Stems several, up to 0.6–2 m tall, angular-striated, strongly branching from the middle, nearly glabrous. Basal leaves densely clustered, pointing upward, long-petiolate; blades 30–70 cm long, 6–12 cm wide, elliptic or oblong in outline, 4–5-pinnate with filiform or narrow-linear segments. Inflorescence an irregular compound umbel, 8–20 rays; umbellets 10–15-flowered. Sepals triangular, acute. Petals obovate, ca.1.5 mm long, yellow. Fruit a schizocarp with 2 mericarps; mericarps oblong-cylindrical, 15–18 mm long, often violet in color with prominent ribs, grooves between ribs narrow, lined with tubercles.

Other distinguishing features: Leaves cause strong burns and photosensitivity when touched. Leaves quickly senescing after which the stem is covered with leaf remnants. Flowers along the outer margin of the umbellets are bisexual; flowers in the center are male.

Phenology: Flowers in May-June, fruits in June-July.

Reproduction: By seeds.

Distribution: Toshkent, Samarqand, Jizzax, Qashqadaryo, and Surxondaryo provinces of Uzbekistan; Osh and Jalal-Abad provinces of Kyrgyzstan.

Habitat: The tau and yailau zones. Clayey and clayey-stony mountain slopes.

Population status: Common, often found in large populations.

Traditional use: Root decoctions and root tinctures, sometimes mixed with tinctures of iodine and St. John's wort (*Hypericum*), are used to treat scabies in humans (Ogolevitz 1951). A decoction of the roots is used to disinfect the mouth and to kill ticks, fleas, and bed bugs on farm animals. The roots are put on hot ashes and after 2–3 h are then put on surface wounds. The above and underground parts are used in a bath to treat skin diseases (scabies, fungal, etc.). A decoction of the aboveground parts is used as a mouth wash to treat toothaches (Khodzhimatov 1989).

Documented effects: In experiments the coumarin osthol increased blood pressure, pulse rhythm, stimulated respiratory activity, weakened acetylcholine effect, and had vermifugal activity. (Ogolevitz 1951; Jamwal et al. 1962). A butenyl coumarin isolated from the plant had analeptic activity on respiration and the heart, stimulated brain functions, and exhibited antiacetylcholinic and antihistaminic action (Chicco 1966). It is also used as antidote in the poisoning due to hypnotics. The coumarin osthol showed antibiotic activity against *Staphylococcus aureus* (MRSA) and *Pseudomonas aeruginosa*. Twelve compounds isolated from this plant had immunosuppressive activity (Tada et al. 2002).

Phytochemistry: Leaves, fruits, roots and root resin contain coumarins. The coumarin osthol and the furocoumarins oxypeucedanin, imperatorin, prangenin, prangenidin, and others have been isolated from the total coumarins. Seeds contain 0.2–0.3 % alkaloids from which the alkaloid prangosine has been isolated (Ogolevitz 1951; Yunusov 1981; Khodzhimatov 1989). This plant is a rich source of coumarins, coumarin derivatives, and terpenoids with 29 different compounds being identified (Tada et al. 2002). One hundred twenty-eight compounds were characterized from the volatile constituents of the fruits. The major constituents of the essential oil were α-humulene, bicyclogermacrene, spathulenol, germacrene D, and α-pinene (Ozek et al. 2007).

Prunus sogdiana Vassilcz. – Rosaceae

Synonyms: *Prunus cerasifera* ssp. *sogdiana* (Vassilcz.) Cinovskis, *Prunus cerasifera* var. *orientalis* Popov, *Prunus mirabilis* Sumner, *Prunus orientalis* (Popov) Kudr.

English name: Sogdian plum

Russian name: Слива согдийская (Sliva sogdiyskaya)

Uzbek name: Togolcha

Kyrgyz name: Жапайы алча (Zhapayy alcha)

Description: Tree or shrub from 2.5 to 7 m tall, with multiple trunks. Older bark dark-gray, cracked; young branches brownish-green to red-brown. Leaves alternate, petiolate, elliptic, ovate or obovate, 4.5–5.6 cm long, 2.2–4 cm wide, glabrous above, light in color and pubescent along midvein below, margins serrate or serrate-crenate. Flowers ca. 2 cm in diameter, pedicillate. Sepals 5, glabrous. Petals 5, ovate, white or with purple base. Fruit a dark purple drupe, spherical to slightly elongated, 1–2 cm in diameter, often glaucous.

Other distinguishing features: Stamens 15–30, in 2 whorls, filaments unequal in height.

Phenology: Flowers in May, fruits in July-October.

Reproduction: By seeds.

Distribution: Jalal-Abad, Osh, and Chuy provinces of Kyrgyzstan; Toshkent and Surxondaryo provinces of Uzbekistan.

Habitat: On the edges, and in the understory of deciduous forests, and among bushes.

Population status: Common, forming dense groups.

Traditional use: The fruits, leaves, flowers, bark and gum are used in folk medicine. An infusion of the leaves and flowers is used as a light laxative. A decoction of the dried fruits is used to increase appetite, to aid in digestion, and as an expectorant. The gum is used as a treatment for coughs. A water extract of the bark and roots is used as a diaphoretic, antipyretic, and anti-inflammatory (Nuraliev 1989). In Pamir-Alai it is used to treat acute respiratory diseases (Zapryagaeva 1964).

Documented effects: No data.

Phytochemistry: Fruits contain sugars, organic acids (malic and citric), vitamin C, provitamin A, pectins, tannins, minerals, and fatty oil (Nuraliev 1989).

Pseudosophora alopecuroides (L.) Sweet – Fabaceae

Synonyms: *Goebelia alopecuroides* (L.) Bunge, *Sophora alopecuroides* L., *Vexibia alopecuroides* (L.) Yakovl.

English name: Unknown

Russian Name: Вексибия лисохвостная, Талхак обыкновенный (Veksibiya lisokhvostnaya, Talkhak obyknovennyy)

Uzbek name: Akmia, Achikmia

Kyrgyz name: Ак мыя (Ak myya)

Description: Herbaceous perennial. Stems 40–70 cm tall. Leaves alternate, odd-pinnate, with 5–12 pairs of oval or elongate-ovate leaflets; leaflets 1.2–3 cm long, 4–12 mm wide. Inflorescence an densely-flowered, apical raceme. Calyx widely campanulate with 5 uneven teeth. Corolla papilionaceous, white or slightly yellowish. Fruit a legume, 5–12 cm long, constricted between the seeds, with extended tip at the end. Seeds spherical, light-brown, smooth.

Other distinguishing features: The whole plant is gray-green hairy.

Phenology: Flowers in May-June, fruits in June-August.

Reproduction: By seeds and rhizomes.

Distribution: All provinces of Kyrgyzstan; Toshkent, Farg'ona, Samarqand and Buxoro provinces of Uzbekistan.

Habitat: In foothills and in abandoned and cultivated fields.

Population status: Common, found in dense groups.

Traditional use: In Tibetan and Mongolian medicine the roots are used to treat diseases of the heart, aorta, and vascular system, diphtheria, and rheumatism, and are used as an antipyretic and restorative as well (Khaidav 1965). The ground seeds are used to treat poor digestion and loss of appetite (Khalmatov 1964).

Documented effects: In modern medicine preparations of this plant are used to increase respiratory function (Khalmatov 1964). In experiments, large doses of sophocarpine act as a weak ganglioblocker and cause contraction of the myometrium. Matrine, sophoridine, sophocarpine, and aloperine have stimulating activities, but aloperine causes short-term hypotension. Sophocarpine and sophoridine cause narrowing of the peripheral vessels, and in small doses strengthen peristalsis and intestinal tonus, paralyze skeletal muscles, and have gangloblocking properties (Georgadze 1938; Kruglikova-Livova 1952). Quinolizidine alkaloids isolated from the plant have very weak antiviral activities (Ma et al. 2002a).

Phytochemistry: The roots contain alkaloids (spartein, sophoridine, and sophocarpine), flavonoids (quercetin and rutin), flavanones (sophoraflavanone G and leachianone A), flavonostilbenes (alopecurones A-F), and anthraquinones (aloemodin, anthraquinone sennosides, etc.). The aboveground parts contain alkaloids (sophoridine, cytisine, neosophoramine, sophoramine, sophocarpine and aloperine; Yusupova et al. 1984; Plant Resources of the USSR 1987; Iinuma et al. 1990, 1995). The alkaloids oxymatrin, oxysophocarpine, cytisine, matrine, sophocarpine, sophoridine, and nicotine have been isolated from the seeds (Zhang et al. 1997).

▲ **Potentilla canescens Bess.**
Photo: Andrei Lubchenko

▲ **Pseudosophora alopecuroides (L.) Sweet**
Photos: Lina Valdshmit

▲ **Prunus sogdiana Vassilcz.** Photos: *left*: John B. Taft; *right*: Vladimir Epiktetov

▼ **Prangos pabularia Lindl.** Photos: Evgeny Davkaev

Psoralea drupacea Bunge – Fabaceae

Synonyms: *Cullen drupacea* (Bunge) C.H. Stirt., *Lotodes drupaceum* (Bunge) Kuntze.

English name: Scurfy-pea

Russian name: Псоралея костянковая (Psoraleya kostyankovaya)

Uzbek name: Ok kuraiy

Kyrgyz name: Соокчёлуу ак куурай (Sookchyoluu ak kuuray)

Description: Herbaceous perennial to 40–150 cm tall, with vigorous, thick, woody roots. Stems erect, branched, densely hairy, glandular. Leaves alternate, simple or ternate, short-petiolate; leaflets nearly round, 1.5–5 cm long, 2–6 cm wide, densely hairy beneath with glands on both sides, margins coarsely dentate; stipules linear-lanceolate, 0.5–1.5 cm long, hairy, glandular. Inflorescences in loose axillary racemes. Flowers 4–7 mm long, on very short pedicels. Calyx tubular-campanulate with unequal teeth, densely hairy with glands. Corolla white-lilac. Fruit a 1-seeded legume, suborbicular, densely hairy, ca. 5 mm long, 2.5–3.5 mm wide. Seed very small, adnate to the fruit wall.

Other distinguishing features: The root has a yellow color inside. The fruit is indehiscent and beakless.

Phenology: Flowers in May-June-July, fruits in June-September.

Reproduction: By seeds.

Distribution: Toshkent, Farg'ona, Samarqand, Qashqadaryo, and Surxondaryo provinces of Uzbekistan; Chuy, Talas and Jalal-Abad provinces of Kyrgyzstan.

Habitat: The chul and adyr zones. Found in combination with ephemeroid vegetation in shallow, loamy, sierozem soil, rocky-clay loams, and light-clay loams.

Population status: Common, found in large populations.

Traditional use: Leaf powder is used in folk medicine to treat abscesses (furuncles and carbuncles), vitiligo, eczema, and hair loss (Shimanov 1972). The essential oil from the fruits and galenical preparations of the legumes and roots are used to treat skin problems (Mamedov et al. 2004).

Documented effects: The chemical psoralen has photosensitizing, estrogenic, contraceptive, and embryotoxic actions (Shimanov 1972; Kurmukov and Akhmedkhodzhaeva 1975; Kurmukov et al. 1976, 1977). A medical preparation *Psoralen* (a mixture of furocoumarins from the fruits) is used to treat vitiligo and patchy, irregular hair loss. Treatment with the preparation *Drupanol*, at the dose of 10 mg/kg for 10 days, has anabolic and androgenic effects. In tests on chickens, the androgenic effect of drupanol caused a significant growth stimulation of the crest. The testing groups' crest increased by 1.6 times compared to control group (Syrov et al. 1976; Akopov 1990).

Phytochemistry: The aboveground parts contain the antibiotic bakuchiol and the coumarin umbelliferol. The fruits and roots contain psoralen, isopsoralen, tannins, and semi-solid essential oil, and the fruits contain fatty oil, the alkaloid drupacine, and the phenol drupanol (Golovina and Nikonov 1973). The mature fruits have the highest coumarin content (0.1 %) and the roots contains tannins (Akopov 1990).

Pulicaria salviifolia Bunge – Asteraceae

Synonyms: *Pulicaria afghanica* Kitam., *Pulicaria lachnophylla* C. Winkl., *Pulicaria olivascens* Rech. f., *Pulicaria sublepidota* Rech. f.

English name: Sage-leaf fleabane

Russian name: Блошница шалфеелистная (Bloshnitsa shalfeelistnaya)

Uzbek name: Gulband

Kyrgyz name: Шалфей жалбырактуу пуликария (Shalfey zhalbyraktuu pulikariya)

Description: Herbaceous perennial, 40–60 cm tall, densely covered with woolly hairs. Stem single or few, each with many straight branches from near the base. Leaves simple, spatulate, apex obtuse, gradually tapering to petiole, margins sinuate; lower leaves much larger than upper. Inflorescences capitula, numerous, arranged in loose racemes. Involucre 9–15 mm in diameter; involucral bracts usually in 2 rows, lanceolate, acute with membranaceous margin, grayish-hairy; inner bracts much more narrow and membranaceous. Flowers yellow; ray flowers with short, obovate ligules; disc flowers narrow, 5–8 mm long. Fruits achenes, 2–2.7 mm long, sparsely hairy, glandular; inner row of pappus plumose-barbed, whitish hairs, 5–8 mm long.

Other distinguishing features: Plant strongly sweet-scented, smelling like honey. Varieties of this species differ in the color and amount of pubescence.

Phenology: Flowers in July-August, fruits in August-September.

Reproduction: By seeds.

Distribution: Toshkent, Farg'ona, Samarqand, and Surxondaryo provinces of Uzbekistan; Osh and Jalal-Abad provinces of Kyrgyzstan.

Habitat: The adyr zone. Dry stony, slopes with rocky debris and pebbly, gypsum-soiled foothills.

Population status: Common, often forming dense groups.

Traditional use: A decoction of the herb is used for decreasing blood sugar content for treatment of diabetes (Abdunazarov 2000).

Documented effects: The flavonoid pulicarin exhibited hypolipidemic effects in experiments with rats (Sagitdinova et al. 1992). When administered orally, salvin, salvicin and salvifolin, show significant hypoglycemic activity in rats (Tashmukhamedova et al. 1992).

Phytochemistry: The plant contains terpenoids and diterpenoids (salvin, salvifolin, salvicin, salvicinin, salvicinolide, and salvicinolin, etc.), as well as flavonoids (rutin, etc.), triterpenoids, and sterols (Nurmukhamedova et al. 1985, 1986; Sagitdinova et al. 1992, 1994; Eshbakova et al. 1997; Eshbakova and Saidkhodzhaev 2001).

Reseda luteola L. – Resedaceae

Synonyms: None
English name: Weld, Yellow dye, Dyer's rocket
Russian name: Резеда жёлтенькая (Rezeda zhyolten'kaya)
Uzbek name: Sayok
Kyrgyz name: Сары резеда (Sary rezeda)
Description: Herbaceous biennial, up to 80 cm high. Stem single, erect, glabrous, densely foliaceous. Leaves alternate, oblanceolate to linear, 3–9 cm long, 5–12 mm wide, glabrous, sessile, margins entire. Inflorescences spiciform racemes, 15–45 cm long, erect. Flowers with 4 rounded sepals and 4 yellowish, irregularly lobed petals. Stamens 20–25. Fruit a subglobose capsule, 3-parted. Seeds ca.1 mm long, brown-black, glabrous, shiny, smooth.
Other distinguishing features: The bracts, calyx, and filaments are persistent during fruiting.
Phenology: Flowers and fruits in May-August.
Reproduction: Only by seeds.
Distribution: Toshkent, Samarqand, and Surxondaryo provinces of Uzbekistan; Jalal-Abad province of Kyrgyzstan.
Habitat: The adyr and tau zones. Dry hills, along roads, river valleys, mountain slopes and near field crops.
Population status: Uncommon, found in small populations of 4–8 individuals.
Traditional use: A decoction of the root is taken as a vermifuge (Khalmatov 1964).
Documented effects: The glycoside glucobarbarin has antithyroid action (Khalmatov 1964). In an inhibition assay, an extract of the plant inhibited the activity of trypsin by 97 % (Johansson et al. 2002). Luteolin showed anti-inflammatory activity in a variety of different in vivo assays and has also exhibited anti-cancer activity (Chowdhury et al. 2002; Ziyan et al. 2007).
Phytochemistry: The aboveground parts of the herb contain 1–3 % dyeing substances (luteolin), mustard essential oil (mostly in roots). 32–34 % fatty oil was extracted from the seeds. The leaves, inflorescence and the seeds include the glycosides glucocapparin and glucobarbarin (Khalmatov 1964). The aboveground parts contain cinnamamide and alkaloids (Lutfullin et al. 1976, 1977). The plant was also found to contain phenyl-β-naphthylamine (Sultankhodzhaev and Tadzhibaev 1976).

Rhamnus cathartica L. – Rhamnaceae

Synonyms: None

English name: Common buckthorn

Russian name: Жостер слабительный (Zhoster slabitel'nyy)

Uzbek name: Togzhumroot

Kyrgyz name: Ич алдыргыч карк моюл (Ich aldyrgych kark moyul)

Description: Bush or small tree, up to 3–8 m tall, usually dioecious. Old bark nearly black, rough, exfoliating; young bark red-brown; some branches ending in short spines. Leaves opposite, ovate to elliptic, 2–8 cm long, 1.5–5 cm wide, base round-cuneate, margins crenate-serrate. Flowers perfect or unisexual, in leaf axils. Sepals 4, twice as long as petals. Petals erect, lanceolate, 1–1.5 mm long in staminate flowers, ca. 0.5 mm in pistillate flowers. Fruits black drupes, 6–8 mm in diameter, juicy, round, shiny.

Other distinguishing features: Lateral leaf veins strongly upcurved. Fruits 4-seeded.

Phenology: Flowers in May, fruits in August.

Reproduction: By seeds.

Distribution: All provinces of Kyrgyzstan; Toshkent, Farg'ona and Samarqand provinces of Uzbekistan.

Habitat: On mountain slopes, along rivers, among bushes and in forest plantations.

Population status: Common, found in small groups.

Traditional use: Thin branches, bark and fruits are used in folk medicine. Fruits are used as a laxative to treat chronic constipation, and the treatment starts to take effect 8–10 h after ingestion. A decoction of the fruits, with the addition of dairy whey, oil, and baking soda is used as a laxative for atonic and spastic constipation, and is used as a stool softener for cases of hemorrhoids and wounds to the colon. An infusion of the fruits in vodka is used externally to treat rheumatism. A decoction of the branches is used internally to treat ulcers and externally, as a compress to heal wounds. A decoction of the bark is used to treat stomach catarrh, low acidity of the stomach, and Polish plait (Maznev 2004).

Documented effects: Preparations of this species have laxative properties that are associated with the presence of anthraglycosides and related compounds (mainly emodin), which act by stimulating the walls of the large intestines and moderately strengthen wave-and pendulum-like movements (Maznev 2004). An ethanolic extract of this species had high antimycobacterial activity but also purgative effects (Newton et al. 2000).

Phytochemistry: Fruits contain anthraglycosides (glucofrangulin, frangulin, frangula-emodin, etc.), flavonol glycosides, sugars, organic acids, pectins, etc. Leaves contain ascorbic acid. Bark contains chrysophanic acid, anthraglycosides, and high amounts of tannins (Khalmatov et al. 1984).

▲**Rhamnus cathartica L.** Photos: Dmitri Oreshkin

▲**Reseda luteola L.** Photos: Bazar Dovletov

►**Psoralea drupacea Bunge** Photos: *left*: Evgeny Davkaev; *right*: Alim Gaziev

▼**Pulicaria salviifolia Bunge** Photos: Alim Gaziev

Rheum maximowiczii Losinsk. – Polygonaceae

Synonyms: *Rheum emodi* Wall., *Rheum megalocarpon* Maxim.

English name: Unknown

Russian name: Ревень Максимовича (Reven' Maksimovicha)

Uzbek name: Rovach, Chukhra

Kyrgyz name: Чукурук (Chukuruk)

Description: Herbaceous perennial with a thick rhizome. Stem 40–100 cm tall, up to 2 cm thick, leafless, branched, tough, reddish, rough due to tiny papillae, rarely smooth. Ocreae rust-colored, tightly surrounding the stem above each leaf axil. Leaves basal, round to kidney-shaped, 18–50 cm long, 20–60 cm wide, with 3 veins and short, flat petioles. Inflorescence paniculiform, pyramidal; flowers clustered on long peduncles. Tepals 6, each 3 mm long, 1 mm wide, greenish. Fruits achenes, 2 cm long, 1.5 cm wide, winged, lilac-reddish.

Other distinguishing features: The adaxial sides of the leaves are glabrous, but the abaxial sides are rough due to papillae near the veins.

Phenology: Flowers and fruits in May-June.

Reproduction: By seeds.

Distribution: Toshkent, Samarqand, Qashqadaryo, and Andijon provinces of Uzbekistan; Osh and Jalal-Abad provinces of Kyrgyzstan.

Habitat: The tau zone. Grassy slopes or slopes with rocky debris.

Population status: Common.

Traditional use: A decoction of the root and fresh juice from the leaves are used to treat diarrhea. Juice from the petioles is recommended as a treatment for malaria (Khalmatov 1964). The young petioles and stems, and the fresh juice or compote made from them are used as a tonic, antipyretic and hypotensive to prevent anemia and to detoxify. The plant is used to increase the appetite and to treat gastritis with low acidity, liver (hepatitis) and gallbladder diseases, tuberculosis, hemorrhoids, chronic constipation, polyarthritis, and fevers (Nuraliev 1989).

Documented effects: The powdered root has astringent properties (Khalmatov 1964). This species improves the liver's ability to detoxify, helps patients with moderately high blood pressure, has diuretic actions, and is good for treatment of constipation and fevers (Nuraliev 1989). Compounds isolated from plants collected in Uzbekistan exhibited antioxidative activity (Kogure et al. 2004). (+)-rhododendrol and epi-rhododendrin isolated from *Acer nikoense* Maxim. suppressed nitric oxide (NO) production in mouse peritoneal macrophages in vivo (Fushiya et al. 1998).

Phytochemistry: The roots contains tannins (catechins, gallic acid, pyrogallol, and pyrocatechin), carbohydrates, and glycosides (Khalmatov 1964). The aboveground parts contain vitamins C, A, E, B_1, B_2, B_6, B_{15}, organic acids (malic and oxalic), sugars, fibers, hemicellulose, pectin, and macroelements (Nuraliev 1989). Roots collected in Uzbekistan contained new phenylbutanoid and stilbene derivatives as well as the known compounds rhododendrol, epi-rhododendrin, lindleyin, torachrysone, etc. (Shikishima et al. 2001).

Rhodiola linearifolia Boriss. – Crassulaceae

Synonyms: *Rhodiola kirilowii* (Regel) Maxim., *Rhodiola longicaulis* (Praeger) S.H. Fu, *Rhodiola macrolepis* (Franch.) S.H. Fu, *Rhodiola robusta* (Praeger) S.H. Fu, *Sedum kirilowii* Regel, *Sedum longicaule* Praeger, *Sedum macrolepis* Franch., *Sedum robustum* Praeger.

English name: Unknown

Russian name: Родиола линейнолистная (Rodiola lineynolistnaya)

Uzbek name: Unknown

Kyrgyz name: Сызгыч чегендир (Syzgych chegendir)

Description: Herbaceous perennial, with a thick caudex. Stems 10–60 cm tall, densely leafy. Caudex leaves scale like; stem leaves alternate, sessile, linear-lanceolate, 2–6 cm long, 3–15 mm wide, slightly serrate. Inflorescences cymose, dense, compact. Flowers unisexual or occasionally bisexual. Sepals linear or triangular, 1.5–3 mm long. Petals 3–4 mm long, brownish-red, pink or yellow. Fruits paired elongate follicles with curved apical beaks.

Other distinguishing features: Stem leaves linear-lanceolate. Stamens 8–10, yellow.

Phenology: Flowers in May-July, fruits in July-September.

Reproduction: By seeds.

Distribution: Jalal-Abad, Naryn, Osh, Ysyk-Kol, and Chuy provinces of Kyrgyzstan; not found in the flora of Uzbekistan.

Habitat: In wet places, forest meadows, and high mountains.

Population status: Common, found growing as individual plants.

Traditional use: An infusion of the underground parts is used to treat weariness, neurotic conditions, and decreased ability to work (Krasnov and Demidenko 1981).

Documented effects: The total flavonoids isolated from the underground parts show antitumor activity (Krasnov and Demidenko 1981). Salidroside has protective effects against oxidative stress-induced cell apoptosis and has been shown to enhance the ability of hemoglobin to carry oxygen and protect neuronal cells against hypoxia/reoxygenation injury in vitro (Chen et al. 2007; Yang et al. 2007; Zhang et al. 2007). Daucosterol was shown to have an inhibitory effect on the viral enzyme reverse transcriptase (Kimura et al. 2003).

Phytochemistry: Underground parts contain organic acids (citric and malic), triterpene glycoside derivatives of oleanolic acids, phenols, phenolcarbonic acids, coumarins, flavonoids, tannins, tyrosol, daucosterol, lotaustralin, salidroside, and sucrose. The aboveground parts contain coumarins and tannins (Krasnov et al. 1979; Kurlin and Zapesochnaya 1986; Peng et al. 1994; Kolesnikov and Gins 2001).

Ribes nigrum L. – Grossulariaceae (formerly in Saxifragaceae)

Synonyms: *Botrycarpum nigrum* (L.) A. Rich., *Grossularia nigra* (L.) Rupr., *Ribes cyathiforme* Pojark., *Ribes olidum* Moench, *Ribes pauciflorum* Turcz. ex Ledeb.

English name: Black currant

Russian name: Смородина чёрная (Smorodina chyornaya)

Uzbek name: Kora smorodina, korakat

Kyrgyz name: Чыны карагат (Chyny karagat)

Description: Shrub up to 1.5 m tall. Young branches dull yellow, hairy; older branches brownish, almost glabrous. Leaves 3–5-lobed, up to 12 cm wide, dotted with yellow glands beneath; lobes wide-triangular, margins serrate-dentate. Inflorescences drooping racemes, 3–8 cm long, 4–12-flowered. Flowers perfect, usually 5-merous, 5–7 mm wide, pedicellate. Hypanthium campanulate, pink, greenish-red or greenish-yellow, pubescent, glandular. Calyx lobes reflexed, 3–4 mm long. Petals ovate, 2–3 mm long. Fruit a many-seeded, black berry, ca. 10 mm in diameter.

Other distinguishing features: Ovary inferior. Stamens inserted below rim of hypanthium, alternating with petals.

Phenology: Flowers in May-July, fruits in July.

Reproduction: By seeds and vegetatively.

Distribution: Ysyk-Kol and Naryn provinces of Kyrgyzstan; cultivated in Uzbekistan.

Habitat: In gorges and mountain river valleys up to 3,000 m elevation.

The Population status: Common, forming dense groups.

Traditional use: Fruits and leaves are used to treat anemia and edema, and as a light laxative. Fresh fruits are used to decrease blood pressure, to treat heart and liver diseases and atherosclerosis. A decocotion of the young branches is drunk to treat children's diabetes and skin tuberculosis (Poludenny and Zhuravlev 2000). A decoction of the fruits is used as a diaphoretic, anti-inflammatory, and diuretic. The fresh juice is used to treat stomach and duodenum ulcers and gastritis with low stomach acidity; mixed with honey it is used to treat respiratory diseases. Leaves are used in a tea to treat skin and bladder diseases, kidney stones, rheumatism, common colds, and also as a diuretic (Khalmatov et al. 1984).

Documented effects: This species is used to treat infectious diseases, hemorrhagic diathesis, gastritis (particularly with low acidity), and is used as a tonic to treat the cardio-vascular system (Akopov 1990). An extract of the fruits has shown antiviral activity against herpes and influenza A and B viruses (Knox et al. 2001, 2003; Suzutani et al. 2003). Proanthocyanidins isolated from the leaves exhibit anti-inflammatory effects in rats (Garbacki et al. 2004).

Phytochemistry: The fruits contain vitamins (ascorbic acid, B_1, and carotene), sugars, organic acids (citric and malic), pectins, anthocyanin compounds (cyanidin and delphinindin) and their glycosides, as well as quercetin and isoquercetin. Buds contain essential oil with d-pinene, l- and d-sabinene, d-caryophyllene, alcohol, and phenols. The leaves contain essential oil and ascorbic acid (Akopov 1990; Knox et al. 2001).

Roemeria refracta DC. – Papaveraceae

Synonyms: *Glaucium refractum* Steven ex DC., *Papaver refractum* (DC.) K.F. Gunther.

English name: Spotted Asian poppy

Russian name: Рёмерия отогнутая (Ryomeriya otognutaya)

Uzbek name: Kizgaldok

Kyrgyz name: Ийилген кызгалдак (Iyilgen kyzgaldak)

Description: Annual herb, slightly hairy. Stem usually branched, rarely simple, 8–60 cm tall. Leaves bi- or tripinnatisect; basal and lower stem leaves petiolate; upper leaves alternate, sessile, pinnatisect. Flowers solitary, axillary and terminal. Petals 4, bright red, with a black spot at the base, broadly obovate, 2–4 cm long, 1.5–3.5 cm wide. Fruit a capsule, 4–6 cm long, 2–3 mm wide, glabrous. Seeds kidney-shaped, gray, pitted or reticulated.

Other distinguishing features: Pedicel 10–12 cm long when in fruit. Fruits have 3–4 awn-like projections on the top, each 3–5 mm long.

Phenology: Flowers in April-May, fruits in May-June.

Reproduction: By seeds.

Distribution: All regions of Uzbekistan and Kyrgyzstan.

Habitat: The chul and adyr zones. On clay slopes of foothills and as a weed in crop fields and orchards.

Population status: Common.

Traditional use: The juice from the petals are used in a drink to treat children with sunstroke and as a wash to treat eye problems. A decoction of the plant is used to treat smallpox and fevers, and is applied externally to treat skin rashes (Khalmatov 1964). The dried petals are used as a sedative to treat cardiac and digestive organ pains (Khodzhimatov 1989).

Documented effects: Pharmacological studies of the alkaloid roemerine showed that it has curare-like actions and an overdose can cause convulsions. The derivatives of roemerine also have curare-like and ganglio-blocking actions, but only for a short time. Roemerine has the ability to potentiate the effects of the analeptics corazol, cardiamine, caffeine, and strychnine. The alkaloid, and one of its derivatives, have strong antibacterial action against pathogenic microorganisms (Khodzhimatov 1989). (−)-roemerine isolated from the leaves of *Annona senegalensis*, was found to enhance the cytotoxic response mediated by vinblastine with multidrug-resistant human cancer cells in vitro (You et al. 1995).

Phytochemistry: At the time of flowering the plants contain 0.2 % total alkaloids (roemerine, l-isoremerin, anonaine, liriodenine, remrefidine, remrefine, l-ephedrine, d-pseudoephedrine, and l-mecambroline). The plant has also been found to contain a variety of additional alkaloids (Gozler et al. 1988, 1990a, b). The aboveground parts also contain tannins, organic acids, vitamin C, and sugar. The seeds contain a significant amount of fatty oils (Yunusov 1981; Khodzhimatov 1989).

▲**Ribes nigrum L.** Photos: *left*: Sergey Mayorov; *center* and *right*: Denis Mirin

◄**Rheum maximowiczii Losinsk.** Photos: Evgeny Davkaev

◄**Rhodiola linearifolia Boriss.** Photos: *left*: John B. Taft; *center* and *right*: Vladimir Epiktetov

▼**Roemeria refracta DC.** Photos: *left*: Evgeny Davkaev; *center*: Alim Gaziev; *right*: Authors

Rosa canina L. – Rosaceae

Synonyms: *Rosa ciliatosepala* Blocki, *Rosa sosnovskyi* Chrshan.

English name: Dog rose

Russian name: Шиповник собачий, Роза собачья (Shipovnik sobachiy, Roza sobach'ya)

Uzbek name: Itburun

Kyrgyz name: Ит мурун (It murun)

Description: Shrub, up to 3 m tall. Stems arching with stout, flattened, hooked or rarely straight prickles. Leaves alternate, stipulate, pinnately compound with 5–7 leaflets; leaflets glabrous, elliptic, apex acute, margins sharply serrate. Inflorescence a corymb or rarely single flowered. Flowers 2–8 cm wide. Sepals 5, usually glabrous, reflexed, deciduous. Petals 5, bright pink, pale pink or white. Stamens many. Fruit a large hip (1.5–2.6 cm long), wide-ovoid or elongate-ovoid, smooth, bright or light-red, containing stony achenes.

Other distinguishing features: Stipules adnate to petiole for more than half their length. Outer sepals pinnatifid.

Phenology: Flowers in June, fruits in August.

Reproduction: By seeds.

Distribution: Jalal-Abad and Osh provinces of Kyrgyzstan; Toshkent, Farg'ona, Samarqand, Qashqadaryo and Surxondaryo provinces of Uzbekistan.

Habitat: Along rivers and streams, on edges of deciduous forests, and in juniper stands.

Population status: Common, found in small groups.

Traditional use: A decoction of the petals, leaves, branches and roots is used to treat rheumatism, radiculitis, and stomach and heart ailments (Poludenny and Zhuravlev 2000). Decoction or tea of the fruits is used to treat scurvy, common colds, and as a diuretic. A decoction of the roots is used to treat liver and gastrointestinal tract diseases (Khalmatov et al. 1984; Khodzhimatov 1989). A decoction and infusion of the fruits is taken as an astringent (particularly for regular and bloody diarrhea), to treat fevers, intestinal infections, as a hemostatic for uterine bleeding, to improve the metabolism, and as a mouth wash for gum disease. The seeds are used as a diuretic and to treat kidney diseases. The powdered leaves are used to treat wounds and skin ulcers (Khalmatov et al. 1984).

Documented effects: Fruits are used as raw material for the preparation *Kholosas*, which has choleretic activity and is used to treat cholecystitis and hepatitis (Khalmatov et al. 1984). Extracts of the fresh fruits exhibit high anti-ulcerogenic activity in rats (Gurbuz et al. 2003). A galactolipid, which is found in this species, has been shown to possess antitumor-promoting properties, as well as anti-inflammatory effects (Larsen et al. 2003). In a clinical trial, treatment with a standardized rose-hip powder showed significant reduction of symptoms associated with osteoarthritis (Warholm et al. 2003).

Phytochemistry: Fruits contain vitamin C, sugars, tannins, flavonoids (cyanidin-3-*O*-glucoside, phloridzin, isoquercitrin and glycosides of kaempferol, quercetin, taxifolin, and eriodictyol), conjugates of methyl gallate, pigments (carotene, lycopene, xanthophyll, etc.), pectins, pentosan and vitamins K_1, B_2, P and E. The seeds contain fatty oils and the flowers contain essential oil (Tolmachev 1976; Khalmatov et al. 1984; Hvattum 2002).

Rosa fedtschenkoana Regel – Rosaceae

Synonyms: *Rosa caraganifolia* Sumn., *Rosa coeruleifolia* Sumn., *Rosa epipsila* Sumn., *Rosa lavrenkoi* Sumn., *Rosa lipschitzii* Sumn., *Rosa minusculifolia* Sumn., *Rosa oligosperma* Sumn.

English name: Fedtschenko's rose

Russian name: Шиповник Федченко, Роза Федченко (Shipovnik Fedchenko, Roza Fedchenko)

Uzbek name: Namatak

Kyrgyz name: Федченко ит мурун (Fedchenko it murun)

Description: Shrub, 2–3(–6) m tall. Branches prickly; prickles yellowish, firm, straight, expanded at the base, up to 13 mm long. Leaves alternate, stipulate, pinnately compound with 5–9 leaflets, 3–4.5 cm long; leaflets 1–2.5 cm long, ovate to elongate-ovate, glabrous, margins serrate. Flowers 3–9 cm in diameter, solitary or in groups of 3–4. Sepals 5, lanceolate, pubescent above, glandular below. Petals 5, white or pink, broad-obovate. Fruit a fleshy, red hip, 2–5 cm long, elongate-ovoid, glandular-bristly, with persistent sepals, and containing stony achenes.

Other distinguishing features: Leaflets glabrous. Hip to 5 cm long, densely glandular-bristly.

Phenology: Flowers in June-August, fruits in July-September.

Reproduction: By seeds.

Distribution: All provinces of Kyrgyzstan; Toshkent, Farg'ona, Samarqand, Qashqadaryo and Surxondaryo provinces of Uzbekistan.

Habitat: In forest glades, among bushes.

Population status: Common, found in small groups.

Traditional use: Fruits are widely used as a tonic, choleretic, and as a remedy for scurvy. An infusion of the fruits with honey is used to treat upper respiratory colds and coughs. A decoction of the roots is drunk to treat diarrhea. A decoction of the leaves is taken to treat dysentery and as a diuretic (Khodzhimatov 1989). The hips from this and related species are used to prevent scurvy and avitaminosis, to treat arteriosclerosis, cholecystitis, hepatitis, and gastrointestinal diseases, particularly with reduced bile production (Altimishev 1991). Oil of rose is used externally to treat cracked and injured nipples of breast feeding women, bedsores, trophic ulcers of the shins, and dermatosis (Muravyova 1978). The fruits are used to treat lung tuberculosis, diphtheria, scarlet fever, flu, and sore throat (Zakordonets 1953).

Documented effects: The fruits of this species are an official source of polyvitamins. Preparations (extracts, syrups, candies, pills, etc.) are used treat hypo- and avitaminosis (particularly vitamin C deficiency) as well to treat diseases related to vitamin deficiency. The fruits are used as a component in an anti-asthmatic mixture. Oil from the seeds is used to treat burns, dermatosis, and radiation exposure. Ascorbic acid and an oil extract *Karotolin* (containing carotenoids, vitamin E, and linolic acid) are isolated from the pericarp. *Karotolin* is used to treat trophic skin ulcers, eczema, erythrodermia, and other skin diseases (Khalmatov et al. 1984).

Phytochemistry: Fruits contain vitamins C, E, P, B_2, K_1 carotene, organic acids (malic and citric), sugars, flavonoids, pectins, and tannins. Seeds contain up to 37 % fatty oil (Tolmachev 1976; Khodzhimatov 1989). The flowers were found to contain glycosides of quercetin, kaempferol, cyanidin, and peonidin (Mikanagi et al. 1995).

Rubia tinctorum L. – Rubiaceae

Synonyms: *Rubia iberica* (Fisch. ex DC.) K. Koch.

English name: Madder, common madder

Russian name: Марена красильная (Marena krasil'naya)

Uzbek name: Ruyan

Kyrgyz name: Боечу марена (Boyechu marena)

Description: Herbaceous perennial, with taproot and horizontal rhizomes. Stems 0.5–2 m tall, prostrate or climbing, 4-sided, with curved prickles. Leaves in whorls of 4 or 6, up to 9 cm long, up to 3 cm wide, narrow-ovate, apex acute. Inflorescences spreading complex panicles. Flowers small. Corollas yellow, 1–1.5 mm in diameter, 5-lobed. Stamens 5. Fruits berry-like, juicy, black with 2 hemispherical seeds.

Other distinguishing features: Abaxial midvein and margins of leaves with curved prickles.

Phenology: Flowers in June-August, fruits in August-September.

Reproduction: By seeds and rhizomes.

Distribution: The Osh province of Kyrgyzstan; Toshkent, Andijon, Farg'ona, Samarqand and Surxondaryo provinces of Uzbekistan.

Habitat: Along canals, near springs, in orchards, and near rivers in tree-shrub forests.

Population status: Common, found in small groups.

Traditional use: In the past the underground parts of this species were used to treat rickets, constipation, jaundice, joint ailments, rheumatic back aches, and other diseases. Avicenna used a water extract as a strong diuretic to purify the liver and spleen and to treat spleen tumors. The extract, mixed with honey, was drank to treat the sciatic nerves and paralysis. The underground parts, mixed with vinegar, were applied to treat fungal skin diseases and to remove skin spots. In recent times the underground parts have been used to treat kidney stones, gallstones, and gout, and also as a diuretic and laxative. In Central Asia, the roots mixed with honey is used to treat jaundice, to improve memory, and as a diuretic (Khalmatov et al. 1984; Grinkevich 1991).

Documented effects: Alcohol and water extracts of the roots inhibited the growth of *Aeromonas hydrophila*, *Bacillus megaterium*, *Corynebacterium xenosis*, *Pseudomonas aeruginosa*, *Micrococcus luteus*, *Enterococcus faecalis*, and *Staphylococcus aureus*, but was not an effective inhibitor of *Escherichia coli* (Golcu et al. 2002). In experiments with rats that ate fresh roots decreased bladder and kidney stone formation was observed, but increased death rates were exhibited. In experiments with rabbits that were given root extracts orally, decreased calcium oxalate crystal formation in the kidneys and hepatotoxicity was observed. Genotoxic effects were observed in bacterial and mammalian cell systems (Blumenthal 1998).

Phytochemistry: The underground parts contain anthraglycosides and anthraquinone derivatives (ruberythric acid, galiosin, purpurin, rubiadin, mollugin, lucidin, etc.), organic acids (citric, malic, and tartaric), sugars, and traces of alkaloids. The young shoots contain the glycoside asperuloside. (Khalmatov et al. 1984; Gammerman et al. 1990; Kawasaki et al. 1992; Derksen et al. 2002).

Rubus caesius L. – Rosaceae

Synonyms: *Rubus psilophyllus* Nevski, *Rubus turkestanicus* (Regel) Pavlov.

English name: European dewberry

Russian name: Ежевика сизая (Ezhevika sizaya)

Uzbek name: Parmanchak, Maimunzhon

Kyrgyz name: Когултур кара булдуркон (Kogultur kara buldurkon)

Description: Shrub, up to 1 m tall. Primocanes arching, glaucous, with stout, hooked prickles, rooting at the tip. Leaves trifoliate (basal leaves sometimes 5-foliate), stipulate, petiole prickly; leaflets broad-ovate, margins unevenly dentate, soft-pubescent beneath. Inflorescence racemiform or paniculiform. Flowers with 5 sepals and 5 white petals. Stamens and pistils many. Fruit an aggregate of drupelets, black or red, glaucous.

Other distinguishing features: Fruits separating from the stem with receptacle.

Phenology: Flowers in June-July, fruits in July-August.

Reproduction: By seeds and vegetatively.

Distribution: All provinces of Kyrgyzstan; Toshkent, Namangan, Farg'ona, Qashqadaryo and Surxondaryo provinces of Uzbekistan.

Habitat: Among shrubs, in forests and deforested areas, and along rivers and canals.

Population status: Common, found in dense groups.

Traditional use: Fresh fruits, infusion of the dried fruits, syrup or jam, or taken with tea, are widely used to quench the thirst, as a tonic, diaphoretic, diuretic, laxative, and sedative, as a remedy to increase the appetite, and to treat chronic gastritis and enterocolitis, stomach and duodenum ulcers, liver diseases, the flu, sore throats, pneumonia, stomatitis, dysentery, typhoid and fever. Water extracts, infusions or tea of the leaves and roots, is commonly used to treat stomach ulcers, chronic gastritis, and kidney stones (Nuraliev 1989). A decoction of the fruits, leaves, and branches is taken to treat cystitis, pyelitis, bronchitis, diabetes, urinary incontinence, eczema, vitiligo, psoriasis, fungal skin diseases, hair loss, and during menopause (Kurochkin 1998).

Documented effects: A decoction of the fruits is used as a source of vitamins, to improve digestion, and as a laxative and diaphoretic (Nuraliev 1989).

Phytochemistry: Fruits contain sugars, pectins, organic acids (citric, tartaric, malic, and salicylic), fiber, tannins, rutin, nicotinic acid, flavonoids, and vitamins C, P, B_1, A, PP, E, and K. The leaves and branches contain flavonoids, tannins and ascorbic, malic, oxalic and lactic acids (Nuraliev 1989; Kurochkin 1998; Gudej and Tomczyk 2004).

▲**Rosa canina L.** Photos: *top left*: Alim Gaziev; *top right*, *center* and *bottom*: Maxim Kucherov

▲**Rosa fedtschenkoana Regel** Photos: *top*: Alim Gaziev; *bottom*: Vladimir Epiktetov

▲**Rubia tinctorum L.** Photos: Maxim Kucherov

▼**Rubus caesius L.** Photos: *left* and *right*: Maxim Kucherov; *center*: Sergey Appolonov

Rubus idaeus L. – Rosaceae

Synonyms: None

English name: Red raspberry

Russian name: Малина обыкновенная (Malina obyknovennaya)

Uzbek name: Parmanchak, Malina

Kyrgyz name: Кадимки дан куурай (Kadimki dan kuuray)

Description: Shrub, up to 80–200 cm tall. Primocanes green, glaucous, with thin, straight prickles. Floricanes yellowish or green, slightly woody. Leaves odd-pinnate with 3–5(–7) leaflets, stipules thread-like; leaflets white tomentose below, margin unevenly serrate. Flowers in few-flowered racemes in corymbiform-paniculate inflorescences. Sepals 5, reflexed, grayish-green. Petals 5, white. Stamens and pistils many. Fruit a red (raspberry) aggregate of drupelets .

Other distinguishing features: Fruits separating from the receptacle.

Phenology: Flowers in June, fruits in July-August.

Reproduction: By seeds and vegetatively.

Distribution: Jalal-Abad, Ysyk-Kol, Osh, and Chuy provinces of Kyrgyzstan; cultivated in Uzbekistan.

Habitat: In meadows, along rivers, and in deforested areas in the shrub and forest belt of mountains.

Population status: Common, found in dense groups.

Traditional use: Fruits are used as a diaphoretic and antipyretic. Leaves are used as an astringent and hemostatic, and to treat diarrhea. A decoction and infusion of the leaves is recommended as a cough remedy, and is gargled to treat sore throats. An infusion of the leaves and flowers is used to treat hemorrhoids and gynecological conditions. A paste of the fresh leaves is used to treat acne and rashes. A decoction of the flowers is used as a wash for acne, erysipelas, and conjunctivitis (Khalmatov et al. 1984). The fresh fruits are considered to have sobering effects for drunkenness (Kurochkin 1998).

Documented effects: Preparations from raspberries improve stomach and intestine function, have antiseptic, analgesic, antipyretic, expectorant, anti-inflammatory, and anti-emetic properties, and improve metabolism (Maznev 2004). Extracts of the fruits have antioxidant effects and exhibit antimicrobial properties (Kahkonen et al. 1999; Rauha et al. 2000; Puupponen-Pimia et al. 2001). Components of raspberry leaf extract exhibited relaxant activity in an in vitro gastrointestinal tissue (Rojas-Vera et al. 2002).

Phytochemistry: Fruits contain vitamins (C, B_1, B_2, B_6, PP, E, and A), organic acids (citric, malic, salicylic, tartaric, formic, and capronic), ellagic acid and its derivatives, sugars, pectins, minerals, essential oil, anthocyans, flavonoids, and tannins. Seeds contain fatty oils, sitosterin, tocopherols, neutral lipids, phospholipids, and free fatty acids. The main fatty acids of crude oil were 18:2 (54.5 %), 18:3 (29.1 %), 18:1 (12 %), and 16:0 (2.7 %; Tolmachev 1976; Khalmatov et al. 1984; Kurochkin 1998; Oomah et al. 2000; Zafrilla et al. 2001).

Rumex confertus Willd. – Polygonaceae

Synonyms: *Rumex alpinus* L. var. *subcalligerus* Boiss.

English name: Russian dock

Russian name: Щавель конский (Shchavel' konskiy)

Uzbek name: Ot quloq

Kyrgyz name: Ат кулак (At kulak)

Description: Herbaceous perennial, 60–150 cm tall, with a thick root. Stems erect, striated. Blades of basal leaves elongated triangular-oval, 15–25 cm long, 6–12 cm wide, apex obtuse, cordate, margins sinuate, abaxial side with stiff hairs towards veins; petiole equal to or exceeding the length of the blade. Cauline leaves smaller, acute, oval-lanceolate, short-lanceolate. Inflorescence terminal, narrow-cylindrical or wide-paniculiform, composed of pedicellate flowers densely arranged in multiflorous whorls. Perianth with 6 tepals, 6–9 mm long, 6–11 mm wide. Fruit a triquetrous achene, 3–5 mm long, 1.5–2.5 mm wide.

Other distinguishing features: Ocreae mostly deciduous.

Phenology: Flowers and fruits in May-June.

Reproduction: By seeds.

Distribution: Toshkent and Qashqadaryo provinces of Uzbekistan; not found in Kyrgyzstan.

Habitat: The adyr and tau zones. River banks and grassy slopes and as a weed in cultivated fields.

Population status: Common, sometimes makes dense populations.

Traditional use: This plant has been used for treatment of multiple diseases such as scabies and scurvy, and as an astringent for diarrhea. A decoction of roots and leaves is used to treat skin disorders (fungal skin diseases and rashes), ulcers, and wounds (Seredin and Sokolov 1969; Khalmatov et al. 1984).

Documented effects: Small doses of preparations (infusions and extracts) have astringent effects and in big doses, purgative effects. Currently they are recommended to improve intestinal function. They are also used for anemia with simultaneous gastrointestinal tract dysfunction, and for colitis, hemorrhagic enterocolitis, hemorrhagic colitis, and child's diarrhea (Seredin and Sokolov 1969). Experiments show that a preparation of this species acts as a vermifuge and has hemostatic and hypotensive ability (Sokolov and Zamotaev 1985). An extract of the plant exhibited cytotoxic effects against human lymphoblastoid cells in vitro (Spiridonov et al. 2005). Chrysophanic acid, isolated from *Dianella longifolia*, has been found to inhibit the replication of poliovirus types 2 and 3 in vitro (Semple et al. 2001). In vitro, emodin inhibits tyrosine kinase, an enzyme overexpressed in certain breast cancer cells. The combination of emodin and paclitaxel synergistically inhibited tumor growth and prolonged survival in mice (Zhang et al. 1999).

Phytochemistry: Underground organs contain tannins (ellagic acid, phloroglucinol, and caffeic acid), flavonoids (nepodin, chrysophanic acid, emodin, etc.), resins, essential oils, and calcium oxalate. Leaves contain flavone glycosides (hyperoside and rutin), carotene, vitamin C, and calcium oxalate (Seredin and Sokolov 1969; Mukhamed'yarova and Chumbalov 1979).

Rumex tianschanicus Losinsk. – Polygonaceae

Synonyms: None

English name: Unknown

Russian name: Щавель тяньшанский (Shchavel' tyan'shanskiy)

Uzbek name: Unknown

Kyrgyz name: Ат кулак (At kulak)

Description: Herbaceous perennial. Stem single, up to 2 m tall, thick, branched, hollow, largely striated. Basal leaves wide-ovate, 17–25 cm long, up to 15 cm wide, apex acute, base cordate, margin undulate, short-petiolate; stem leaves smaller; ocreae membranous, falling off early. Inflorescence paniculate. Flowers with 6 tepals arranged in 2 whorls. Fruits 3-sided achenes, pointed, light-brown, 2 mm long.

Other distinguishing features: Each tepal with a prominent vein.

Phenology: Flowering and fruits in May-June.

Reproduction: By seeds.

Distribution: Chuy province of Kyrgyzstan; not found in the flora of Uzbekistan.

Habitat: In rivers valleys and orchards.

Population status: Common, found in small groups.

Traditional use: In Uzbekistan, a fresh leaf is applied externally to an abscess to provoke maturation (Sezik et al. 2004).

Documented effects: The underground parts have slight antitumor activity. An infusion and alcohol extract are used to treat pellagra and dyspepsia. Fruits are used to treat dyspepsia in children (Belodubrovskaya et al. 2002).

Phytochemistry: All parts of the plant contain phenolcarbonic acids, flavonoids, and catechins. The seeds contain fatty oil (Plant Resources of the USSR 1985). The roots also contain sugars, inulin, organic acids, tannins, anthraquinones, and leucoanthocyanides. The leaves contain vitamins (C, P, K), carotenoids, and tannins (Belodubrovskaya et al. 2002; Kharlamova 2007).

Salvia deserta Schangin – Lamiaceae

Synonyms: *Salvia jailicola* Klokov, *Salvia moldavica* Klokov, some considered *S. deserta* a synonym of *Salvia nemorosa* L.

English name: Unknown

Russian name: Шалфей пустынный (Shalfey pustynnyy)

Uzbek name: Mavrak

Kyrgyz name: Чол шалфейи (Chol shalfeyi)

Description: Perennial subshrub. Stems erect, simple or branched, densely curly pubescent, 60–80 cm tall. Leaves opposite, ovate to ovate-lanceolate, apex acute to acuminate, base cordate, adaxial side dark green, abaxial side gray-pubescent, margin crenate-serrate, petiolate. Inflorescences terminal, racemiform, composed of verticillasters with 4–6 flowers, pubescent. Bracts broadly ovate, 4–6 mm long, purple-red. Flowers short pedicellate. Calyx 5–6 mm long, 2-lipped. Corolla 9–10 mm long, 2-lipped, blue-purple to violet. Nutlets rounded-triangular, black, 1.5 mm long.

Other distinguishing features: Upper lip of calyx shorter than lower lip.

Phenology: Flowers in May-August, fruits in June-September.

Reproduction: Only by seeds.

Distribution: Toshkent, Farg'ona, and Surxondaryo provinces of Uzbekistan; all of Kyrgyzstan.

Habitat: The adyr and tau zones. As a weed in orchards, on grassy slopes and as a weed in oases.

Population status: Common, often found as single individuals.

Traditional use: A decoction of leaves and flowers is used for cardiac neurosis and neurasthenia, to increase appetite, as a gargle to treat sore throat, and for intestinal infections and fever. A powder of roasted seeds is recommended for dysentery and heart palpitations. Ground fruits mixed with oil are used to heal wounds. Preparations of *Salvia deserta* are used in the same way as the preparations of the aboveground parts of *Salvia sclarea* (Khalmatov and Kosimov 1992; Gammerman et al. 1990).

Documented effects: Essential oils from this species are used in the pharmaceutical industry to add an aroma to drugs and in the fragrance industry as an aroma fixative (Khalmatov and Kosimov 1992; Gammerman et al. 1990). Water and MeOH extracts of the plant strongly inhibited aldose reductase activity, an enzyme associated with diabetic complications (Kasimu et al. 1998). Compounds from the plant were found to inhibit prolyl endopeptidase (PEP), an enzyme thought to be involved with learning and memory processes, and the inhibition of which may produce anti-amnesic effects (Tezuka et al. 1999).

Phytochemistry: Flowering plant tops contain 0.01–0.04 % essential oil (similar to *Salvia sclarea* in composition) with a pleasant aroma. In leaves there were 47 mg% vitamin C, and seeds contained up to 19 % oil (Khalmatov 1964). Triterpenoids, including ursane, oleanane, and lupane derivatives were isolated from the aboveground parts (Savona et al. 1987). The roots were found to contain a number of caffeic acid derivatives (rosmarinic acid, lithospermic acid B, etc.), diterpenes (royleanone, ferruginol, taxodione, etc.), and the steroid daucosterol (Tezuka et al. 1998).

▼**Salvia deserta Schangin** Photos: Evgeny Davkaev

▼**Rumex tianschanicus Losinsk.** Photos: Vladimir Epiktetov

▼**Rubus idaeus L.** Photos: *top* and *bottom*: Dmitri Oreshkin; *center*: Sergey Mayorov

▼**Rumex confertus Willd.** Photos: *top*: Sergey Appolonov; *bottom*: Anatoly Lisitzyn

Salvia sclarea L. – Lamiaceae

Synonyms: *Salvia asperata* Falc. ex Benth., *Salvia pamirica* Gand.

English name: Clary, Clary sage

Russian name: Шалфей мускатный (Shalfey muskatnyy)

Uzbek name: Mavrak, Marmarak, Khutan

Kyrgyz name: Мускат шалфейи (Muskat shalfeyi)

Description: Herbaceous perennial, with taproot. Stems few, erect, 20–150 cm tall, 4-sided, hairy, branched, upper portions glandular. Leaves opposite, simple, 7–30 cm long, 3–22 cm wide, rugose, ovate or oblong-ovate, margins unevenly dentate. Inflorescences verticillasters in panicles. Bracts round-ovate, 1–3 cm long, often whitish with red-purple tips. Calyx tubular, 2-lipped, upper lip 3-lobed, lower lip 2-lobed. Corolla 2-lipped, pink, lilac or white. Fruits are ellipsoid nutlets, brown, 2–3 mm long.

Other distinguishing features: Upper lip of corolla arching, longer than tube and extending past lower lip.

Phenology: Flowers in June-August, fruits in August-September.

Reproduction: By seeds.

Distribution: Osh, Jalal-Abad, Talas, and Chuy provinces of Kyrgyzstan; Toshkent, Andijon, Farg'ona, Samarqand and Surxondaryo provinces of Uzbekistan.

Habitat: On the slopes of mountains, along high mountain rivers, and in fallow fields and orchards.

Population status: Common, forming dense groups.

Traditional use: The aboveground parts are used to treat fevers, stomach ulcers, headaches, epilepsy, to improvement digestion, and as an antiseptic. It is used in bathes to treat bladder diseases, polyarthritis, osteomyelitis, deforming arthrosis, and trophic ulcers. The leaves are used as a antispasmodic and anti-inflammatory. A decoction of the leaves is used as a mouth wash for acute respiratory diseases and throat illnesses, periostitis and is applied externally to purulent wounds and furuncles. The decoction of the leaves and inflorescences are used to treat tachycardia and asthenia (Plant Resources of the USSR 1991).

Documented effects: Clinical studies showed that an ointment (with 5–20 % plant extract) was highly effective in treating psoriasis (Khalmatov et al. 1984). An emulsion of the oil was successfully used to treat osteomylitis, varicose veins, paronychia, burns, and other diseases (Sklarovsky 1972). Extracts of the roots show antibacterial activity and are used in antibacterial preparations (Gammerman et al. 1990). In experiments, a tincture of the herb increased respiration and arterial pressure and had diuretic properties. The tincture affected an isolated frog heart in a similar manner as camphor. An infusion of the herb is used in stomatology to treat caries, pulpitis, periodontitis, and catarrhal gingivitis (Plant Resources of the USSR 1991). A number of the diterpenoids and sesquiterpenes isolated from the plant were found to be active against *Staphylococcus aureus* and *Candida albicans* and caryophyllene oxide showed activity against *Proteus mirabilis* (Ulubelen et al. 1994).

Phytochemistry: Aboveground parts contain essential oil (linalyl-acetate, linalool, ocimene, myrcene, cedrene, nerolidol, sclareol, etc.), coumarins, flavonoids, saponins, and trace alkaloids. Seeds contains drying fatty oil which contains oleanolic, linoleic, linolenic, arachidic, behenic, lignoceric, and cerotinic acids, pigments (carotene and chlorophyll), and stearins. The roots contains quinones (tanshinone, isotanshinone, oxytanshinone, etc.; Khalmatov et al. 1984; Khodzhimatov 1989; Gammerman et al. 1990). An extract of the whole plant contained flavonoids (apigenin, luteolin and their derivatives, etc.), diterpenes (sclareol, manool, ferruginol, etc.), sesquiterpenes (caryophyllene oxide and spathulenol), alpha-amyrin, and β-sitosterol (Ulubelen et al. 1994).

Sanguisorba officinalis L. – Rosaceae

Synonyms: *Sanguisorba glandulosa* Kom.

English name: Great burnet, Official burnet

Russian name: Кровохлёбка аптечная (Krovokhlyobka aptechnaya)

Uzbek name: Sangvizorba, Dorivor kukat, Dorivor krovoklebka

Kyrgyz name: Дары кансоргуч (Dary kansorguch)

Description: Herbaceous perennial, with thick rhizome. Stems up to 1 m tall, single or few, hollow, ribbed, branched above. Leaves alternate, odd-pinnately compound, glabrous, stipulate; leaflets elongate-ovate, margins serrate; lower leaves large, long-petiolate; upper leaves sessile. Inflorescences ellipsoid to cylindrical heads, 1–3 cm long. Sepals 4, petaloid, purple-brown. Petals lacking. Fruit a brown achene.

Other distinguishing features: Stamens 4, equal in length to sepals.

Phenology: Flowers in June, fruits in August.

Reproduction: By seeds.

Distribution: Chuy province of Kyrgyzstan; not found in the flora of Uzbekistan.

Habitat: Along rivers, in the forest-meadow mountain belt, among shrubs, on grassy slopes, and in bogs.

Population status: Rare, found in small groups.

Traditional use: Used in folk medicine as an astringent and hemostatic. Used to treat gastrointestinal diseases, tuberculosis, hemoptysis, and uterine bleeding. Used externally to heal wounds (Khalmatov et al. 1984; Akopov 1990; Grinkevitch 1991).

Documented effects: The species is used to treat upper respiratory illnesses, hemorrhoids, scurvy and gingivitis, and as an expectorant and astringent (Kovaleva 1971). The decoction of this species is used externally to treat wounds and skin ulcers and as a douche to treat cervical erosion (Akopov 1990). A decoction of the roots has antimicrobial effects against *Trichomonas*, *Candida* sp., and *Giardia lamblia* (Zavrazhanov et al. 1977). Two triterpene glycosides isolated from the roots were found to have cytotoxic activity against human carcinoma cells in vitro (Mimaki et al. 2001). Both in vitro and in vivo, a triterpene glycoside isolated from the roots diminished tumor necrosis factor-alpha production (Cho et al. 2006).

Phytochemistry: Underground parts contain tannins (pyrogallic groups), saponins, stearins, acids (gallic, ellagic, oxalic, and ascorbic), a number of triterpenes and triterpene glycosides, gallotannins, carotene, starch, pigments, phytoncides, essential oil, and micro- and macroelements (Kurochkin 1998; Mimaki et al. 2001; Liu et al. 2005; Cho et al. 2006).

Scabiosa songarica Schrenk – Dipsacaceae

Synonyms: *Trochocephalus songaricus* (Schrenk) Á. Löve & D. Löve.

English name: Unknown

Russian name: Скабиоза джунгарская (Skabioza dzhungarskaya)

Uzbek name: Zhoongor scabiozasi

Kyrgyz name: Жунгар бешилик чобу (Zhungar beshilik chobu)

Description: Herbaceous perennial, with woody roots. Stems 20–65(−100) cm tall, with short hairs. Basal and lower stem leaves petiolate; lower stem leaves lanceolate, entire or slightly pinnate; upper stem leaves opposite, lanceolate, hairy, pinnatifid with a larger apical lobe. Inflorescence a head, 2.5–3 cm in diameter; involucral bracts narrow-lanceolate, densely long-bristled; involucel expanded above into corona. Marginal flowers up to 2 cm long. Corolla yellow-violet, hairy outside. Fruits bristly achenes, adnate to the involucel and crowned by the calyx.

Other distinguishing features: Calyx teeth twice as long as corolla.

Phenology: Flowers in June-July, fruits in August-September.

Reproduction: By seeds and vegetatively.

Distribution: All provinces of Kyrgyzstan; Toshkent, Andijon, Farg'ona, Samarqand, and Surxondaryo provinces of Uzbekistan.

Habitat: On foothills, in steppes with a wide diversity of grass species.

Population status: Common, forming dense groups.

Traditional use: A decoction of the herb is used to treat respiratory infections and common colds (Sinitsin 1959).

Documented effects: In experiments on animals, the total saponins isolated from the roots and the preparation *Zongorozid* caused a significant decrease in arterial pressure, increased resistance to hypoxia, and had sedative effects. In experiments with dogs, the preparation *Zongorozid* increased the sodium in erythrocytes and reduced potassium in blood plasma as well as in erythrocytes. A one time dose of the preparation has blood coagulating effects but multiple applications, over 5–7 days, have better effects. The effects include an increase in tolerance to heparin, reduction of prothrombin time and fibrinolytic activity, increase in fibrinogen content (up to 45 %), and an increase of the adhesion index with an increase in blood coagulation potential (Alimbaeva et al. 1986).

Phytochemistry: Roots contain organic acids, saponins (17 triterpene glycosides and oleanolic acid derivatives), steroids, alkaloids, vitamin C, flavonoids, coumarins, and tannins. The aboveground parts contain organic acids, saponins, alkaloids, phenolcarbonic acids, coumarins, and flavonoids (Alimbaeva and Akimaliev 1975).

Serratula sogdiana Bunge – Asteraceae

Synonyms: *Serratula alatavica* C.A. Mey. ex Rupr., *Serratula dissecta* var. *asperula* Regel & Herder, *Serratula trautvetteriana* Regel & Schmalh.

English name: Unknown

Russian name: Серпуха согдийская (Serpukha sogdiyskaya)

Uzbek name: Unknown

Kyrgyz name: Согдия чогойносу (Sogdiya chogoynosu)

Description: Herbaceous perennial with a thick, woody, branching rhizome and string-like roots. Stems erect, 25–55 cm tall, ribbed, foliaceous, with long, appressed, straight, thin branches. Basal and lower leaves thin-coriaceous, oblong-lyrate, ca. 8 cm long, lower half of blade incised-toothed, upper half entire, petiolate with stipule-like auricles at the base; middle leaves and leaves on branches linear-lancolate, ca. 3 cm long, some deeply incised, toothed; upper leaves becoming very reduced and spinescent. Inflorescences capitula, mostly solitary, 12–15 mm wide, 25 mm long; peduncles with several small spiny leaves; involucral bracts coriaceous, yellowish-green, imbricate, short-hairy on the outside, gradually tapering into pointed tip. Corollas pink or purple, ca.1.5 cm long with linear lobes, protruding well past involucral bracts. Fruits oblong achenes, ca.5 mm long, dentate-edged on top, reddish-brown; pappus with dense, plumose bristles, deciduous.

Other distinguishing features: Receptacle with smooth bristles that are ca. 1 cm long.

Phenology: Flowers in June-August, fruits in August-September.

Reproduction: By seeds.

Distribution: Farg'ona province of Uzbekistan; lower belt of Alai mountain range (mountains-Kiziltau, Galtin, Mashalang, Katrantau, Hurdjuntau); Osh, Chuy and Jalal-Abad provinces of Kyrgyzstan.

Habitat: The lower tau zone. On gentle, stony slopes.

Population status: Uncommon, sometimes found in small populations.

Traditional use: Decoctions and infusions of plants in the genus *Serratula* are used to heal wounds, to treat anemia, as a restorative for weakness due to fever and as a treatment for liver diseases (Zavrazhanov et al. 1972).

Documented effects: Ecdysterone and extracts of *Serratula sogdiana* L. have anabolic activity as well as the ability to keep nitrogenous compounds in the organism and assist in acceleration of protein synthesis (Syrov and Kurmukov 1975a, b, c; Saatov et al. 1999).

Phytochemistry: The phytoecdysteroids ecdysterone, viticosterone, and sogdisterone were identified in extracts of the inflorescences (Zatsny et al. 1971, 1973a, b; Saatov et al. 1999).

▲**Sanguisorba officinalis L.** Photos: *left*: Denis A. Davydov; *center* and *right*: Rostislav Lezhoyev

▲**Scabiosa songarica Schrenk**
Photos: *left* and *center*: Evgeny Davkaev; *right*: Alim Gaziev

◄**Salvia sclarea L.**
Photos: *left* and *center*: Alim Gaziev; *right*: Evgeny Davkaev

▼**Serratula sogdiana Bunge**
Photos: Evgeny Davkaev

Silybum marianum (L.) Gaertn. – Asteraceae

Synonyms: *Carduus marianus* L.

English name: Blessed milk thistle, Milk thistle

Russian name: Расторопша пятнистая (Rastoropsha pyatnistaya)

Uzbek name: Unknown

Kyrgyz name: Unknown

Description: Annual or biennial 0.3–3 m, usually ~1.5 m tall. Stem erect, usually branched, grooved, farinose, thinly arachnoid-hairy, foliaceous. Leaves green with large white spots and veins, oblanceolate to elliptical, pinnately lobed, petiolate; lobes prickly or prickly toothed; basal leaves up to 80 cm long and 30 cm wide, forming a rosette; upper leaves reduced, sessile, clasping, prickly lobed. Inflorescences terminal capitula, nodding, oblong or globose, 3–6 cm in diameter, solitary with slender peduncles. Involuctral bracts imbricate; outer and middle bracts up to 3 cm long, spreading, stiff, erect, spinescent with 4–6 spines on the margin. Flowers discoid, 2.5–3.5 cm long, pink, purple or white, numerous; tube long, slender, throat abruptly wider, corolla lobes linear. Fruits elliptical or obovate achenes, ca. 6 mm long, 3 mm wide, slightly flattened, brownish-black and sometimes white spotted, glabrous; pappus composed of a deciduous ring of minutely barbed bristles, ca. 2 cm long.

Other distinguishing features: Receptacles flat and covered with whitish bristles.

Phenology: Flowers in April-May, fruits May-June.

Reproduction: Only by seeds.

Distribution: Qashqadaryo and Surxondaryo provinces of Uzbekistan; not found in Kyrgyzstan.

Habitat: The chul and adyr zones. A weed growing along roads and edges of agricultural fields.

Population status: Uncommon, sometimes makes dense populations numbering up to 40 individuals.

Traditional use: The seeds are used to treat jaundice, hepatitis, chronic coughing and hemoptysis, gall-stones and inflammation of the gall bladder and bile duct, liver and spleen diseases, fevers, hemorrhoids, and other diseases. Juice from the leaves is drunk as a choleretic and diuretic and to treat colitis and constipation. A decoction of the root is drunk to treat stomach catarrh. Currently, an alcohol-water extraction of the seeds is used to treat liver diseases (Khalmatov 1964; Khodzhimatov 1989).

Documented effects: Preparations such as *Karsil*, *Legalon*, and *Silimarin* are used in modern medicine to restore liver membranes and to treat bile-duct and gall-bladder diseases (Gammerman et al. 1990). A variety of experiments have shown that silymarin increases liver regeneration after damage caused by liver diseases. Similar effects were found in kidney cells in vitro (Sonnenbichler et al. 1999).

Phytochemistry: The aboveground parts contain flavonoids and fumaric acid. Seeds contain 0.08 % essential oil, vitamin K, mucilage, resins, biogenic amines (thiramine and histamine), trace alkaloids, saponins, flavonoids, and flavolignans (isosilibinin, silibinin, silicristin and silidianin; Khodzhimatov 1989; Gammerman et al. 1990; Kurochkin 1998; Sonnenbichler et al. 1999). The seed oil is rich in linoleic and oleic acids and contains 5 major triacylglycerols. Campesterol, 5-stigmasterol, β-sitosterol, 7-stigmasterol, avenasterol, and spinasterol were also detected in the seed oil (El-Mallah et al. 2003).

Sorbus tianschanica Rupr. – Rosaceae

Synonyms: *Pyrus tianschanica* (Rupr.) Franch.

English name: Tian Shan mountain ash, Tian Shan rowan

Russian name: Рябина тяньшанская (Ryabina tyan'shanskaya)

Uzbek name: Kizilchetan

Kyrgyz name: Тяньшань четини (Tyan'shan' chetini)

Description: Tree, 3–5 m tall. Branches brown, with lenticels; young shoots reddish- brown. Leaves alternate, odd-pinnately compound with 13–15 leaflets, 12–18 cm long (including rachis), stipules membranaceous; leaflets ovate-lanceolate, glabrous, margins serrate. Inflorescences loose clusters, many-flowered. Flowers 1.5–2 cm wide, hypanthium campanulate. Sepals 5, triangular. Petals 5, ovate or elliptic, white. Stamens 15–20. Styles 3–5. Fruits nearly round pomes, 10–12 mm wide, scarlet to dark-red, glaucous.

Other distinguishing features: Buds white, pubescent.

Phenology: Flowers in June-July, fruits in August-September.

Reproduction: By seeds.

Distribution: All provinces of Kyrgyzstan; Toshkent and Samarqand provinces of Uzbekistan.

Habitat: In the upper forest-shrub belt of mountains (2,000–3,200 m).

Population status: Common, found growing as single plants.

Traditional use: The fruits of this species are used to treat hepatitis and cholecystitis (Sumnevich 1942).

Documented effects: Fruits and seeds have antibacterial properties. An alcohol extract and fatty oil are used to treat paratyphoid fever (Aitbaeva 1972).

Phytochemistry: All parts of plant contain phenolcarbonic acids, flavonoids, and catechins. The seeds contain fatty oil. Fruits contain ascorbic acid, vitamin A, tannins, and carotene (Zapesochnaya et al. 1973; Dzhangaliev et al. 2003).

Sphaerophysa salsula (Pall.) DC. – Fabaceae

Synonyms: *Phaca salsula* Pall.

English name: Alkali swainsonpea, Austrian Peaweed

Russian name: Сферофиза солонцовая, Круглоплодник солончаковый (Sferofiza solontsovaya, Krugloplodnik solonchakovyy)

Uzbek name: Shildir bosh

Kyrgyz name: Шорчул сферофиза (Shorchul sferofiza)

Description: Herbaceous perennial with scattered short, appressed hairs. Stems erect, 30–70 cm tall, with appressed branches. Leaves alternate, odd-pinnate, 4–9.5 cm long; leaflets in 6–10 pairs, elliptic to oblong-elliptic, mucronulate. Inflorescences racemes, 4–10 cm long. Flowers numerous, short-pedicellate. Calyx campanulate, 4–5 mm long with 5 short teeth. Corolla papilionaceous, brick-red. Stamens diadelphous. Fruits swollen legumes, wide-oblong, 2.5–3.5 cm long, 1.8–2 cm wide, papery-membranous, glabrous or with scattered hairs, stipitate. Seeds ~1.5 mm long, round to kidney-shaped, brown, dull.

Other distinguishing features: Legume many-seeded, indehiscent.

Phenology: Flowers in May-June, fruits in July-August.

Reproduction: By seeds and vegetatively by rhizomes.

Distribution: Nearly all provinces of Uzbekistan; Talas and Jalal-Abad provinces of Kyrgyzstan.

Habitat: The chul zone. Wet, salty places, river banks, and tugais.

Population status: Uncommon, usually found as small populations.

Traditional use: An infusion of the herb is used as a hemostatic after childbirth and to treat uterine atonia (Akopov 1981). It is used for the treatment of hypertension in China (Ma et al. 2002b).

Documented effects: The alkaloid spherophysine, which was isolated from the aboveground plant parts, has hypotensive activity and effects uterine action. In the form of a benzoic-acid salt, it is used for essential hypertension of the first and second degrees. Spherophysine is used for arterial hypertension, weak birthing activity (labor difficulties), and post natal bleeding (Sokolov and Zamotaev 1989). A stilbene isolated from the plant was synthesized and tested for antioxidant activity and showed superior antioxidative activity when compared to the well-known antioxidants resveratrol, vitamin C and butylated hydroxyanisole (BHA) (Venkateswarlu et al. 2003). Additionally, synthesized stilbenes based on naturally occurring compounds were active against leukemia and lymphoma cell lines (Tolomeo et al. 2005).

Phytochemistry: Aboveground plant parts contain 0.4 % total alkaloids, from which spherophysine, spherosine, and saponins with hemolytic index of 1:40 have been isolated (Sokolov and Zamotaev 1989). Isoflavans, lignans, coumarins, flavonoids, and sterols have also been isolated from the plant (Ma et al. 2002b, 2003, 2004a, b; Hou et al. 2005).

Spinacia turkestanica Iljin – Amaranthaceae (formerly in Chenopodiaceae)

Synonyms: *Spinachia tetrandra* Steven ex M. Bieb.

English name: Turkestan spinach, Wild spinach

Russian name: Шпинат туркестанский (Shpinat turkestanskiy)

Uzbek name: Chuchka tikan

Kyrgyz name: Туркстан шпинаты (Turkstan shpinaty)

Description: Dioecious, herbaceous annual, glabrous or with slight farinaceous bloom. Stem 10–60 cm tall, unbranched or sometimes with elongated lower branches. Basal leaves and lower stem leaves runcinate, with a large triangular-hastate terminal lobe and oblong or linear lateral lobes, long-petiolate; upper stem leaves triangular-hastate with shorter petioles. Male inflorescences interrupted spikes, axillary and terminal, nearly-leafless. Female flowers clustered in leaf axils. Fruits consist of 4–6 flowers accreted in to a spiny aggregate (3–8 mm long) with thorny horns. Surface of aggregate and horns smooth or wrinkled; horns usually oblong- pyramidal, triangular in cross-section.

Other distinguishing features: Staminate flowers with 4 perianth segments and very exserted stamens. Fresh leaves have an alkaline flavor.

Phenology: Flowers and fruits in April-June.

Reproduction: By seeds.

Distribution: Toshkent, Samarqand, Buxoro, and Surxondaryo provinces of Uzbekistan; Osh and Chuy provinces of Kyrgyzstan.

Habitat: The adyr and tau zones. A weed of irrigated and unirrigated fields and foothill pastures.

Population status: Common.

Traditional use: The leaves are used as a carminative. It is recommended as a poly-vitamin for treatment of anemia and rickets (Khalmatov 1964).

Documented effects: Spinach is a valuable food crop due its high iodine, calcium, iron, vitamin, protein, and fat content. Spinach can compete with milk with its protein content; the protein is mainly contained in the leaves (Bakiev and Makhkamov 1987). The spinach protein, secretin, is used in medicine like pilocarpine, as a therapeutic agent to stimulate the mucus coating of the stomach lining and the pancreatic glands (Khalmatov 1964).

Phytochemistry: The leaves contain 80 mg% carotene, 64 units/100 g of vitamin B1, up to 40 units of vitamin B2, 16 mg% of vitamin C, and peculiar proteins (Khalmatov 1964).

▲**Silybum marianum (L.) Gaertn.** Photos: Evgeny Davkaev

◄**Sphaerophysa salsula (Pall.) DC.** Photos: Alexander Ivanov

▼**Spinacia turkestanica Iljin** Photos: Chris Kik

▼**Sorbus tianschanica Rupr.** Photos: Vladimir Epiktetov

Tanacetum vulgare L. – Asteraceae

Synonyms: *Chrysanthemum tanacetum* Vis., *Chrysanthemum vulgare* (L.) Bernh., *Pyrethrum vulgare* (L.) Boiss., *Tanacetum boreale* Fisch. ex DC., *Tanacetum crispum* Steud., *Tanacetum umbellatum* Gilib.

English name: Common tansy

Russian name: Пижма обыкновенная (Pizhma obyknovennaya)

Uzbek name: Oddi dastarbosh

Kyrgyz name: Кадимки танацетум (Kadimki tanatsetum)

Description: Herbaceous perennial, with long, woody rhizomes. Stems many, erect, 50–150 cm tall, branched in upper part. Leaves alternate, up to 20 cm long, 3–10 cm wide, bipinnatisect, elongate-ovate; basal leaves petiolate; stem leaves sessile; lobes pinnatifid or dentate. Inflorescences capitula in flat-topped corymbs; capitula semispherical, compact, 5–10 cm wide, with up to 200 flowers. Disc flowers yellow, 2–3 mm long, 5-lobed, peripheral flowers ca. 20, 3–4-lobed; ray flowers absent. Fruits elongate achenes, often ribbed.

Other distinguishing features: Leaves nearly or completely glabrous.

Phenology: Flowers and fruits in July-October.

Reproduction: By seeds and vegetatively.

Distribution: Naryn, Ysyk-Kol, Osh, and Chuy provinces of Kyrgyzstan; not found in the flora of Uzbekistan.

Habitat: On steppes, grassy slopes, in meadows, among shrubs, along rivers and roads, and in spruce forests.

Population status: Common, found in small groups.

Traditional use: Inflorescences are used as a vermifuge, carminative, and choleretic, to heal wounds, and to treat intoxication due to lung tuberculosis, fevers, gastrointestinal diseases, and low acidity (Khalmatov et al. 1984). An infusion of the inflorescences is used in Russian folk medicine to increase appetite, bile and sweat production, blood pressure, and decrease heart rates. The infusion is also used as an antipyretic, antispasmodic, anti-inflammatory, analgesic, vermifuge, insecticide, and anti-microbial, and to heal wounds. In the folk medicine of North Caucasus a decoction of the herb is used to treat headaches, and is used externally to treat rheumatism. A decoction of the inflorescences is used to treat skin cancer (Altimishev 1991).

Documented effects: A decoction of the inflorescences is used as a vermifuge (for ascarides and pinworm), to treat liver diseases (hepatitis and angiocholitis), gall bladder diseases, and acute gastrointestinal diseases. A water infusion of the inflorescences has shown to be an effective treatment for enterocolitis and other intestinal diseases. An infusion of the inflorescences and leaves is used externally as a bath and compress as a pain killer, to treat gout, rheumatism, joint pain, sprains, bruises and to heal wounds. The infusion of this plant is prohibited for pregnant women (Altimishev 1991). In experiments with animals, an infusion of the inflorescences increased heart beat amplitude and blood pressure, decreased heart rate, increased choleresis, tonified the gastrointestinal tract, and increased its secretions (Akopov 1990). An extract of the plant and isolated compounds have been shown to have anti-inflammatory properties in vivo (Schinella et al. 1998).

Phytochemistry: Leaves and inflorescences contain essential oil (α-thujone, β-thujone, L-camphor, thujol, borneol, pinene, etc.), flavonoids (luteolin, quercetin, apigenin, diosmetin, etc.), tannins, bitter substances, and alkaloids (Khalmatov et al. 1984; Akopov 1990; Schinella et al. 1998; Williams et al. 1999).

Taraxacum officinale F.H. Wigg. – Asteraceae

Synonyms: *Leontodon taraxacum* L., *Taraxacum dens-leonis* Desf., *Taraxacum retroflexum* H. Lindb., *Taraxacum sylvanicum* R. Doll.

English name: Common dandelion

Russian name: Одуванчик лекарственный (Oduvanchik lekarstvennyy)

Uzbek name: Koki, Momakaimok, Gulkoku

Kyrgyz name: Дары какымы (Dary kakymy)

Description: Herbaceous perennial, with taproot. Leaves in basal rosette, numerous, oblanceolate, 10–25 cm long, 1.5–5 cm wide, pinnifid or wide-triangularly toothed. Inflorescences capitula, with hollow peduncles up to 50 cm tall; involucral bracts in 2 series. Flowers all ligulate, yellow. Fruits light brown achenes, 3–4 mm long, with a long, thin beak, bearing white pappus.

Other distinguishing features: Mature inflorescences with mature fruits look spherical due to large pappi.

Phenology: Flowers in April-May, fruits in May-June.

Reproduction: By seeds.

Distribution: Almost all provinces of Kyrgyzstan and Uzbekistan.

Habitat: In meadows, forest glades, in orchards and parks, near roads and in populated areas as a weed.

Population status: Common, forming dense groups.

Traditional use: Fresh juice from the leaves is recommended as a laxative and to treat anemia and general weakness. Milky juice, from the roots, is used to eliminate warts and a galenical preparation of the roots is used to treat skin conditions (Khalmatov et al. 1984; Mamedov et al. 2004). The roots are collected in autumn and the leaves in spring before flowering. In Chinese medicine all parts of the plant are used as a antipyretic (diaphoretic), and leaves are used to strengthen the function of mammary glands (Akopov 1990). The plant is used as a remedy for jaundice, liver and gallbladder disorders, and as a treatment for water retention and breast and uterus cancer (Koo et al. 2004).

Documented effects: A methanol extract of the flowers inhibited inflammation in induced mouse ear edema experiments (Yasukawa et al. 1998). In scientific medicine a decoction or extract is used to increase the appetite, to aid in function of the digestive tract and is used as a choleretic and laxative. A powder from the roots is used in a complex remedy to treat arteriosclerosis (Khalmatov et al. 1984). Flower extracts have shown antioxidant activity in vitro (Hu and Kitts 2004). An aqueous extract of the plant has exhibited anti-tumor actions and was shown to induce apoptosis of human carcinoma cells in vitro. Taraxasterol has also been shown to have anticarcinogenic activity (Koo et al. 2004).

Phytochemistry: Roots contain sesquiterpene lactones, triterpene compounds (taraxerol, taraxasterol, pseudotaraxasterol, β-sitosterin, and stigmasterin), taraxol, inulin, caoutchouc, and fatty oil, which contains glycerides of palmitic, oleic, linoleic, melissic, and cerotinic acid. The inflorescence and leaves contain coumarins (cichoriin and aesculin), flavonoids and flavonoid glycosides (chrysoeriol [3′-methoxyluteolin], luteolin, luteolin 7-glucoside and its derivatives), carotinoids (taraxanthin, flavoxanthin, and lutein), triterpene alcohols (arnidiol and faradiol) and vitamin B2. The leaves contain ascorbic and chicoric acid. Monocaffeyltartaric and chlorogenic acid have been found throughout the plant (Tolmachev 1976; Akopov 1990; Williams et al. 1996; Kisiel and Barszcz 2000).

Thalictrum foetidum L. – Ranunculaceae

Synonyms: *Thalictrum minus* var. *foetidum* (L.) Hook. f. & Thomson
English name: Foetid meadow rue
Russian name: Василистник вонючий (Vasilistnik vonyuchiy)
Uzbek name: Sassik sanchikoot
Kyrgyz name: Сасык тармал чоп (Sasyk tarmal chop)
Description: Herbaceous perennial. Stems 15–100 cm tall, simple or branched, gray-green, often violet. Leaves petiolate, wide-triangular, tri- or quadripinnate; leaflets 4–15 mm long, 2–15 mm wide, broad-ovate, 3-lobed; lobes entire or with 2–3 rounded teeth. Inflorescence a loose, spreading panicle. Sepals 5, 2.5–4 mm long, ovate, violet. Petals absent. Stamens many, with yellow anthers. Fruits ovoid achenes, 3–5 mm long.
Other distinguishing features: The entire plant, especially the lower surfaces of the leaves, are covered with hairs and small glands.
Phenology: Flowers in June-July, fruits in July-August.
Reproduction: By seeds.
Distribution: Ysyk-Kol, Jalal-Abad, and Chuy provinces of Kyrgyzstan; not found in the flora of Uzbekistan.
Habitat: On stony slopes with rocky debris, in subalpine and alpine mountain belts.
Population status: Common, found in large groups.
Traditional use: The plant is used as an antiemetic. In Tibetan medicine this species is used to treat edema and gynecological diseases (Akopov 1990). Decoction or tea of the aboveground parts is recommended to treat epilepsy, jaundice, edema, lung tuberculosis, nose bleeds, gastrointestinal ailments, common colds, and gynecological diseases, as well as a general tonic. The herb is used in a poultice to treat bruises, wounds, abscesses, and rheumatism. A decoction of the roots is drunk to treat diarrhea, ulcers, and liver and kidney diseases. A tea of the seeds and herb is drank to treat side pains, headache, dizziness, and bronchitis (Khodzhimatov 1989; Mamedov et al. 2004).
Documented effects: An infusion is used as a treatment for early stage hypertension, stenocardia, and poor blood circulation. The alkaloid foetidin has anti-inflammatory and antiedemic action (Gammerman et al. 1990; Grinkevich 1991). Glycosides isolated from the plant reduced cholesterol in the blood serum and showed anti tumor activity as well as contraceptive effects (Khamidullina et al. 2006).
Phytochemistry: The aboveground parts contains alkaloids (foetidine, magnoflorine, thalfine, and thalfinine), flavonoids (rutin, glycosides, etc.), coumarins, triterpene glycosides, tannins, organic acids, and resins. The underground parts contain alkaloids (Ganenko et al. 1986; Rakhimov et al. 1987; Akopov 1990).

Thalictrum isopyroides C.A. Mey. – Ranunculaceae

Синонимы: Unknown

English name: Unknown

Russian name: Василистник изопироидный (Vasilistnik izopiroidnyy)

Unbek name: Sanchikut

Kyrgyz name: Терен кесиктуу тармал чоп (Teren kesiktuu tarmal chop)

Description: Herbaceous perennial with fibrous roots. Stem 8–45 cm tall, simple or branched, glabrous. Leaves tri- or quadripinnatisect, gray, glabrous, with short petioles, concentrated at the base of stem, usually in groups of 2–3; leaflets broadly rhomboid, 3-lobed, thick; lobes lanceolate-linear or oblanceolate; terminal lobe divided up to the middle or to the base into 2–3 lanceolate segments with smooth margins. Inflorescence a very loose panicle. Sepals greenish, ca. 2 mm long. Petals absent. Stamens 5–8. Fruits narrow-ovoid achenes, 4–5 mm long, 1–2 mm wide.

Other distinguishing features: Stigma triangular-winged, persistent in fruit.

Phenology: Flowers in April- early May, fruits in May-June.

Reproduction: By seeds.

Distribution: Toshkent, Andijon, Farg'ona, Samarqand, Qashqadaryo, and Surxondaryo provinces of Uzbekistan; Talas, Osh and Jalal-Abad provinces of Kyrgyzstan.

Habitat: The adyr and tau zones. On shallow-soiled, stony slopes with rocky debris on hills and mountains.

Population status: Rare, found as single individuals.

Traditional use: In Tajikistan a tea made from the aboveground parts is used to treat fever, chest pain, and as an anticonvulsive. A decoction of the herb is drunk to treat epilepsy, jaundice, tachycardia, nose bleeds, lung tuberculosis, gastrointestinal, and feminine diseases. A decoction of the roots is drunk to treat stomach ulcers, liver and kidney disease, and high blood pressure. A tea of the seeds is recommended to treat dizziness, high blood pressure, bronchitis, and edema (Khodzhimatov 1989). An infusion of the plant is used to treat diarrhea, jaundice, malaria, epilepsy and lung tuberculosis, and is used externally to treat skin diseases (Khalmatov 1964).

Documented effects: In experiments with animals, the alkaloid thalisopine acted as a sedative and had pronounced anticonvulsant activity which surpassed that of phenytoin and trimetin (Tashbaev and Sultanov 1962, 1965). When injected intravenously, it had distinct antiarrhythmic action on experimental models (Akbarov et al. 1972). The alkaloid cryptopine stimulated uterine smooth muscles, had vasoconstrictive action, and increased arterial pressure in narcotized animals. The alkaloid magnoflorine reduced blood pressure due to its ganglio-blocking action (Fakhrutdinov 1971; Fakhrutdinov and Sultanov 1972). In anesthetized animals, intravenous injections of the alkaloid thalicminine caused short-term reduction of blood pressure and heart rate (Abdalla et al. 1991).

Phytochemistry: Plants studied were found to contain 3.22 % total alkaloids. Thalisopine, thalisopidine, dehydrothalicmine, thalicmine, thalicminine, cryptopine, magnoflorine, and others were isolated from the total alkaloids (Yunusov 1974; Abduzhabbarova et al. 1978).

▲**Thalictrum isopyroides C.A. Mey.** Photos: Alim Gaziev

▼**Taraxacum officinale F.H. Wigg.** Photos: *top*: Dmitri Oreshkin; *center* and *bottom*: Sergey Appolonov

▲**Thalictrum foetidum L.** Photos: Petr Filippov

◀**Tanacetum vulgare L.** Photos: *left*: Mary Backlund; *right*: Sergey Appolonov

Thalictrum minus L. – Ranunculaceae

Синонимы: Unknown

English name: Small meadow-rue, lesser meadow-rue

Russian name: Василистник малый (Vasilistnik malyy)

Uzbek name: Sanchyq ut

Kyrgyz name: Кичинекей тармал чоп (Kichinekey tarmal chop)

Description: Herbaceous perennial. Stems 30–100 cm high, glabrous, smooth, erect or irregularly bending, evenly foliaceous. Leaves alternate, tri- or quadripinnatisect, wide-triangular in outline, greenish-gray, petiolate (upper leaves sessile); leaflets almost round, 0.8–4 cm long and wide, irregularly lobed. Inflorescence an oval or pyramidal panicle. Sepals ovate, 3–4 mm long, 2 mm wide, yellowish-green. Stamens 10–15. Fruits ovoid achenes, 4–5 mm long, 2 mm wide, ribbed, with an erect or slightly bent tip, sessile.

Other distinguishing features: Inflorescence spreading and much branched. Leaves deflected from the stem.

Phenology: Flowers in June-July, fruits in June-August.

Reproduction: By seeds.

Distribution: Toshkent, Jizzax, Samarqand, Qashqadaryo, and Surxondaryo provinces of Uzbekistan; all of Kyrgyzstan.

Habitat: The adyr and tau zones. In the valleys of lowland and mountain rivers.

Population status: Uncommon, found as single individuals.

Traditional use: An infusion of the herb is used to treat various diseases: skin, diarrhea, hepatitis, malaria, epilepsy, tuberculosis, fevers, and is also used as a hemostatic (Khalmatov et al. 1984).

Documented effects: Extracts are used as a hemostatic, for hypotonia to increase blood pressure, and as an antibacterial against gram-positive bacteria. In various animals using different modes of application, the alkaloids thalicmine and thalicmidine caused depression of the central nervous system and elongated effects of soporifics. In higher doses they produced catalepsy (Zabirov and Kasmaliev 1962; Sadritdinov et al. 1971; Sadritdinov 1973; Sadritdinov and Khamdamov 1975). The alkaloid thalictrimine had ganglion blocking action (cardiac ganglion n. vagus) and inhibited the cough reflex (Sadritdinov and Kurmukov 1980). Thalicminine and thalmine have sedative and short-term hypotensive effects; thalmine also had anti-inflammatory, analgesic, and antipyretic action (Sadritdinov and Sultanov 1971; Fakhrutdinov and Sultanov 1972; Sadritdinov 1971b, 1973; Abdalla et al. 1991). The alkaloids O-methyl-thalicberine, thalisopine, and thalmine had antiarrhythmic action. Thalisopine exceeded the activity of quinidine and procainamide-hydrochloride (Akbarov et al. 1978). Experiments have shown that a number of the alkaloids isolated from the plant have antimicrobial activity against *Mycobacterium smegmatis* (Liao et al. 1978a, b). The alkaloid thaliblastine exhibited activity against various types of cancer (Mircheva and Stoychkov 1976; Ilarionova et al. 1980; Stoychkov and Miloushev 1980; Todorov and Zeller 1992; Chen et al. 1992).

Phytochemistry: The aboveground parts contained up to 1 % total alkaloids (thalmine and thalminine), flavonoids (1.64 %), saponins (3.1 %), vitamin C (175.7–761.7 mg%), organic acids, tannins, bitter, and other substances; the roots contained 1.1 % total alkaloids (thalicmine, thalicmidine, thalicmitrine, tolmetin, argemonine and others). The seeds contained 22.9–28.4 % fatty oil (Yunusov 1981; Khalmatov et al. 1984; Sidjimov et al. 1998).

Thermopsis alterniflora Regel & Schmalh. – Fabaceae

Synonyms: *Thermopsis rigida* Vassilcz.

English name: Unknown

Russian name: Термопсис очередноцветковый (Termopsis ocherednotsvetkovyy)

Uzbek name: Afsonak (Афсонак)

Kyrgyz name: Кезек гулдуу сары мыя (Kezek gulduu sary myya)

Description: Herbaceous perennial with vigorous rhizomes. Stems erect, up to 100 cm tall, branched, middle and upper portion with varying amounts of hairs. Leaves alternate, trifoliate, petiolate with lanceolate stipules; leaflets oblong-elliptic, 2.5–5 cm long, 1–2 cm wide, acuminate, adaxial side glabrous, abaxial side slightly hairy. Inflorescence a loose apical raceme, 9–20 cm long, with oblanceolate bracts. Flowers alternate. Calyx 10–20 mm long, densely covered with silky hairs. Corolla papilionaceous, yellow. Fruits oblong-elliptic legumes, 3–6 cm long, 1–1.6 cm wide, covered with short, appressed hairs, few-seeded. Seeds kidney-shaped, 5–6 mm long, 3–4 mm wide, brownish-red-greenish, glabrous.

Other distinguishing features: Calyx teeth one third to one half as long as tube. All 10 stamens free.

Phenology: Flowers in May-June, fruits in June-August.

Reproduction: By seeds and rhizomes.

Distribution: Toshkent province of Uzbekistan, in the Western Tien Shan; Osh and Jalal-Abad provinces of Kyrgyzstan.

Habitat: The tau zone. Shallow-soiled slopes and mountain brook valleys; as a weed among unirrigated cereal crops.

Population status: Uncommon, sometimes in populations with up to 50 individuals.

Traditional use: A galenical preparation of the stems, leaves, flowers and fruits is used to treat bronchial asthma (Mamedov and Craker 2001).

Documented effects: Used as an expectorant and vermifuge (Khalmatov 1964). The alkaloid cytisine is used to prepare a 0.15 % solution, called *cytion*, which is used to increase respiration in cases of respiratory standstill, such as during operations and traumas, from infectious diseases, shocks, various intoxication (such as poisoning by carbon oxide, prussic acid, and narcotics), asphyxia of newborns, and others. Pachycarpine increases uterine contractility and is used in obstetrical practice to stimulate contractions for weak labors, and also to stop bleeding during the post-natal period (Mashkovskii 1984).

Phytochemistry: At the beginning of flowering, 3.5 % total alkaloids were obtained from the aboveground parts. Cytisine (>50 % of total alkaloids), pachycarpine, methylcytisine, thermopsine, anagirine, argentine, alteramine, dimethamine, and other alkaloids were isolated from the total alkaloids. The flavonoids cinaroside, luteolin, chrysoeriol, thermopsocide, genistein, and genistin were also isolated from the aboveground parts. The aboveground portion also contained 4.88 % titrated organic acids, up to 4.8 % sugars, and 5.08 % resins. Roots contained 0.81 % and seeds contained up to 3.34 % total alkaloids (Khalmatov et al. 1984).

Thermopsis lanceolata R. Br. – Fabaceae

Synonyms: *Sophora lupinoides* L., *Thermopsis dahurica* Czefr., *Thermopsis glabra* Czefr., *Thermopsis lupinoides* (L.) Link., *Thermopsis sibirica* Czefr.

English name: Lanceleaf thermopsis

Russian name: Термопсис ланцетный, Мышатник (Termopsis lantsetnyy, Myshatnik)

Uzbek name: Lantsetcemon termopsis, Lantsetcemon afsonak

Kyrgyz name: Ланцетный сары мыя (Lantsetnyy sary myya)

Description: Herbaceous perennial, with a deep main root and lateral rhizomes. Stems up to 40 cm tall, many, erect, branched, striated, hairy. Leaves alternate, petiolate, trifoliate; leaflets elongate- or oblanceolate, 2.3–7.6 cm long, 0.8–2.3 cm wide, glabrous above, hairy below. Flowers in whorls forming terminal racemose inflorescences. Calyx campanulate, with 5 lanceolate lobes. Corolla papilionaceous, yellow. Fruit a narrow-linear legume, straight or slightly arched, 4–8.8 cm long, 0.7–1.2 cm wide, short hairy. Seeds nearly round, dark olive or nearly black, glaucous.

Other distinguishing features: Stamens 10, all distinct. Legumes not flattened, sharply tapering at the end.

Phenology: Flowers in May-June, fruits in July-August.

Reproduction: By seeds and rhizomes.

Distribution: Ysyk-Kol, Naryn, and Chuy provinces of Kyrgyzstan; not found in the flora of Uzbekistan.

Habitat: Found among *Achnatherum splendens* (Trin.) Nevski on the coast of Lake Ysyk-Kol, on solonetzic soils, and in fallow and cultivated fields. Not found high into the mountains.

Population status: Common, found in dense groups.

Traditional use: The aboveground parts are harvested before flowering. Decoctions of the aboveground parts are used to treat respiratory catarrh, flu, bronchitis, pneumonia, and headaches (Akopov 1990; Mamedov and Craker 2001).

Documented effects: An infusion of the herb is used as an expectorant to treat chronic bronchitis and residual pneumonia. The preparation *Cytiton*, which contains the alkaloid cytosine isolated from the seeds, in is used to stimulate respiratory function and improve blood circulation. The preparation is used to treat asphyxia in newborns and when a person stops breathing during surgical procedures or from trauma (Khalmatov et al. 1984). The alkaloid pachycarpine, isolated from this plant, is used to treat peripheral vessel spasms and to induce labor when necessary (Akopov 1990).

Phytochemistry: The herb contains alkaloids (thermopsine, homothermopsine, methylcytisine, pachycarpine, and anagyrine), saponins, tannins, resins, mucilage, traces of essentail oil, and ascorbic acid. The seeds contain alkaloids, mainly cytisine (Tolmachev 1976; Akopov 1990).

Thermopsis turkestanica Gand. – Fabaceae

Synonyms: *Thermopsis kaxgarica* Chang Y. Yang, *Thermopsis lanceolata* ssp. *turkestanica* (Gand.) Gubanov.

English name: Unknown

Russian name: Термопсис туркестанский (Termopsis turkestanskiy)

Uzbek name: Unknown

Kyrgyz name: Туркестан сары мыясы (Turkestan sary myyasy)

Description: Herbaceous perennial, with a deep main root and lateral rhizomes. Stems many, erect, 30–50 cm tall, striated, branched; branches appressed to main stems. Leaves alternate, petiolate, trifoliate; leaflets 3.5–8 cm long, 0.5–1 cm wide, narrowly lanceolate. Flowers in whorls forming terminal racemose inflorescences. Calyx campanulate; lobes 5, lanceolate. Corolla papilionaceous, yellow. Fruit an elongate-linear legume, 4.5–7 cm long, 0.8–1 cm wide, light-brown, densely covered with short hairs. Seeds ellipsoid, dark green.

Other distinguishing features: Stamens 10, all distinct. Legumes flattened and slowly tapering to the end.

Phenology: Flowers in June-July, fruits in July-August.

Reproduction: By seeds and rhizomes.

Distribution: Ysyk-Kol, Naryn, and Chuy provinces of Kyrgyzstan; not found in the flora of Uzbekistan.

Habitat: In fallow fields and on solonetzic soils among *Achnatherum splendens* (Trin.) Nevski.. Found in high mountain valleys.

Population status: Common, forming dense groups.

Traditional use: A decoction of the herb is used to treat low blood pressure and as an expectorant (Teslov 1960).

Documented effects: An alcoholic extract of the aboveground parts strengthens respiratory function and raises blood pressure (Chefranova 1954).

Phytochemistry: The aboveground parts contain alkaloids (cytisine, thermopsine, N-methylcytisine, anagyrine, and sparteine; Plant Resources of the USSR 1987).

▲**Thermopsis alterniflora Regel & Schmalh.** Photos: Alim Gaziev

▲**Thalictrum minus L.** Photos: Alim Gaziev

▼**Thermopsis lanceolata R. Br.** Photo: Klazina Witteveen

▼**Thermopsis turkestanica Gand.** Photo: Vladimir Epiktetov

Thymus marschallianus Willd. – Lamiaceae

Synonyms: *Thymus amictus* Klok., *Thymus latifolius* (Bess.) Andrz., *Thymus pannonicus* All., *Thymus pannonicus* ssp. *marschallianus* (Willd.) Soó, *Thymus platyphyllus* Klok., *Thymus pseudopannonicus* Klok., *Thymus stepposus* Klok. & Shost.

English name: Unkown

Russian name: Тимьян Маршаллов (Tim'yan Marshallov)

Uzbek name: Kaklikoot, Toshchop

Kyrgyz name: Кадимки кийик оту (Kadimki kiyik otu)

Description: Perennial subshrub. Stems short, much branched, upper-half retrorse-pubescent, flower-bearing branches 12–37 cm tall. Leaves opposite, sessile, oblanceolate or elongate-elliptic, 12.5–30 mm long, 2.5–7.5 mm wide, abaxially glandular, margin entire or slightly serrulate. Inflorescences verticillasters in apical spikes; pedicels densely pubescent. Calyx tubular-campanulate, 2-lipped; upper lip 3-toothed; lower lip 2-toothed. Corolla red-purple, lilac or white, pubescent, 2-lipped; lower lip 3-lobed. Fruits ovoid nutlets.

Other distinguishing features: Plants gynodioecious.

Phenology: Flowers in May, fruits in August.

Reproduction: By seeds.

Distribution: Ysyk-Kol and Chuy provinces of Kyrgyzstan; not found in the flora of Uzbekistan.

Habitat: On foothills, in meadow-steppes, meadow slopes, on the edges of spruce forests, among juniper stands.

Population status: Common.

Traditional use: An infusion of the herb is used to treat stomatitis and toothaches. A decoction in milk is used to treat acute respiratory infections and amenorrhea. In Bulgaria the herb is used to heal wounds and a decoction is used to treat stomach ulcers and bad breath. An infusion is used in the Altai region to treat fevers and headaches and in the Middle Volga region as an expectorant for acute respiratory infections and pertussis (Plant Resources of the USSR 1991).

Documented effects: In modern medicine the herb is used in a similar manner as *Thymus. serpyllum*. A decoction and liquid extract is recommended for internal use as an expectorant to treat bronchitis and other upper respiratory illnesses. It is used externally in compresses and baths as an analgesic to treat radiculitis and neuritis (Tolmachev 1976). The complex preparation *Pertussin*, containing this herb, is used as an expectorant and cough suppressant and to treat bronchitis and other upper respiratory illnesses (Kurochkin 1998). Ethanolic extracts of *Thymus marschallianus* exhibited antioxidative activity (Budincevic et al. 1995). Volatile oils isolated from the plant exhibited antibacterial activity against the gram-positive bacterium *Diplococcus pneumoniae* (Oprean et al. 2007).

Phytochemistry: Aboveground parts contain phenolcarbonic acids and their derivatives (caffeic, rosemarinic, 1-caffeoylquinic, and 5-caffeoylquinic acids), flavonoids (luteolin, apigenin, scutellarein, and anthocyans) and essential oil (containing thymol, carvacrol, α-pinene, camphene, sabinene, n-thymol, isoborneol, boroneol, undecanoic acid, and amyl alcohol; Plant Resources of the USSR 1991; Kolesnikov and Gins 2001; Stahl-Biskup 2002).

Tribulus terrestris L. – Zygophyllaceae

Synonyms: *Tribulus bicornutus* Fisch. & Mey.

English name: Puncturevine, Caltrop

Russian name: Якорцы стелющиеся (Yakortsy stelyushchiyesya)

Uzbek name: Temirtikan

Kyrgyz name: Тошолмо мык тикен (Tosholmo myk tiken)

Description: Herbaceous annual with a thin taproot. Stems 20–80 cm long, branched, spreading, prostrate and rising only at tips, usually hairy. Leaves opposite, even-pinnate, 3–6 cm long, short-petiolate, with small stipules; leaflets in 6–8 pairs, oblong, 4–10 mm long, adaxial side glabrous, abaxial side hairy. Flowers solitary in leaf axils, 1–1.2 cm in diameter, pedicels 4–10 mm long. Sepals 5. Petals 5, yellowish. Stamens 10. Style 1. Fruits schizocarpic, flattened, star-shaped; mericarps 5, dry, angular, tuberculate with 2 or 4 divergent spines.

Other distinguishing features: Plant often appears glaucescent.

Phenology: Flowers and fruits in May-August.

Reproduction: By seeds.

Distribution: All of Uzbekistan and Kyrgyzstan.

Habitat: The adyr zone. Waste places, oases, unirrigated fields, near roads, dry slopes, and slopes along rivers and brooks.

Population status: Common, not forming dense groups.

Traditional use: This plant has been used since ancient times for various diseases. Avicenna recommended caltrop for tumors and ulcers, especially for festering ulcers of the gums, as a diuretic, and to remove kidney and bladder stones. Folk medicine in the East uses decoctions and infusions of the herb as a purgative, diuretic and tonic, for gonorrhea, headaches and eye inflammations, and for strong side pains. Cleaned roots are boiled in milk and used for chronic malaria and as a energizing remedy (Seredin and Sokolov 1969; Khalmatov et al. 1984). In Western countries it is used to increase the libido, and as a tonic, astringent, and diuretic (Gammerman et al. 1990).

Documented effects: A liquid extract of this species (collected during flowering) is used to treat people with low levels of stomach acidity due to hypo- and anacidic gastritis and as a diuretic to treat swelling. Extracts made from the plant (collected during fruiting period) are also used as a diuretic. A preparation from the leaves, *Tribusponin*, which contains steroid glycosides, is used as an antisclerotic treatment (Seredin and Sokolov 1969; Gammerman et al. 1990). Two compounds isolated from the plant, tribulosin and β-sitosterol-D-glucoside, exhibited antihelmintic activity (Deepak et al. 2002). Steroidal saponins, isolated from the plant, exhibited antifungal activity against *Candida albicans* and *Cryptococcus neoformans* and anti-cancer activity against a variety of cancer cell lines (Bedir et al. 2002). Rats that were given an oral extract of the fruits exhibited weight gain and improvement in sexual behavior parameters (Gauthaman et al. 2003). The systolic blood pressure of hypertensive rats that were fed an extract of the fruits was significantly decreased compared to unfed hypertensive rats. The ACE (angiotensin-converting enzyme) activity in all tissues of extract fed hypertensive rats was significantly lower than that of the control rats (Sharifi et al. 2003).

Phytochemistry: The plant contains flavonoids, alkaloids (harman, etc.), amides, and steroidal saponins (diosgenin dehydration products including crystalline diosgenin, gitogenin, ruscogenin, and 25-D-spirosta-3,5-diene), and saponins with a hemolytic index of 1:240. The leaves contain up to 160 mg% vitamin C. The seeds contain alkaloids and the fruits contain around 5 % tannins and fatty drying oil (Seredin and Sokolov 1969; Gammerman et al. 1990; Wang et al. 1997; Wu et al. 1999b; Deepak et al. 2002; De Combarieu et al. 2003).

Trichodesma incanum (Bunge) A. DC. – Boraginaceae

Synonyms: *Friedrichsthalia incana* Bunge.

English name: Unknown

Russian name: Триходесма седая (Trikhodesma sedaya)

Uzbek name: Kampir chopon

Kyrgyz name: Боз триходесма (Boz trikhodesma)

Description: Rhizomatous perennial up to 30–100 cm tall. Stems ascending, branched, herbaceous, densely covered with short, gray pubescence, becoming shiny, woody and glabrescent below. Leaves alternate or subopposite, ovate to oblanceolate, 3–8 cm long, 1.3–2.8 cm wide, apex acute, margins entire, sessile, both sides silky gray-pubescent. Inflorescences loose, narrow-paniculate, composed of terminal scorpioid cymes. Flowers pedicellate, drooping. Calyx ovate-campanulate, gray-tomentose with 5 deeply divided oblanceolate lobes. Corolla ca. 2 cm in diameter, with a short tube and 5 broad, triangular-ovate lobes with tail-like appendages; at the at the beginning of flowering the tube is white and lobes light-blue, later the tube turns pink and lobes dark-blue. Anthers yellow, forming an exserted cone. Fruits ovoid nutlets, 6–8 mm long, grayish-brown, dull and covered with tiny wrinkles and tubercles, edges slightly uneven or toothed.

Other distinguishing features: Anthers with spirally-twisted awn-like appendages. Calyx enlarged in fruit, becoming disk-shaped and membranous.

Phenology: Flowers and fruits from May to November.

Reproduction: By seeds and rhizomes.

Distribution: Karakalpakstan autonomous republic, Toshkent, Andijon, Farg'ona, Samarqand, and Surxondaryo provinces of Uzbekistan; Talas, Batken and Osh provinces of Kyrgyzstan.

Habitat: The adyr and tau zones. Loess slopes of hills, stony slopes with rocky debris, as well as unirrigated and abandoned fields.

Population status: Uncommon, found as single individuals.

Traditional use: The plant roots (as a root-powder plaster) are used to heal persistent wounds and furunculosis. A decoction of the roots and leaves is recommended for scabies and is applied on infected skin areas (Khalmatov 1964).

Documented effects: This plant is highly poisonous. The alkaloids contained in this species act as neurovascular toxins. The alkaloids depress blood production, destroy erythrocytes, induce hypoxia in tissues, and increase vein wall permeability. These toxins have the ability to accumulate in the body (Vilner 1974). The alkaloid incanine (the N-oxide [amine oxide]) and the alkaloid trichodesmine lower arterial pressure and have antispasmodic action (Mashkovskii 1983).

Phytochemistry: All plant parts contain alkaloids. Immature fruits contain up to 1.5 %, mature fruits 2.7 %, and the above-ground parts, before flowering, up to 1 % alkaloids. The flowering herb contains only 0.3 % total alkaloids. The alkaloids incanine (1.5 % in seeds), N-oxide form of incanine, trichodesmine, and N-oxide form of trichodesmine have been isolated from the total alkaloids. At flowering period the plant top contains up to 70 % trichodesmine from the total alkaloid content (Yunusov 1981).

Trifolium pratense L. – Fabaceae

Synonyms: *Trifolium ucrainicum* Opperm. ex Wissjul.

English name: Red clover

Russian name: Клевер луговой, Клевер красный (Klever lugovoy, Klever krasnyy)

Uzbek name: Sebarga

Kyrgyz name: Шалбаа уй бедеси (Shalbaa uy bedesi)

Description: Herbaceous perennial, slightly hairy, with taproot. Stems to 80 cm tall, erect or suberect, simple or branched. Leaves trifoliate, stipulate; lower leaves long-petiolate; upper leaves short-petiolate or sessile; leaflets obovate or elliptical, usually sinuate, rarely serrulate, often with a white triangular blotch. Inflorescence head-like, ovoid or globose. Flowers 1.3–2 cm long, in globose or ovoid heads. Calyx tubular-campanulate, with 5 teeth (1 longer, 4 shorter). Corolla papilionaceous, light-pink to dark-red. Fruits small legumes nearly enclosed by calyx.

Other distinguishing features: Stamens 10 (9 united). Heads on top of stems and lateral branches.

Phenology: Flowers in May-September, fruits in June-October.

Reproduction: By seeds.

Distribution: All provinces of Kyrgyzstan; Toshkent, Farg'ona, Samarqand, Qashqadaryo and Surxondaryo provinces of Uzbekistan.

Habitat: Along rivers, in damp meadows and valleys, in the high-mountain meadow and forest belt, and in tallgrass meadows.

Population status: Common, found in dense groups.

Traditional use: An infusion or tea of the flowers is used as an antiseptic, expectorant, diuretic, anti-inflammatory, and analgesic, and to treat lung and upper respiratory tract diseases, sore throat, bronchial asthma, pertussis, malaria, rheumatism, hypertension, stenocardia, anemia, uterine bleeding, leukorrhea, shortness of breath, coughs, and painful menstruation. Freshly ground leaves or fresh juice from the plant are applied externally to treat infected wounds and skin ulcers, burns, and rubella (Nuraliev 1989; Akopov 1990). The seeds are used to increase the libido and are used to treat prolonged fevers (Khodzhimatov 1989).

Documented effects: A tincture of the plant is used to treat arteriosclerosis in patients with normal blood pressure (Nuraliev 1989). In modern medicine this species is used as an expectorant, diuretic, and antiseptic (Khodzhimatov 1989). Metabolites of isoflavones found in the plant were found to protect against UV radiation-induced inflammation and immunosuppression (Widyarini et al. 2001). Isoflavones found in red clover inhibited COX enzyme activity in certain cancer cell types (Lam et al. 2004). Extracts of red clover and individual flavonoid constituents exhibited estrogenic activity in a variety of in vitro assays (Overk et al. 2005).

Phytochemistry: The herb contains many flavonoids (trifolin, isotrifolin, trifoside, etc.), asparagine, tyrosine, coumarinic and salicylic acids, alkaloids, fatty oil, carotene, B vitamins, and vitamin C. The roots contain coumarins (Khalmatov 1964; Akopov 1990; Lin et al. 2000; Klejdus et al. 2001).

▲**Trifolium pratense L.** Photos: *left*: Dmitri Oreshkin; *center*: Sasha Eisenman; *right*: Sergey Appolonov

▼**Thymus marschallianus Willd.** Photos: Andrei Lubchenko

▼**Trichodesma incanum (Bunge) A. DC.** Photos: Alim Gaziev

◄**Tribulus terrestris L.** Photos: *top*: Sergey Mayorov; *center* and *bottom*: Sasha Eisenman

Tussilago farfara L. – Asteraceae

Synonyms: None

English name: Coltsfoot

Russian name: Мать-и-мачеха обыкновенная (Mat'-i-machekha obyknovennaya)

Uzbek name: Okkaldirmok

Kyrgyz name: Кадимки огой Эне (Kadimki ogoy ene)

Description: Herbaceous perennial, with creeping rhizomes. Flowering stems 7–45 cm tall, felted-hairy, with small, alternate bracts. Leaves basal, long-petiolate, orbicular-cordate, 8–15 cm long, up to 12 cm wide, glabrous above, felted-hairy beneath, shallowly lobed, margins unequally dentate. Stem leaves scale-like, ovate-lanceolate, 0.6–1.5 cm long, 0.3–0.8 cm wide, sessile, purple-violet, appressed to stem. Inflorescences capitula. Ray flowers 100–300, golden yellow; disc flowers 20–40, yellowish. Fruits linear achenes, 3–4.5 mm long; pappus white, longer than achene.

Other distinguishing features: The plant blossoms before the leaves appear.

Phenology: Flowers in May-September, fruits in June-October.

Distribution: All provinces of Kyrgyzstan; Toshkent, Farg'ona, Samarqand and Surxondaryo provinces of Uzbekistan.

Habitat: On damp lawns, along mountain rivers and streams, and in areas with water-eroded soils and alluvial deposits.

Population status: Common, found in dense groups.

Traditional use: An infusion and decoction of the leaves and flowers are used as an expectorant and cough suppressant, to treat bronchial asthma, as well as a diuretic to treat edema and scrofula. It is applied externally as a poultice or wash to treat tumors, abscesses, and furuncles. Juice from fresh leaves and roots is used to treat tuberculosis and malaria, and as a choleretic and diaphoretic (Khalmatov et al. 1984). Leaves are used to treat acute and chronic bronchitis, catarrh of the upper respiratory system, pneumonia, laryngitisis, bronchial pneumonia, and a hoarse voice. Preparations of coltsfoot are used to treat tracheitis, kidney and bladder diseases, the gastrointestinal tract, loss of appetite, fever, erysipelatous skin inflammation, scrofula, hair loss, and abscesses. Fresh juice from the leaves is inhaled into the nostrils to eliminate sinus colds. The juice of leaves is also mixed with powdered sugar to treat tuberculosis (Maznev 2004).

Documented effects: An infusion of the leaves and flowers is used as an expectorant and demulcent. A tea made from the leaves is used to treat bronchitis, laryngitis, bronchiectasis, abscesses, and gangrene of the lungs. A poultice is used externally as a demulcent, disinfectant, and anti-inflammatory (Tolmachev 1976; Khalmatov et al. 1984). A sesquiterpene isolated from extracts of the buds was found to have anti-inflammatory effects in vitro and reduced induced rat foot edema (Hwang et al. 1987). Extracts of both the aboveground parts and rhizomes showed antimicrobial activity against *Bacillus cereus* and *Staphylococcus aureus* (Kokoska et al. 2002). Flavonoids isolated from the flower buds exhibited antioxidative activity (Kim et al. 2006). Various compounds isolated from the plant induced cardiovascular and respiratory stimulation and have been shown to have anti-inflammatory activities by inhibiting arachidonic acid metabolism, platelet-activating factor receptors, and the activity of nitric oxide synthesis (Ryu et al. 1999). The ethyl acetate fraction of the plant extract had neuroprotective and antioxidant effects in vitro (Cho et al. 2005).

Phytochemistry: Leaves contain bitter glycosides, carotenoids, alkaloids, flavonoids, coumarins, saponins, mucilage, tannins, organic acids, cytosterin, inulin, tussilagin, and vitamin C. Flowers contain rutin, arnidiol, faradiol, taraxanthin, stigmasterin, cytosterin, phytosterins, n-heptacosane, tannins, etc. (Tolmachev 1976; Khodzhimatov 1989; Ryu et al. 1999).

Ungernia victoris Vved. ex Artjushenko – Amaryllidaceae

Synonyms: None
English name: Unknown
Russian name: Унгерния Виктора (Ungerniya Viktora)
Uzbek name: Omonqora
Kyrgyz name: Unknown
Description: Herbaceous perennial up to 20 cm tall. Bulbs ovoid, small, 4–7 cm wide, with multiple black-brown, papery coats. Stem bearing inflorescence flattened, 5–10 cm long. Leaves 7–10 in number, in 2 rows, linear, 20–25 cm long and 2–3 cm wide, light blue-gray, smooth. Inflorescence an umbel bearing 4–7 flowers. Flowers funnelform, 5–6 mm wide, yellowish to yellow-pink. Stamens 6. Fruit a capsule, 2–3 cm wide, with 3 wide heart-shaped valves. Seeds flat, black.
Other distinguishing features: Flowers after leaves have senesced.
Phenology: Flowers in August, fruits in September.
Reproduction: By seeds and vegetatively by bulbs.
Distribution: Gissar mountain range, Chulbair mountains in Surxondaryo province of Uzbekistan; not found in Kyrgyzstan.
Habitat: The tau zone. Shallow-soiled slopes and ravines.
Population status: Rare endemic of southwest Pamiro-Alai; listed in the Red Book of Rare and Endangered Species of Uzbekistan.
Traditional use: Baked bulbs are used to heal wounds; they are also applied to furuncles to remove pus. According to Avicenna the herb and its seeds are the best treatment for diarrhea. If the seeds are taken with water or wine it helps to heal stomach ulcers and improve digestion. Wine infused with the seeds are used to treat kidney stones (Karimov and Shomakhmudov 1993).
Documented effects: This species is recommended as the raw material to obtain the alkaloid galanthamine. Hydrobromic salt of galanthamine is widely used in medical practice to treat myasthenia, myopathia, and for post-poliomyelitis, radiculitis, and polyneuritis palsies, as well as traumatic injuries of sensory and motor nerves. The alkaloid narwedine has anti-narcotic action and facilitates transfer of nervous excitation to H- and M-cholinergic synapses. A preparation was recommended for clinical trials as an anti-narcotic drug. The alkaloid pancratine lowers blood pressure, has sedative action, and increases activity of soporifics. In acute tests, hordenine shows adrenomimetic action. It is used for intestinal peristalsis inhibition in diarrhea. Licorine has anti-inflammatory, analgesic, and antipyretic action and strengthens the hypothermic action of amidopyrine. This alkaloid also strengthens the secretion of intestines and lung-bronchial glands in dogs and cats (Sadritdinov and Kurmukov 1980). An extract derived from cultured plant cells exhibited antimutagenic properties (Dvornyk et al. 2002).
Phytochemistry: Leaves contain 0.35–1 %, bulbs 0.8–0.9 %, and roots 2.5 % total alkaloids. Galanthamine, pancratine, narwedine, hordenine, and licorine have been isolated from the leaves. Similarly, galanthamine, licorine, pancratine, tatsetine, and hippeastrine have been isolated from the bulbs (Yunusov 1981).

Urtica dioica L. – Urticaceae

Synonyms: None

English name: Stinging nettle

Russian name: Крапива двудомная (Krapiva dvudomnaya)

Uzbek name: Gazanda, Kichitki oot, Chayan oot

Kyrgyz name: Чалкан (Chalkan)

Description: Dioecious, herbaceous perennial plant, with creeping rhizomes. Stems and leaves covered with stinging hairs. Stems erect, 30–170 cm tall. Leaves opposite, simple, ovate-lanceolate, 8–17 cm long, 2–8 cm wide, apex acuminate, margins large dentate. Inflorescences axillary panicles. Flowers unisexual, small, green. Staminate flowers with 4 equal tepals; stamens 4. Pistillate flowers with 4 tepals, inner 2 equal to achene, outer 2 smaller. Fruits ovoid or elliptic achenes, 1–1.5 mm long.

Other distinguishing features: Staminate inflorescences ascending, pistillate lax or recurved in fruit. Achenes smooth.

Phenology: Flowers and fruits in May-September.

Reproduction: By seeds and rhizomes.

Distribution: All provinces of Kyrgyzstan; Karakalpakstan autonomous republic, Toshkent, Andijon, Namangan, Farg'ona, Samarqand, Qashqadaryo and Xorazm provinces of Uzbekistan.

Habitat: Found as a weed in settlements, along canals, in woods, and among bushes.

Population status: Common, forming dense groups.

Traditional use: A decoction and powder of leaves is used as a diuretic, laxative, expectorant, vasoconstrictor, and hemostatic to treat internal bleeding and hemorrhoids, and also to treat rheumatism, stomach diseases, diabetes, and chronic ulcers. It is used in a bath to treat various types of swelling. The roots and fruits are used to treat diarrhea. The leaves are used in a wash to treat hair loss. A water extract of the leaves, along with extracts of different plant species are used to prepare a cream with cow bone marrow which is used to wash and encourage hair growth (Khalmatov et al. 1984; Khodzhimatov 1989).

Documented effects: Preparations from this plant species are used internally as a hemostatic, to increase uterine contractions and to increase blood coagulation. The preparations are effective in treating climacteric menopause and hemorrhoids. Extracts of this species are used to normalize the menstrual cycle. A preparation of the herb has pressor action on internal organ vessels (Tolmachev 1976). Preparations of this species decrease blood cholesterol content and have choleretic and anti-inflammatory activities. In studies with diabetic patients, a decoction of leaves decreased blood and urine sugar levels (Kurochkin 1998). Preparations of nettle normalize metabolism and blood sugar content, increase blood coagulability, increase milk production in feeding mothers, normalize lipid metabolism, increase hemoglobin and erythrocyte content, increase intestine and cardiovascular tonus and stimulate epithelization of wounded tissues. This species helps to treat liver illnesses, joint rheumatism, and gastrointestinal and bladder diseases (Maznev 2004). A water extract of the plant had antioxidant and analgesic activity, showed antimicrobial activity against 9 microorganisms, and exhibited antiulcer activity against ethanol-induced ulcerogenesis (Gulcin et al. 2004). A methanolic extract of the roots exhibited antiproliferative effects on human prostate cancer cells in vivo and in vitro (Konrad et al. 2000). A fraction from the extract of the leaves caused a marked increase in insulin secretion by the pancreatic islets of Langerhans in normal and induced diabetic rats (Farzami et al. 2003).

Phytochemistry: Leaves contain carotene and other carotenoids, organic acids (formic, pantothenic, caffeic, p-coumaric, and ferulic), glycosides (urticin), sitosterin, phytoncides, quercetin, acetylcholine, histamine, tannins, mineral salts, vitamins C, K and group B, resin, protoporphyrin, koproporphyrin, and 5-hydroxytryptamine (Tolmachev 1976; Chikov 1989; Kurochkin 1998).

Vaccaria hispanica (Mill.) Rauschert – Caryophyllaceae

Synonyms: *Saponaria segetalis* Neck, *Saponaria vaccaria* L., *Vaccaria parviflora* Moench., *Vaccaria pyramidata* Medik., *Vaccaria segetalis* (Neck) Garke ex. Asch., *Vaccaria vulgaris* Host.

English name: Cow cockle, Cow herb

Russian name: Тысячеголов пирамидальный, Тысячеголов посевной (Tysyachegolov piramidal'nyy, Tysyachegolov posevnoy)

Uzbek name: Qora mug

Kyrgyz name: Айдама мин баш (Aydama min bash)

Description: Herbaceous annual, glabrous, glaucous. Stem erect, 30–70 cm tall, heavily branched towards top. Leaves opposite, simple, sessile, ovate-lanceolate to oblong-ovate, 2–9 cm long, blue-gray, apex acute, base almost cordate and slightly connate. Inflorescence a paniculiform-cyme, pedicels 1–6 cm long. Calyx 1.3–1.5 cm long, consists of 5 connate sepals, yellowish-green. Petals 5, with linear claws, pink. Stamens 10. Styles 2. Fruit a capsule, wide-ovoid, shorter than the calyx. Seeds black, globose, tuberculate, 1.5 mm wide.

Other distinguishing features: Calyx with 5 raised longitudinal ribs. When fruiting, calyx swollen at the base, the top very narrowed.

Phenology: Flowers and fruits in April-July.

Reproduction: By seeds.

Distribution: All provinces of Uzbekistan; in agricultural zones of all provinces of Kyrgyzstan.

Habitat: The chul and adyr zones. A weed of cultivated fields, especially in unirrigated wheat fields.

Population status: Common.

Traditional use: In Chinese medicine, the seeds are used as an analgesic, to stimulate milk let-down, to promote diuresis, to activate blood circulation, relieve carbuncles, and to treat amenorrhea and breast infections. It is also used in ointments, which are used for treating skin diseases (eczema and psoriasis). In Central Asia, a plaster of the herb is used to treat tumors and as an analgesic (Khalmatov 1964; Morita et al. 1997b; Sang et al. 2000).

Documented effects: Hemolytic index of the herb is equal to 1:1450, of the roots 1:4000, and the seeds contain 3.18 % of a poisonous saponin with a high hemolytic index (1:50,000 in human blood and 1:25,000 in dog's blood). Convolvine and convolamine act as local anesthetics. However, because they are highly toxic and not very effective, they are not used for this purpose. After modification, a derivative of convolamine, convocaine, was introduced for use in hospitals (Ogolevitz 1951). Peptides isolated from the seeds exhibited estrogen-like activity and caused uterine contractions in vitro (Morita et al. 1997a, b).

Phytochemistry: A wide assortment of chemical compounds have been isolated from the seeds including triterpene saponins, alkaloids (up to 0.5 %, convolvine and convolamine), cyclic peptides, phenolic acid, flavonoids, and steroids. Roots contain 5 % saponins, sugars, saporubin, and saporubinic acid. Leaves contain the glycoside saponarin (Ogolevitz 1951; Morita et al. 1997a, b; Sang et al. 2000, 2003).

▲**Tussilago farfara L.** Photos: *left* and *right*: Alexander Naumenko; *center*: Evgeny Davkaev

▼**Ungernia victoris Vved. ex Artjushenko** Photos: Alim Gaziev

▲**Urtica dioica L.** Photos: Sergey Appolonov

▼**Vaccaria hispanica (Mill.) Rauschert** Photos: Andrei Lubchenko

Valeriana officinalis L. – Valerianaceae

Synonyms: *Valeriana baltica* Pleijel, *Valeriana exaltata* Mikan fil., *Valeriana palustris* Kreyer.

English name: Valerian, Garden valerian, Garden heliotrope

Russian name: Валериана лекарственная (Valeriana lekarstvennaya)

Uzbek name: Asaroon

Kyrgyz name: Дары мышык тамыр (Dary myshyk tamyr)

Description: Herbaceous perennial, with short rhizomes. Stems single or few, 50–150 cm tall, hollow, furrowed. Leaves opposite, 7–25 cm long, odd-pinnately compound with 6–8 pairs of leaflets, lower leaves petiolate; leaflets ovate-lanceolate or almost linear, entire to dentate. Inflorescence corymbiform or paniculiform, apical. Flowers perfect. Corolla funnelform with 5 lobes, white or pale-lilac. Stamens 3. Fruits flattened achenes, 2–3 mm long.

Other distinguishing features: Calyx initially small, later enlarged with plumose, pappus-like segments. Roots have a strong, specific smell.

Phenology: Flowers in May, fruits in July.

Reproduction: By seeds and division of rhizomes.

Distribution: All provinces of Kyrgyzstan; cultivated in Uzbekistan.

Habitat: Cultivated.

Population status: Common.

Traditional use: Valerian is used as a sedative, carminative, and vermifuge, as an aid in digestion, and to treat hypercondria, psychological traumas, hysteria, migraines, convulsive pains, heart pains, heart failure, epilepsy, insomnia, and anxiety (Turova and Sapozhnikova 1984; Altimishev 1991). A decoction or tincture is used as a heart remedy, a sedative to treat nervous disorders, as well as to treat headaches, and cancer, and to improve the appetite. It is used in a bath to relax hyperactive children so they sleep well and to treat hysteria, convulsions, acute typhus, epilepsy, and internal aches (Kurochkin 1998).

Documented effects: Preparations of valerian influence the nervous system and have sedative effects as well as antispasmodic actions (Kurochkin 1998). They are used to treat insomnia, neurosis of the cardiovascular system and to treat spasms of the gastrointestinal tract. In acute and chronic experiments with dogs given valerian infusion intravenously and orally, arterial pressure was decreased (only when applied intravenously) and the speed of blood coagulation was increased (Akopov 1990). Valepotriates suppress aggression, have anticonvulsant effects against pentylenetetrazol- and strychnine-induced seizures, increase thiopental-induced sleeping time, reduce motility and have dose-dependent sedative effects. The sesquiterpenes reduce locomotion and increase pentobarbital and hexobarbital-induced sleeping time of mice. Some sesquiterpenes, especially valerenic acid, influence serotonin and noradrenaline levels (Ortiz et al. 1999).

Phytochemistry: Underground parts contain essential oil with sesquiterpenes, iridoids, etc. (including bornyl-isovalerianate, didrovaltrate, valtrate, acevaltrate, isovaltrate, valerenic and iso-valerianic acid, borneol, myrtenol, myrtenyl isovalerianate, camphene, α-pinene, d-terpeneol, limonene, alcohols, etc.), alkaloids (valerine, chatinene, etc.), glycosides (valeride), tannins, sugars, acids (formic, acetic, malic, stearic, palmitinic, etc.) and macro- and micro-elements (Akopov 1990; Bos et al. 1998; Kurochkin 1998; Ortiz et al. 1999).

Veratrum lobelianum Bernh. – Melanthiaceae

Synonyms: *Veratrum album* ssp. *lobelianum* (Bernh.) Schuebl. & Martens, *Veratrum album* ssp. *virescens* (Gaudin) Jav. & Soo, *Veratrum album* var. *lobelianum* (Bernh.) Koch, *Veratrum album* var. *virescens* Gaudin.

English name: Unknown

Russian name: Чемерица Лобеля (Chemeritsa Lobelya)

Uzbek name: Maralkulok

Kyrgyz name: Лобел марал кулагы (Lobel maral kulagy)

Description: Herbaceous perennial, with short rhizomes. Stem single, erect, 2–3 cm in diameter, 70–170 cm tall. Leaves cauline, alternate, simple, sheathing the stem, prominently veined, margins entire; lower leaves wide-elliptic, 15–25 cm long, 10–15 cm wide; upper leaves smaller, lanceolate. Inflorescence an apical panicle, 20–60 cm tall. Flowers with 6 white-green tepals and 6 stamens. Fruit an ovoid capsule, 3-lobed. Seeds flat, elliptical, broad-winged, 6–10 mm long.

Other distinguishing features: Ovary superior.

Phenology: Flowers in June-July, fruits in July-September.

Reproduction: By seeds and rhizomes.

Distribution: Ysyk-Kol province of Kyrgyzstan; not found in the flora of Uzbekistan.

Habitat: In wet meadows, bogs and along rivers.

Population status: Common, forming dense groups.

Traditional use: The underground parts infused in cream is recommended to treat eczema (Khalmatov et al. 1984). This plant is used to treat mental illness and is used externally to treat joint rheumatism and neuralgia (Altimishev 1991). A tincture of the rhizome is used as a hypotensive in Bulgarian folk medicine (Ivancheva and Stantcheva 2000).

Documented effects: A preparation of this species has insecticidal activity. In medicine an alcoholic or water infusion is used externally on skin parasites and scabies. An alcohol infusion, decoction, and ointment prepared with the underground parts are used as an analgesic to treat neuralgia, arthritis, rheumatism, and common colds. Alkaloids isolated from this species have hypotensive and analgesic abilities. Because of the high toxicity of the alkaloids they are not widely used in medicine (Khalmatov et al. 1984; Kurochkin 1998). The alkaloid jervine isolated from this plant effected fibroblasts and isolated animal organs in vitro in a similar manner as seratonin (Suladze et al. 2006).

Phytochemistry: The plant contains alkaloids (jervine, pseudojervine, rubijervine, isorubijervine, etc.), tannins, resins, sugars, and pigments (Bondarenko 1972; Khashimov et al. 1970; Tolmachev 1976; Shakirov et al. 1995; Suladze et al. 2006).

Verbascum songaricum Schrenk – Scrophulariaceae

Synonyms: *Verbascum khorassanicum* Boiss., *Verbascum lychnitis* L., *Verbascum polystachyum* Kar. & Kir.

English name: Songar mullein

Russian name: Коровяк джунгарский (Korovyak dzhungarskiy)

Uzbek name: Sigir kuyruq

Kyrgyz name: Жунгар аюу кулагы (Zhungar ayuu kulagy)

Description: Herbaceous biennial, densely stellate hairy. Stem 40–150 cm high, erect, foliaceous, branched on top. Basal leaves lanceolate to oblanceolate, 15–40 cm long, 4–12 cm wide, base attenuate, margin nearly entire, grayish-hairy on both sides, nearly sessile to petiolate; stem leaves lanceolate to oblong, sessile; base of upper leaves subcordate. Inflorescence pyramidal-paniculate, 20–40 cm long. Flowers in bunches of 4–7, pedicellate. Calyx 4–10 mm long with 5 deep linear-lanceolated lobes, whitish-hairy. Corolla yellow, 1.5–3 cm in diameter, 5-lobed. Stamens 5. Fruit a wide-ovoid capsule, 5–8 mm long, densely hairy. Seeds tiny, obconic-prismatic, 0.7–0.9 mm long, 0.6 mm wide, linearly pitted.

Other distinguishing features: Staminal filaments are coated with whitish hairs.

Phenology: Flowers in June-August, fruits in July-September.

Reproduction: By seeds.

Distribution: Toshkent, Farg'ona, Samarqand, and Surxondaryo provinces of Uzbekistan; all of Kyrgyzstan.

Habitat: The adyr and tau zones. Dry slopes of foothills.

Population status: Common, found as single individuals.

Traditional use: The plant is used to heal wounds. Thoroughly boiled leaves are put on burns, tumors and wounds, and the fresh leaf juice is applied on the surface of wounds. It is also used for toothaches, eye inflammations, and as an expectorant to relieve chronic cough. A decoction of the flowers is used to treat stomach and intestinal catarrh and gall bladder and liver inflammation (Seredin and Sokolov 1969).

Documented effects: An infusion of the flowers in water is used as an expectorant. Decoctions of the leaves and flowers of this species, as well as the related species *Verbascum thapsus*, *V. phlomoides*, and *V. thapsiforme* are used as an expectorant and to coat and sooth the mouth and throat to reduce effects of catarrh and coughs (Seredin and Sokolov 1969). Phenylethanoid glycosides isolated from a methanolic extract of the plant inhibited mammalian DNA polymerases (Iida et al. 2003).

Phytochemistry: The whole plant contains alkaloids, including anabasine and plantagonine, saponins, triterpenoid saponins, and vitamin C (Khodzhimatov 1989; Seifert et al. 1991; Hartleb and Seifert 1995). The aboveground parts contains saponins with a hemolytic index of 1:250 (Khalmatov 1964).

Verbascum thapsus L. – Scrophulariaceae

Synonyms: None

English name: Common mullein

Russian name: Коровяк обыкновенный (Korovyak obyknovennyy)

Uzbek name: Unknown

Kyrgyz name: Аю кулак (Ayu kulak)

Description: Herbaceous biennial, densely felted-hairy. Stem thick, leafy, up to 2 m tall. Lower leaves in a basal rosette, petiolate, oblong or oblanceolate, up to 30 cm long, up to 5–10 cm wide, usually entire; upper leaves alternate, becoming smaller, sessile, decurrect on stem to next leaf below. Inflorescence a dense, apical, spiciform raceme, appearing in the second year. Calyx deeply 5-lobed. Corolla yellow, 1–2.5 cm in diameter, 5-lobed, the lower 3 lobes slightly longer than the upper 2. Fruit a septicidal capsule with 2 valves. Seeds small, furrowed.

Other distinguishing features: Stamens 5, upper 3 shorter than the lower 2.

Phenology: Flowers in May-June, fruits in July-August.

Reproduction: By seeds.

Distribution: Jalal-Abad, Osh, and Chuy Provinces of Kyrgyzstan; not found in the flora of Uzbekistan.

Habitat: In fallow fields and pastures and along canals.

Population status: Common, found in loosely arranged groups.

Traditional use: A decoction of the herb is used to treat neurosis and epilepsy, as a diuretic to treat kidney stones, and gout and swelling due to kidney and heart problems. It is used externally to treat throat diseases, neuralgia of facial nerves, in a bath to treat hemorrhoids, scrofula, and rickets, and as a compress or lotion to heal wounds and treat eye diseases. A decoction of the roots and leaves is used to treat diarrhea. An infusion and decoction of the leaves and flowers is used as an expectorant, anti-inflammatory, demulcent and coating to treat acute respiratory diseases, pneumonia, bronchial asthma, gastritis and liver and gall bladder diseases (Plant Resources of the USSR 1990).

Documented effects: Extracts of the plant exhibited varying antibacterial activity against *Klebsiella pneumonia*, *Staphylococcus aureus*, *Staphylococcus epidermidis*, and *Escherichia coli*, inhibited *Agrobacterium tumefaciens*-induced tumors in vitro, and had antiviral activity (McCutcheon et al. 1995; Turker and Camper 2002).

Phytochemistry: The plant contain polysaccharides, iridoid glycosides (harpagoside, harpagide, and aucubin) flavonoids (3-methylquercetin, hesperidin, and verbascoside) saponins, essential oil, steroids, mucilage, etc. (Turker and Gurel 2005).

▲**Valeriana officinalis L.**
Photos: Vadim Prokhorov

▲**Veratrum lobelianum Bernh.**
Photos: *top*, *upper center* and *lower center*: Vadim Prokhorov; *bottom*: Sergey Appolonov

▲**Verbascum songaricum Schrenk** Photos: Alim Gaziev
▼**Verbascum thapsus L.**
Photo: Sasha Eisenman

Vexibia pachycarpa (Schrenk ex C.A. Mey.) Yakovlev – Fabaceae

Synonyms: *Goebelia pachycarpa* (Schrenk ex C.A. Mey.) Bunge ex Boiss., *Sophora pachycarpa* Schrenk ex C.A. Mey.

English name: Unknown

Russian name: Вексибия толстоплодная (Veksibiya tolstoplodnaya)

Uzbek name: Achykmiya

Kyrgyz name: Unknown

Description: Herbaceous perennial, 30–60 cm tall, stems branching from the base, densely covered with short, white hairs. Leaves alternate, compound, odd-pinnate, 10–18 cm long; leaflets in 6–12 pairs, elliptic or oblong, 1.5–2 cm long, 3–8 mm wide, both sides white-hairy. Inflorescences cylindrical, apical racemes. Calyx wide-campanulate, with wide-triangular teeth, densely hairy. Corolla papilionaceous, white to creamy-yellow colored, up to 1.5 cm long. Fruits club-shaped legumes, 3–6 cm long, 7–9 mm wide, with varying amounts of pubescence, legumes oriented vertically. Seeds slightly kidney-shaped to oval, deep-brown, glabrous.

Other distinguishing features: Legumes slightly constricted between seeds, with elongated, conical tip.

Phenology: Flowers in April-May, fruits in June-July.

Reproduction: By seeds and rhizomes.

Distribution: Toshkent, Farg'ona, Samarqand, and Buxoro provinces of Uzbekistan; found in some provinces of Kyrgyzstan.

Habitat: The chul and adyr zones. On river banks, in long-fallow fields, on loess hills, sandy soils, and as a weed in unirrigated wheat fields.

Population status: Common, usually occurs in small populations or as single individuals.

Traditional use: The ground seeds are recommended for loss of appetite. An decoction of the aboveground plant parts is used to treat skin diseases (eczema, fungal, and scabies) and as a spasmolytic, analgesic, and vermifuge (Khalmatov et al. 1984; Khodzhimatov 1989; Mamedov et al. 2004).

Documented effects: Only pachycarpine is used in medical practice. Pachycarpine is a ganglionic blocking agent and is used internally for hypertension strokes, peripheral vessels spasms (endarteritis, intermittent claudication), for myopathy and to stimulate labor during child birth. Dermatologists use pachycarpine preparations for scleroderma, idiopathic skin atrophy, and to treat chronic eczema (Mashkovskii 1984).

Phytochemistry: The aboveground parts contain up to 3 %, and seeds up to 2.2 %, total alkaloids. Plants from Kenimess massif (Buxoro province, Uzbekistan) contained 3.90–6.4 % (aboveground) and 1.5–2.98 % (roots) total alkaloids. The main alkaloids are pachycarpine, sophocarpine, matrine, and sophoramine. Pachycarpidine, quercetin, kaempferol, and genistein and its xyloglucoside have also been isolated. The roots contained 9–12 % (and the root bark 22–25 %) phenolic pigments, flavonoids, steroid glucosides, etc. (Yunusov 1981; Botirov et al. 2006; Muminova et al. 2006; Emami et al. 2007).

Vicia cracca L. – Fabaceae

Synonyms: *Vicia hiteropus* Freyn, *Vicia lilacina* sensu B. Fedtsch., *Vicia macrophylla* (Maxim.) B. Fedtsch.

English name: Bird vetch, cow vetch, tufted vetch

Russian name: Горошек мышиный (Goposhek myshinyy)

Uzbek name: Unknown

Kyrgyz name: Жапайы жер буурчак (Zhapayy zher buurchak)

Description: Herbaceous perennial vine. Stems trailing or climbing, pubescent. Leaves pinnate, usually with 5–11 pairs of leaflets, a tendril replacing the terminal leaflet; leaflets linear-lanceolate to narrowly oblong, 1.5–3 cm long, 4–8 mm wide, apex mucronate. Inflorescence a long-peduncled one-sided raceme, many-flowered. Calyx campanulate, pink or bluish, lobes unequal. Corolla papilionaceous, blue-violet or rarely white. Fruits elongate-lanceolate legumes, 15–20 mm long. Seeds dark-brown, spherical.

Other distinguishing features: Legumes glabrous.

Phenology: Flowers in June-July, fruits in August-September.

Reproduction: By seeds.

Distribution: Ysyk-Kol, Naryn, Talas, and Chuy provinces of Kyrgyzstan; not found in the flora of Uzbekistan.

Habitat: In tall-grass meadows, among bushes, in forests, along canals and in floodplains.

Population status: Common, found in small groups.

Traditional use: The aboveground parts are used as a demulcent, hemostatic and to heal wounds. A tincture is used to treat diarrhea and as a diuretic. A poultice is used to treat rectal prolapse and prolapsed hemorrhoids. The crushed, dry or fresh herb is applied as a compress to treat abscesses. In the Bryansk area of Russia a decoction of the roots is used to treat hepatitis. In the Tibetan medicine, the aboveground parts are used to treat swelling, ascites, and as a hemostatic (Plant Resources of the USSR 1987).

Documented effects: Lectins, isolated from this species, show bonding specificity with human blood type A (Sharon and Lis 2004). In experiment on animals, an infusion and decoction of the plant had antibacterial activity (Plant Resources of the USSR 1987).

Phytochemistry: The aboveground parts contains the non-protein, amino acid canavanine (Enneking 1995), cyanogenic glycosides (vicianine), hydrocyanic acid, and vitamin C. The leaves contain vitamins C, P and carotene. The flowers contain vitamins, flavonoids and anthocyans (Savoskin et al. 1971; Shreter 1975).

Vinca erecta Regel & Schmalh. – Apocynaceae

Synonyms: None

English name: Unknown

Russian name: Барвинок прямостоящий, Барвинок прямой (Barvinok pryamostoyashchiy, Barvinok pryamoy)

Uzbek name: Burygul

Kyrgyz name: Туз бору гул (Tuz boru gul)

Description: Herbaceous perennial with horizontal, woody, scale-covered rhizomes. Stems many, erect, 15–50 cm high, unbranched, glabrous or hairy. Leaves opposite, densely arranged, sessile; lower leaves simple, glabrous or pubescent, 1–2 cm long, up to 7 mm wide, apex obtuse or rounded; upper leaves ovate to wide-lanceolate, 2.5–5.5 cm long, 1.2–3 cm wide, apex acute. Flowers single, axillary, pedicellate. Corolla 2–2.5 cm long, pale lilac outside, white inside with dark-violet tube, glabrous. Fruits composed of 2 linear-cylindrical follicles, 3–6.5 cm long, brown, densely covered with large tubercles. Seeds 1.1–1.6 cm long, 2–3 mm wide, light-brown.

Other distinguishing features: Follicles have 1–7 seeds, usually 3–4.

Phenology: Flowers and fruits in May-August.

Reproduction: By seeds and rhizomes.

Distribution: Toshkent, Farg'ona, and Surxondaryo provinces of Uzbekistan; Osh and Jalal-Abad provinces of Kyrgyzstan.

Habitat: The adyr and tau zones. Stony slopes, shale taluses in foothills, on rocks.

Population status: Uncommon, found in small populations.

Traditional use: In mountain zones where these plants are growing, local populations use decoctions and infusions of the aboveground parts to treat febrile diseases. A decoction of the roots is used as an emetic (Kurmukov 1970). The leaves are used in a tea to treat diarrhea and gastrointestinal disorders, headaches and dizziness, and as a mouthwash for toothaches. The fresh leaves are applied to wounds. A powder is used externally as an astringent and to heal wounds (Khalmatov et al. 1984; Khodzhimatov 1989).

Documented effects: The total alkaloids of the aboveground parts have different actions at different doses. In low and middle doses they act as a sedative and at major doses they have a stimulating effect on the central nervous system. The preparation, *Vinsumine*, has antispasmodic, adrenolytic and ganglion blocking (ganglions of cardiac branches of vagus nerves) actions, it changes signals from the carotid and sciatic nerves which influence arterial pressure, and it releases and prevents cardiac arrhythmia caused by electric heart stimulation and by intravenous introduction of 10 % calcium chloride solution. All effects of *Vinsumine* are due to the alkaloids it contains (Kurmukov and Sultanov 1965; Kurmukov 1967a, b). The alkaloids akuamidine, tombozine, and ervine have α-adrenolytic action. Ervine shows pronounced anti-fibrillating action for cardiac arrhythmia (Kurmukov 1968b, 1970, 1975, 1978). Ervamine had tranquilizing and hemostatic effects. Intravenous injection had a brief hypotensive effect, increased the amplitude and decreased the frequency of heart contractions, and also increased the coronary blood flow. Ervamine also had a substantial effect on the smooth muscle of the uterus both in vivo and in vitro (Sultanov and Kurmukov 1965). The alkaloid ervinine is an analeptic of the central nervous system with primary influence on respiration and stimulates reticular formations of the medulla oblongata and midbrain due to stimulation of adrenergic structures (Saidkasimov 1960; Kurmukov and Saidkasymov 1968; Kurmukov and Saidkasimov 1969; Kurmukov 1970). The alkaloid vincamine has stimulatory action on uterine unstriped muscles and stimulates contractions in weak labors. It was used in obstetrics under the preparation name *Vikametrin*. The alkaloid vincarine has anti-arrhythmic action and is not inferior to aimaline (Khanov et al. 1968, 1972; Kurmukov 1968a, 1970; Kurmukov and Sultanov 1971). The alkaloid vincanine is a strychnine-like spasmodic and analeptic of the central nervous system (Sultanov 1959b; Shamansurov and Sultanov 1967). The main effect of the alkaloid vincanidine is an apomorphine-like emetic action (Sultanov 1959a, 1960). Vinervinine suppressed central nervous system activity in mice and its effect on blood pressure and respiration was similar to that of acetylcholine (Kurmukov 1967b).

Phytochemistry: The following alkaloids have been obtained from the aboveground plant parts collected in different areas of Uzbekistan: vincamine, ervamine, ervinine, ervine, vinervine, vinervinine, akuamine, akuamidine, reserpinine, isoreserpiline, and vincamine. Alkaloids, including vincanine and vincanidine, have been isolated from the roots. More than 60 other alkaloids have been isolated from this species (Yunusov 1981; Yagudaev et al. 1983).

Viola suavis M. Bieb. – Violaceae

Synonyms: *Viola pontica* W. Beck.

English name: Russian violet

Russian name: Фиалка приятная (Fialka priyatnaya)

Uzbek name: Gunafsha

Kyrgyz name: Жагымдуу ала гул (Zhagymduu ala gul)

Description: Herbaceous perennial, with a short rhizome and short, stout stolons. Leaves in a rosette, obovate to broad-ovate, base cordate, long-petiolate; spring leaves 3–8 cm long; summer leaves up to 20 cm long, margins dentate. Flowers solitary. Sepals 5. Sepals 5, violet with white throat; lower petal with spur. Fruit a spherical capsule, glabrous or pubescent. Seeds with conspicuous elaiosomes.

Other distinguishing features: Stipules free, lanceolate, long-fimbriate. This species also produces cleistogamous flowers.

Phenology: Flowers in April, fruits in May.

Reproduction: By seeds and stolons.

Distribution: Cultivated in Kyrgyzstan and Uzbekistan.

Habitat: Found escaped into the wild.

Population status: Common.

Traditional use: A syrup made from the aboveground parts is used as a diuretic, anti-inflammatory, expectorant, diaphoretic, and choleretic. A decoction is used to treat coughs, sinus colds and illnesses of the eyes, throat, and stomach. The roots are used as an emetic and laxative. In Turkmenistan and the Caucasus a decoction of the flowers with sugar is used to treat heart illnesses (Plant Resources of the USSR 1986).

Documented effects: Unknown.

Phytochemistry: Aboveground parts have essential oil and vitamin C (Plant Resources of the USSR 1986).

▲**Vexibia pachycarpa (Schrenk ex C.A. Mey.) Yakovlev**
Photos: Evgeny Davkaev

▲**Vicia cracca L.**
Photos: *top*: Evgeny Davkaev; *center*: Rostislav Lezhoyev; *bottom*: Radu Chibzii

▲**Viola suavis M. Bieb.**
Photos: Maxim Zaitsev

▼**Vinca erecta Regel & Schmalh.** Photo: Authors

Xanthium strumarium L. – Asteraceae

Synonyms: *Xanthium americanum* Walter, *Xanthium cavanillesii* Schouw, *Xanthium chasei* Fernald, *Xanthium chinense* Mill., *Xanthium curvescens* Millsp. & Sherff, *Xanthium echinatum* Murray, *Xanthium echinellum* Greene ex Rydb., *Xanthium globosum* C. Shull, *Xanthium inflexum* Mack. & Bush, *Xanthium italicum* Moretti, *Xanthium natalense* Widder, *Xanthium orientale* L., *Xanthium oviforme* Wallr., *Xanthium pensylvanicum* Wallr., *Xanthium pungens* Wallr., *Xanthium speciosum* Kearney, *Xanthium varians* Greene, *Xanthium wootonii* Cockerell.

English name: Common cocklebur, Rough cocklebur

Russian name: Дурнишник обыкновенный (Durnishnik obyknovennyy)

Uzbek name: Guzatkon, Patanak

Kyrgyz name: Кадимки манкоо (Kadimki mankoo)

Description: Herbaceous annual, with taproot. Stems 20–200 cm tall, branched, appressed hairy or subglabrous. Leaves alternate, long-petiolate, broadly ovate to suborbicular, shallowly 3–5-lobed, irregularly dentate. Inflorescences unisexual heads. Staminate heads many-flowered with highly reduced involucre, heads in a terminal cluster. Pistillate heads in short axillary clusters, heads cylindric to ovoid, 1–3.5 cm long, 2-flowered, enclosed by involucre forming a bur (false-fruit) with curved prickles. Fruits thick achenes with no pappus.

Other distinguishing features: Leaves broad, no spines in the axils.

Phenology: Flowers in June-July, fruits in July-September.

Reproduction: By seeds.

Distribution: All provinces of Kyrgyzstan and Uzbekistan.

Habitat: Near roads, canals, waste places, sandy riverbanks, and in agricultural fields.

Population status: Common.

Traditional use: In folk medicine a decoction of the seeds and roots is used to treat dysentery, scrofula, and bladder diseases. A tincture of the entire plant in vodka is drunk to treat goiters, rheumatism and common colds, and inflammatory diseases and is also used as a diaphoretic, antipyretic, and sedative. A tea made from the entire plant is used to treat cancer. An infusion is used to stimulate digestion, to treat intestinal atonia, stomach spasms, liver inflammation, jaundice, acute and chronic bronchitis, pertussis, painful menstruation, kidney stones, goiters, cancer, and to reduce sexual excitability. It is used externally in dry and damp compresses and aromatic baths. Fruits and seeds are used to treat eczema, itchy dermatosis, insect stings, and paralysis. A decoction of the root is used externally to treat skin diseases and furunculosis. A decoction of the entire plant is applied to the face after shaving, especially on pimples and fungal skin diseases (Maznev 2004).

Documented effects: An extract of the leaves exhibited trypanocidal activity in vitro and in vivo (Talakal et al. 1995). Rodents treated with an extract of the plant exhibited alterations in behavior patterns that suggested the extract had significant depressing activity on the central nervous system (Mandal et al. 2001). An extract of the plant showed slight activity against *Candida albicans* (Murillo-Alvarez et al. 2001). Caffeic acid isolated from the fruits induced a dose-dependent decrease of plasma glucose in streptozotocin-induced and insulin-resistant diabetic rats (Hsu et al. 2000).

Phytochemistry: The entire plant contains iodine. The leaves contain alkaloids, ascorbic acid, essential oil (with limonene, carveol, and α-ionone being the major constituents), sesquiterpenoids (xanthanine, xanthanol, xanthosine, xanthamine, xanthinine, xanthumanol, and xanthinosin), phenolic acids (caffeic), chalcones, tannins, steroids, (β- and ε-sitosterin) saponins and carotenoids. The fruits contain drying fatty oil, resins, flavonoids, alkaloids and the glycoside xanthostrumarin (Khodzhimatov 1989; Marco et al. 1993; Belodubrovskaya et al. 2002).

Ziziphora bungeana Juz. – Lamiaceae

Synonyms: Some authors consider this species synonymous with *Ziziphora clinopodioides* Lam., *Ziziphora clinopodioides* ssp. *bungeana* (Juz.) Rech. f.

English name: Unknown

Russian name: Зизифора Бунге (Zizifora Bunge)

Uzbek name: Kiyik ut

Kyrgyz name: Кокомерен (Kokomeren)

Description: Perennial subshrub, with woody roots. Stems many, 12–30 cm tall, branched, bases woody, densely retrorse pubescent towards apex. Leaves opposite, simple, short-petiolate, 5–15 mm long, 1.5–6 mm wide, narrowly lanceolate to ovate-lanceolate, glandular, margins entire. Inflorescences verticillasters, crowded into semiglobose, terminal heads. Calyx tubular. Corolla pink, 2-lipped; upper lip entire; lower lip 3-lobed. Fruits smooth, ovoid nutlets.

Other distinguishing features: Plant has a strong smell. Two longer, fertile stamens, reaching upper corolla lip, and two reduced or absent stamens.

Phenology: Flowers in July, fruits in August.

Reproduction: By seeds.

Distribution: Jalal-Abad, Ysyk-Kol, Naryn, and Chuy provinces of Kyrgyzstan; Toshkent province of Uzbekistan.

Habitat: On stony slopes.

Population status: Common, found in small groups.

Traditional use: Extracts and infusions of the aboveground parts are recommended for hypertonia, for cardiac and climacteric neurosis, rheumacarditis with poor blood circulation, and rheumatic endomyocarditis of children in the active phase of illness. A decoction of the leaves is used to treat gastric colic, nausea, to stimulate the appetite, and as a diuretic. It is used externally to treat throat illnesses in children. An extract of the flowers is used to treat gastritisis, frequent vomiting, and meteorism (Dobrokhotova and Chudinov 1966; Plant Resources of the USSR 1991).

Documented effects: In experiments on animals, an extract, infusion and decoction possessed hemostatic properties, raised the activity of respiratory enzymes during hypoxia, had positive influence on collateral coronary blood flow and showed prophylactic activity for, and effective treatment of, myocardial infarctions and myocarditis. In experiments, the total alkaloids showed cardiotonic properties. The preparation *Ziziphorine* has antiarrhythmic properties on model ventricular arrhythmia in dogs, and has cardiotonic and hypotensive actions (Plant Resources of the USSR 1991). The essential oil exhibited antibacterial activity against *Staphylococcus epidermidis*, *S. aureus*, *Escherichia coli*, and *Bacillus subtilis* (Sonboli et al. 2006).

Phytochemistry: The roots contain organic acids, essential oils, saponins, alkaloids, vitamin C, flavonoids, and tannins. Aboveground parts contain essential oils, triterpenoids, alkaloids, flavonoids, and tannins (Dobrokhotova and Chudinov 1966). The essential oil contains over 32 components with pulegone, isomenthone, 1,8-cineole and piperitenone as the main constituents (Sonboli et al. 2006).

Ziziphora clinopodioides Lam. – Lamiaceae

Synonyms: *Ziziphora afghanica* Rech. f., *Ziziphora borzhomica* Juz., *Ziziphora brevicalyx* Juz., *Ziziphora bungeana* Juz., *Ziziphora clinopodioides* ssp. *afghanica* (Rech. f.) Rech. f., *Ziziphora clinopodioides* ssp. *bungeana* (Juz.) Rech. f., *Ziziphora denticulata* Juz., *Ziziphora dzhavakhishvilii* Juz., *Ziziphora turcomaica* Juz.

English name: Unknown

Russian name: Зизифора пахучковидная (Zizifora pakhuchkovidnaya)

Uzbek name: Kiyik ut

Kyrgyz name: Кокомерен (Kokomeren)

Description: Perennial subshrub, with woody roots. Stems many, 8–40 cm tall, bases woody, rarely branched, densely retrorse pubescent towards top. Leaves opposite, simple, petiolate, 6–25 mm long, 3–12 mm wide, broadly elliptic, ovate or elongate-ovate, glandular, margins entire or slightly toothed. Inflorescences are verticillasters, crowded into semiglobose, terminal heads. Calyx tubular. Corolla lilac, 2-lipped; upper lip entire; lower lip 3-lobed. Fruits smooth, ovoid nutlets.

Other distinguishing features: Plant has a strong smell. Two longer, fertile stamens, reaching upper corolla lip, and two reduced or absent stamens.

Phenology: Flowers in June, fruits in August.

Reproduction: By seeds.

Distribution: Naryn and Chuy provinces of Kyrgyzstan; Toshkent, Jizzax, Samarqand, Qashqadaryo, and Surxondaryo provinces of Uzbekistan.

Habitat: On stony slopes of mountains and gorges in spruce forests and the subalpine zone.

Population status: Common, found in small groups.

Traditional use: In Kyrgyzstan, an infusion and decoction is used to treat tachycardia, gastralgia, and heart illnesses with swelling. Juice from the plant is used as a vermifuge for pinworm in children (Alimbaeva and Goncharova 1971). In the Altai region of Russia, a tincture is used to treat common colds, rheumatism, and scrofula and it is used externally to treat toothaches. In Indian medicine an infusion of the leaves is used as an antipyretic and a decoction is used to treat typhoid fever (Plant Resources of the USSR 1991).

Documented effects: A tincture of the herb possesses hypotensive, cardiotonic, and antihelminthic properties. An 8 and 10 % water solution of the total flavonoids possesses hypotensive properties (Alimbaeva and Goncharova 1971). The essential oil shows antibacterial and fungicidal activity (Delova and Guskova 1974). In experiments with mice, pretreatments with extracts of the plant reduced the biochemical, macro-, and microscopic effects of induced inflammatory bowel disease (Ghafari et al. 2006). Extracts the plant showed significant antibacterial activity against *Staphylococcus aureus* and *Pseudomonas aeruginosa*. The essential oil showed antibacterial activity against *Staphylococcus epidermidis*, *S. saprophyticus*, *Escherichia coli*, *Shigella flexneri*, and *Salmonella typhi* (Fazly Bazzaz and Haririzadeh 2003; Tabatabaei-Anaraki et al. 2007).

Phytochemistry: The aboveground parts, collected during flowering stage, contained essential oil with limonene, menthone, isomethone, isomenthol, and pulegone as the main constituents (Korolyuk et al. 2002). Twenty-six compounds were isolated from the essential oil of plant material collected in Iran. The major components were pulegone and piperitenone (Mohammadreza 2008). In other collections from Iran the main compounds were: thymol, p-cymene and carvacrol, or 1,8-cineole and terpinen-4-ol (Tabatabaei-Anaraki et al. 2007). The aboveground parts also contain saponins, coumarins, and flavonoids. The seeds contain fatty oil (palmitic, oleic, stearic, linoleic, and linolenic; Plant Resources of the USSR 1991).

Ziziphora pedicellata Pazij & Vved. – Lamiaceae

Synonyms: None

English name: Unknown

Russian name: Зизифора цветоножечная (Zizifora tsvetonozhechnaya)

Uzbek name: Kiik ut

Kyrgyz name: Гулсапчалуу кокомерен (Gulsapchaluu kokomeren)

Description: Perennial herb. Stems 20–40 cm tall, numerous, slightly winding. Leaves opposite, short-petiolate, lanceolate, glabrous or with short, spreading hairs. Flowers with long, hairy pedicels, in verticillasters crowded into head-like inflorescences. Calyx tubular, hairy; slightly 2-lipped, upper lip with 3 teeth, lower lip with 2 teeth. Corolla light-violet, 2-lipped, upper lip entire, lower lip 3-lobed, tube surpassing the calyx. Fruits smooth nutlets, almost brown.

Other distinguishing features: When rubbed, leaves and flowers produce a strong menthol aroma.

Phenology: Flowers in June-August and fruit July-September.

Reproduction: By seeds.

Distribution: Toshkent province of Uzbekistan; Western Tien-Shan; not found in Kyrgyzstan.

Habitat: The tau zone. Stony slopes with rocky debris.

Population status: Common.

Traditional use: In Central Asia and Kazakhstan a tincture and decoction of the aboveground parts are used as a diuretic and the fresh ground plant is used to heal wounds. In Uzbekistan an infusion of the herb, taken as a tea, is used as a hypotensive and to treat headaches (Khalmatov 1964; Gusakova and Khomova 1997; Sezik et al. 2004).

Documented effects: In pharmacological studies, infusions, tinctures and liquid extracts of this plant had positive effects on myocarditis and myocardial infarction. The same preparations acted as a cardiotonic, decreased arterial pressure, and increased diuresis (Khalmatov 1964).

Phytochemistry: The plant contains essential oil composed of pulegone, pinene, menthol, menthone, isomenthone, alcohols, and other substances. The seeds and leaves contain carotenoids and lipids. The flowers contain terpenes (Khalmatov 1964; Gusakova and Khomova 1997).

►Xanthium strumarium L.
Photos: *center*: Stan Shebs; *left* and *right*: Clinton Shock

▼Ziziphora clinopodioides Lam.
Photos: Evgeny Davkaev

▼Ziziphora pedicellata Pazij & Vved.
Photos: Alim Gaziev

Ziziphora tenuior L. – Lamiaceae

Synonyms: *Faldermannia parviflora* Trautv.

English name: Unknown

Russian name: Зизифора тонкая (Zizifora tonkaya)

Uzbek name: Chul yalpiz

Kyrgyz name: Ичке кокомерен (Ichke kokomeren)

Description: Annual herb. Stems erect, unbranched or branching from the base, 5–30 cm tall, curly-hairy. Leaves opposite, linear-lanceolate to lanceolate, apex acuminate, base attenuate, the edges and abaxial side along veins curly-hairy, margin entire, short-petiolate; upper leaves ciliate. Inflorescences axillary verticillasters, usually 2–6-flowered, arranged into spikes. Calyx tubular, slightly curving downwards, obscurely 2-lipped, spreading-hairy, upper lip 3-toothed, lower lip 2-toothed. Corolla light violet, 2-lipped; upper lip entire; lower lip 3-lobed, spreading; tube noticeably protruding past the calyx. Fruits oblong-linear nutlets, 1.5 mm long, 3-edged, brown.

Other distinguishing features: Leaves in the inflorescences much longer than the flowers. Plant produces a strong menthol aroma when crushed.

Phenology: Flowers in May-June, fruits in June-August.

Reproduction: Only by seeds.

Distribution: All provinces of Uzbekistan and Kyrgyzstan.

Habitat: The chul, adyr, and tau zones. Clay and stony soils.

Population status: Common in *Artemsia*-ephemeral complexes, sometimes makes small populations.

Traditional use: A decoction of the herb is recommended for intestinal diseases, diarrhea, children's colitis, neurasthenia and for maintaining cardiac activity (Khalmatov 1964).

Documented effects: Pulegone, which is isolated from the essential oil, is reduced to produce menthol (Ogolevitz 1951). Extracts of the plant exhibited antifungal and antibacterial activity (Sardari et al. 1998; Tajadod and Majd 2007).

Phytochemistry: Plants contain 0.3–1 % essential oils which consist of 75–87.1 % pulegone (Khalmatov 1964; Salehi et al. 2005).

Ziziphus jujuba Mill. – Rhamnaceae

Synonyms: *Rhamnus zizyphus* L., *Ziziphus sativa* Gaertn., *Ziziphus vulgaris* Lam.

English name: Jujube, Chinese date

Russian name: Унаби (Unabi)

Uzbek name: Unaby, Chylon jiida

Kyrgyz name: Кадимки унаби (Kadimki unabi)

Description: Shrub or small tree usually to 3–4(–10) m high, with or without spines. Bark brown or gray-brown. New branches purple-red or gray-brown, flexuose, with 2 stipular spines or not; long spines erect, stout, to 3 cm; short spines recurved; annual branchlets pendulous, green, resembling compound leaves, solitary or 2–7-fascicled on short shoots. Leaves alternate, short-petiolate with small spinose stipules at the base, oblong-ovate to broadly-lanceolate, rounded or slightly cordate and unequal at the base, prominently 3-veined, coriaceous, glabrous, dark green and shiny above, pale green below, margins crenate-serrate. Inflorescences axillary cymes on very short peduncles. Flowers 3–4 mm in diameter, with fleshy disk. Sepals 5, ovate-triangular. Petals 5, greenish-yellow, obovate, clawed at base. Stamens 5. Fruit a drupe, globular or oblong, reddish-orange to red-purple.

Other distinguishing features: Leaves have anesthetic effect when chewed, and causes inability to taste sugar, salt and pepper for 1–2 min.

Phenology: Flowers and fruits in July-September.

Reproduction: By seeds and rhizomes.

Distribution: Naturally occurs in Surxondaryo province, but is cultivated throughout Uzbekistan; Jalal-Abad province of Kyrgyzstan.

Habitat: The tau zone. Dry slopes with rocky debris.

Population status: Uncommon, occurs in small populations and as solitary individuals.

Traditional use: Fruits are used for catarrh of the upper airways, fevers, and to treat intestinal infections. The root bark is used as a stimulant and the fruits have antibacterial action. In Central Asia, a decoction of the fruit is used for anemia, chest pains, asthma, coughs, smallpox, diarrhea, and as an analgesic for diseases of the liver, kidneys, and intestines and also as hypotensive drug. In China, the preparation *landutzao* is made by processing the fruits of this species with steam from water in which *Aconitum leucostomum* L. has been boiled. This preparation is used to treat tuberculosis, lymph nodes, bones, skin, eyes, and lungs (Sakhobiddinov 1948; Gammerman et al. 1990).

Documented effects: As a result of pharmacological studies, fruits and leaves in a 10 % infusion were recommended as a medical treatment for its hypotensive and diuretic effects. In the therapeutic clinic of Samarqand Medical Institute, preparations of jujube fruits have shown positive results for the treatment of hypertensive patients (Akopov 1981; Gammerman et al. 1990). Betulinic acid and a fatty acid mixture of linoleic, oleic and stearic acids isolated from extracts of the seeds showed moderate and significant levels of cyclooxygenase-2 inhibition respectively (Su et al. 2002). The flavonoids spinosin and swertish, isolated from the seeds, exhibited significant sedative effects (Cheng et al. 2000). In vivo experiments with mice showed that an extract of the seeds possessed anxiolytic effects at lower dose and sedative effects at higher dose (Peng et al. 2000). Triterpenoids isolated from the fruit exhibited high cytotoxic activity against a number of different tumor cell lines (Lee et al. 2003).

Phytochemistry: Triterpenoid saponins, triterpenoids, flavonoids, and alkaloids have been isolated from species in this genus (Li et al. 2005). Leaves contain 27–30 % tannins (including 15 % pure tannin), tetra-saccharide, methyl ether of gallic acid, and free gallic acid. The leaves also contain myricitrin and other flavonoids, up to 0.01 % of essential oils, up to 122 mg% of vitamin C, and pigments. The fruit coat contains tannins. The fruits contain micro and macro-elements (iron, iodine, zinc, copper, cobalt, and others), as well as triterpenoids (Akopov 1981; Gammerman et al. 1990; Lee et al. 2003). Over 22 different compounds have been isolated from the seeds including flavonoids, phenyl glycosides, triterpenes, and alkaloids (Cheng et al. 2000; Li et al. 2005).

Zygophyllum oxianum Boriss. – Zygophyllaceae

Synonyms: *Zygophyllum fabago* L. var. *oxianum* (Boriss.) Kitam.

English name: Beancaper

Russian name: Парнолистник амударьинский (Parnolistnik amudar'inskiy)

Uzbek name: Tujatovan, It tovon

Kyrgyz name: Unknown

Description: Herbaceous perennial with a thick, woody, vertical root. Stems few, erect, 30–70 cm tall, divaricate-branched above, thick, striated, glabrous. Leaves opposite, compound, with 1 pair of leaflets; stipules 4–7 mm long; leaflets obliquely ovate to orbicular, flat, fleshy, up to 3–4 cm long. Flowers singular or paired in upper leaf axils, pedicels 1–1.2 cm long. Sepals 5. Petals 5, oblong, ca. 1 cm long, lower half orange-red, top white, apex rounded. Stamens 10, orange. Fruits oblong-cylindrical capsules, 1.5–2 cm long, sharp-angular with 5 ribs and 5 grooves, erect. Seeds 5–8 mm long, glabrous, gray.

Other distinguishing features: Differs from *Zygophyllum fabago* which has longer, drooping fruits.

Phenology: Flowers and fruits in May-August.

Reproduction: Most often by seeds, and rarely by rhizomes.

Distribution: All provinces of Uzbekistan; not found in Kyrgyzstan.

Habitat: The chul and adyr zones. Primary habitats are river floodplains, on slightly salty soils, and in oases of desert and semi-desert zones.

Population status: Common in typical habitats, mostly as a solitary individuals.

Traditional use: Plasters made of fresh leaves are used to treat abscesses, as well as to heal wounds. An infusion of the leaves is used as a vermifuge and to treat fatigue and weak heart function. An extract made of this species has bacteriacidal properties. A decoction of the root is used as a wash to treat rheumatism, wounds and carbuncles. An ointment, prepared by mixing powdered root with sheep fat, is used to treat wounds (Seredin and Sokolov 1969). In China the plant is used as a cough suppressant, expectorant, anti-inflammatory, and analgesic (Feng et al. 2007).

Documented effects: An extract of the closely related species *Zygophyllum fabago* exhibited low activity as an acetylcholinesterase inhibitor and exhibited much higher inhibitory activity against butyrylcholinesterase (Orhan et al. 2004). Extracts also showed very strong antifungal activity against *Candida albicans* and significant antibacterial activity against *Escherichia coli* and *Bacillus subtilis* (Zaidi and Crow 2005). Compounds isolated from a bark extract exhibited anti-tumor activity (Feng et al. 2007).

Phytochemistry: The whole plant contains up to 2 % alkaloids, the main ones being zygofabagine, harmine, and others. Leaves contain 15.7–70 mg% and fruits up to 10 mg% of vitamin C (Seredin and Sokolov 1969). The bark contains tritepenoid glycosides, quinovic acid and its derivatives as well as a cincholic acid derivative (Feng et al. 2007).

▲**Ziziphora tenuior L.** Photos: Evgeny Davkaev

◀**Ziziphus jujuba Mill.** Photos: Alim Gaziev

▼**Zygophyllum oxianum Boriss.**
Photos: Alim Gaziev

Appendix 1

English-Russian Translations of Botanical and Ecological Terms

Abandoned field	заброшенная пашня (залеж)
Abundant	обильный
Achene	семянка
Acuminate	заострённый
Acute	острый, заострённый
Adnate	сросшийся
Aggregate fruit	сложный плод
Alluvial deposits	аллювиальные наносы
Along	вдоль
Alpine	альпийский
Alternate	очерёдный
Amplexicaul	стеблеобъемлющий (лист)
Angled	гранистый
Angular	угловатый
Annual	однолетний
Anther	пыльник
Apex	верхушка
Apical	верхушечный
Apiculate	с коротким узкозаострённым концом
Appendage	придаток
Appressed	прижатый
Arachnoid-hairy	паутинисто-опушённый
Arching, arcuate	дуговидный
Arcuate, arching	дуговидный
Aril	присемянник
Artemisia-grass complex	полынно-разнотравный фитоценоз
Ascending	приподнимающиеся
Attenuate	суженный
Auricles	ушки (листа)
Auriculate	ушковидный
Awl-shaped	шиловидный
Awn	ость
Axil	пазуха
Axillary	пазушные
Banner petal	флаг (часть цветка бобовых)
Barb	шип
Bark	кора
Basal leaves	прикорневые листья
Beak	носик (клюв)
Belt-like	ремневидный
Beneath	снизу
Berry	ягода
Biennial	двулетний
Bifurcating	раздвоенный
Bipinnate	дваперистый
Bipinnatipartite	двуперистораздельный
Bipinnatisect	дважды перисторассечённый
Bisexual	обоеполый
Biternate	двоякотройчатый
Bitter	горький
Blade	пластинка
Blunt	тупой
Bog	болото
Bract	прицветник
Bracteate	с прицветниками
Bracteole	прицветник на вторичной оси
Bractlet	прицветничек
Branch	ветка
Branched	ветвистый
Branchlet	веточка
Bristly	щетинистый
Bristles, setae	щетинки
Broadly-oblong	широко-продолговатый
Brook, stream	сай, ручей, маленькая речка
Brook, stream	ручей, сай, маленькая речка
Buds (flower)	бутоны
Bulb	луковица
Bulblet	луковичка
Burst	лопаться
Bush	кустарник
Calyx	чашечка (цветка)
Campanulate	колокольчатый
Canyon bottoms	дно ущелий
Capitate	головчатый
Capitulum	корзинка (форма соцветия)
Capsule	коробочка (форма соцветия)
Carpel	плодолистик

S. Eisenman et al. (eds.), *Medicinal Plants of Central Asia: Uzbekistan and Kyrgyzstan*,
DOI 10.1007/978-1-4614-3912-7, © The Editor 2013

Cartilaginous	хрящеватый
Catkin	серёжка (форма соцветия)
Caudex	каудекс
Cemetary	кладбище
Chain	цепочка
Chamber	камера
Ciliate	бахромчатый, ресничатый
Circumscissile	открывется по круговой линии
City	город
Cladodes	кладодий
Clasping	охватывающий, стеблеобъемлющий
Clasping	стеблеобъемлющий, охватывающий
Claw	ноготок
Clay bluff	глинистый обрыв
Clayey	глинистый
Clay-soiled	глинистая почва
Cleistogamous	клейстогамный
Climbing	цепляющиеся
Club-shaped	булавовидный
Cluster	гроздь
Cluster (flowers)	кисть (соцветие)
Coarse-dentate	крупно-зубчатый
Coat	оболочка
Compact	скученный
Compacted (soil)	хрящеватая (почва)
Compound leaf	сложный лист
Compound umbel	сложный зонтик
Compressed	сжатая
Cone (berry-like)	шишкоягода (плод)
Cone [pine type]	шишка
Cone [shape]	конус
Conical	конический
Conjoined	многоглавый
Connate	сросшийся
Constricted	перетянутый
Convex	выпуклый
Cordate	сердцевидный
Coriaceous	кожистый
Corolla	венчик
Corona	корона
Corymb	щиток
Corymbiform	щитковидный
Corymbiform cyme	щитковидный полузотик
Cotton field	хлопковое поле
Crack	трещина
Cracked	трещиноватый
Creeping	ползучий
Creeping roots	корнеотпрысковый
Crenate	городчатый (лист)
Crescent-shaped	серповидный
Crowded	скученный
Crown	крона
Cultivated	культивируемый
Cultivated fields	возделываемые поля, посевы
Cuneate	клиновидный
Cup-shaped	бокальчатый
Curling	вьющийся
Curly	курчавый
Curly-hairy	курчаво-опушённый
Curved	изогнутый
Cyathium	циатий
Cylindrical	цилиндрический
Cyme	полузонтик (соцветие)
Deciduous	опадающий
Deciduous	теряющий на зиму листву
Decumbent	приподнимающийся
Decurrent	нисходящий
Deep	глубокий
Deeply dissected	глубоко рассечённый
Dehiscent (fruits)	растрескивающийся плод
Densely	густой
Dentate	зубчатый
Desert	пустыня
Diadelphous	двубратственный
Diameter	диаметр
Dichasium	дихазий
Dimorphic	диморфный
Dioecious	двудомное (растение)
Disc flowers	дисковые цветки
Disk-shaped	дискообразный
Dispersal of seeds	распространение семян
Dissected	рассечённый
Divaricate-branched	вильчато-ветвистый
Dots	точки
Drooping	поникающий
Drupe	костянка (плод)
Drupelet	костяночка
Dry	сухой
Dry river-bed	сухое русло реки
Dull	матовый
Elaiosome	элайосома
Elliptic	эллиптический
Elongate	удлинённый, продолговатый
Emarginate	выемчатый
Embankment	насыпь
Endocarp	эндокарпий
Entire (margin)	цельный
Ephemeral	эфемерный
Ephemeroid	эфемероидный
Epicalyx	наружная чашечка
Erect	прямостоячий
Even-pinnate	парноперистый
Evergreen	вечнозелёный
Explosively dehiscing	вскрывается
Exserted	выступающий
Farinose	покрыт мучнистым налётом
Fascicle	пучок
Feather-grass steppes	ковыльные степи
Felted	войлочный
Felted-hairy	войлочно опушённый
Female flower	женский цветок
Fibrous roots	мочковатые корни

Field	поле
Filiform	нитевидный
Fimbriate, fringed	бахромчатый
Finger-like	пальчатообразный
Fir	пихта
Fissure	трещина
Flat	плоский
Flattened	сплющенный
Fleshy	мясистый
Flexuose	извилистый
Floodplain	пойма (реки)
Fluted	желобчатый
Foliaceous	листовидный
Follicle	листовка (плод)
Foothills	предгорье
Forest	лес
Forest edges	опушка леса
Forked	вильчатый
Four-sided (4-sided)	четырёхгранный
Fragrant	душистый
Fringed, fimbriate	бахромчатый
From the base	от основания
Fruit	плод
Funnelform	воронковидный
Furrowed	бороздчатый
Fusiform	веретеновидный
Flabrous	голый
Glade	поляна
Glands	железки
Glandular	железистый
Glandular prickles	шиповидные железки
Glaucous	покрытый налётом
Globular	шаровидный, сферический
Glossy, shiny	блестящий
Gorge	ущелье
Gradually	постепенно
Granular	гранулярный
Groove	борозда
Gum, resin, pitch	камедь, смола
Gypsum	гипс
Habitat	местообитание
Hairs	волоски
Hairy	опушённый
Hanging	пониклый
Hastate	стреловидный
Head (inflorescence)	головка (соцветия)
Head-like	головчатовидный
Helicoid	спиралеобразный
Herbaceous plant	травянистое растение
Heterogamous	гетерогамный
Hill	холм
Hollow	полый
Honey	мёд
Hood	шлем (часть цветка)
Hooked	крюкообразный
Horizontal	горизонтальный
Horn	рожок
Horn-like	роговидный
Hypanthium	гипантий
Imbricate	чешуйчатый
Incised	надрезаный
Incrassate	утолщённый
Incurved	внутрь изогнутый
Indehiscent	нераскрывающиеся (плоды)
Inflated	вздутый
Inflorescence	соцветие
Inflorescence axis	ось соцветия
Inserted	расположенный
Interrupted	прерывистый
Involucel	вторичная обёртка (цветка)
Involucral bract	листочек обёртки
Involucre	обёртка соцветия
Irregular	неправильный
Irrigated	орошаемый
Irrigation canal	арык
Juicy	сочный
Juniper	арча
Juniper stand	арчёвник
Keel	лодочка, киль
Keel	киль, лодочка
Keeled	килевидный
Kidney-shaped	почковидный
Lake	озеро
Lanceolate	ланцетный
Large-dentate	крупнозубчатый
Lateral	боковой
Lawn	лужайка
Leaf	лист
Leaflet	листочек
Leafy	облиственный
Legume	боб
Lenticel	чечевичка
Lenticular	чечевицеобразный
Limestone talus	известняковая осыпь
Linear	линейный
Lip	губа
Loam	суглинок
Lobe	долья, лопасть
Lobe	лопасть, долья
Lobed	лопастной, дольчатый
Lobule	долька
Loess	лёсс
Long	длина
Long-fallow field	перелог
Longitudinal	продольный
Loose	рыхлый
Lyrate	лировидный
Male flower	мужской цветок
Male inflorescence	мужское соцветие
Many-flowered	многочветковый
Many-lobed	многолопастной
Margin entire	цельнокройный

Marginal flowers	краевые цветки
Meadow	луг
Melon	дыня
Melon field	бахчёвое поле
Membranaceous	плёнчатый
Membranous	перепончатый
Mericarp	мерикарп
Milky	молочный
Milky sap	млечный сок
Moniliform	чёткообразный
Monocarpic	монокарпический
Monoecious	однодомное (растение)
Mountain	гора
Mucilage (plant)	слизи (растений)
Mucronulate	маленькое острое окончание (листа)
Narrow	узкий
Nectary	нектарник
Needle-like	игловидный
Nerve, vein	жилка (растения)
Nodding	пониклый
Nutlet	орешек (плод)
Oasis	оазис
Obconical	обратноконический
Oblanceolate	обратноланцетный
Obliquely descending	скошенный
Oblong, elongate	продолговатый
Obovate	обратнояйцевидный
Obtuse	притупленный
Ocrea	раструб
Odd-pinnate	непарноперистый
Odorous	пахучий
Opposite	супротивный
Orchard, garden	сад
Oriented	ориентированый
Ovary	завязь
Ovate	яйцевидный, овальный
Ovate	овальный, яйцевидный
Ovule	семяпочка
Paleaceous	чешуйчатый, плёнчатый
Paleaceous	плёнчатый, чешуйчатый
Palmate	пальчатый
Palmately compound	пальчатосложный
Palmatifid	дланевидно-надрезный
Palmatilobate	пальмовидно-лопастный
Palmatipartite	пальчато-лопастный
Palmatisect	палчато-рассечённый
Panicle	метёлка (соцветие)
Paniculate	метёльчатый
Paniculiform	метёлковидное
Paper-like, papery	бумагообразный
Papilionaceous	мотыльковый (цветок)
Papilla	сосочек
Papillate	бородавчатый
Pappus	хохолок
Parallel	паралельный
Pasture	пастбище
Pebbly	галечниковый
Pedicel	плодоножка, цветоножка
Pedicel	цветоножка, плодоножка
Peduncle	цветонос
Pellucid dots	исколотый
Pendulous	плакучий
Pepo	тыквина (плод)
Perennial	многолетний
Perfect (flower)	обоеполый (цветок)
Perianth	околоцветник
Pericarp	околоплодник
Petal	лепесток
Petaloid	лепестковидный
Petiolate	черешковый
Petiole	черешок
Pinnate	перистый (лист)
Pinnatifid	перистонадрезанный (лист)
Pinnatilobate	перистолопастной (лист)
Pinnatipartite	перистораздельный (лист)
Pinnatisect	перисторассечённый (лист)
Pistil	пестик (цветка)
Pistillate flower	женский цветок
Pitch, gum, resin	камедь, смола
Pith	сердцевина (стебля)
Pitted	ямчатый
Placenta	плацента
Plain	равнина
Plate	пластинка
Plicate	складчатый
Plowed field	пашня
Plumose	перистый
Plumose-barbed	перисто-зазубренный
Pod	стручок
Pomaceous	яблокообразный
Pore	дырочка
Prickle	шип
Prickly	шиповатый
Prismatic	призматический
Projection	выпрост
Prominent	выдающийся
Prostrate	стелящийся
Pubescent	волосистый
Pulp	мякоть (плодов)
Punctate glandular	точечные железки
Pyramidal	пирамидальный
Quadrangular	четырёхгранный
Quadripinnate	четыреждыперистый
Raceme (cluster)	кисть (соцветие)
Racemiform	кистевидный
Rachis	ось
Raised	выступающий
Raised gland	железистый шипик
Ray	луч
Ray flower	язычковый цветок
Receptacle	цветоложе

Reclining	приподнимающийся
Recurved	отогнутый вниз
Red sandstone	краснопесчаник
Reduced	редуцированый, уменьшенный
Reduced	уменьшенный, редуцированый
Reflexed	отогнутый вниз
Resin, gum, pitch	камедь, смола
Resin, gum, pitch	смола, камедь
Reticulate	сетчатый
Reticulate veined	сетка жилок
Retrorse	направленный вниз
Rhizome	корневище, корневой отпрыск
Rhombic	ромбический
Rib	ребро
Ribbed	ребристый
Ridge	рубчик
Rind	корка
Ring	кольцо
Ripe	созревший, спелый
Ripe	спелый, созревший
River	река
River valley	долина реки
Road	дорога
Rocky debris	щебнистый
Root	корень
Root crown	кореневая шейка
Root system	корневая система
Rosette	розетка
Rotate	колесовидный
Rough	шероховатый
Round	круглый
Row	ряд
Ruderal	рудеральный
Rugose	морщинистый
Runcinate	обращённые назад доли (листа)
Sagittate	стреловидное основание (листа)
Salty area (very)	солончак
Sandstone	песчаник
Sandy	песчаный
Sap	сок
Scabrid	шершавый
Scale	чешуя
Scale-like	чешуевидный
Scarious	пластинчатый
Scattered	разбросанный
Schizocarp	распадающийся плод, дробный
Schizocarp	дробный, распадающийся плод
Scorpioid cyme	завиток (соцветие)
Seam	шов
Segment	сегмент
Segmented	сегментированный
Semi-desert	полупустыня
Semispherical	полушаровидный, полусферический
Semi-woody	полудеревянистый
Senescing	скороувядающий (лист)

Sepal	чашелистик
Septicidal	растрескивающийся по перегородкам (плода)
Septum	перегородка
Serrate	пильчатый
Serrulate	мелкозубчатый
Sessile	сидячий
Setae, bristl s	щетинки
Shady	тенистый
Shale, slate	сланец
Shallow	неглубокий
Shallow soil	мелкоземистая почва
Sheath	влагалище (листа)
Shell	скорлупа
Shiny, glossy	блестящий
Short	короткий
Short shoot	короткая веточка
Short-petiolate	короткий черешок
Shrub	кустарник
Silicle	короткий стручок
Silique	стручок
Silique-like	стручковидный
Silky	шелковистый
Simple	простой
Single, solitary	одиночный
Sinuate	выемчатый
Slightly	слегка
Slope	склон
Small groves	небольшая роща
Smell	запах
Smooth	гладкий
Soft	мягкий
Soil	почва
Solitary, single	одиночный
Solonetzic	солонцеватый
Sour	кислый
Spadix	початок
Spathe	обвёртка
Spatulate	лопатчатый, лопатовидный
Spatulate	лопатовидный, лопатчатый
Spear-shaped	копьевидный
Spherical	сферический, шаровидный
Spherical	шаровидный, сферический
Spiciform	колосовидный
Spike	колос (соцветие)
Spine	колючка
Spine, thorn	колючка
Spinescent	колючий
Spine-tipped	оканчивающийся колючкой
Spiny-toothed	колюче-зубчатый
Spirally	винтообразный
Spirally-twisted	спиралезакрученный
Spongy	мочалистый
Spore	спора
Spot	пятно
Spreading branchy	оттопыренно-ветвистый

Spring	родник
Spruce forest	еловый лес
Spur	шпора
Stamen	тычинка
Staminal column	тычиночная колонка
Staminal filament	тычиночная нить
Staminate flowers	мужские (тычиночные) цветки
Staminode	бесплодая тычинка
Stellate	звёзчатый
Stellate-hairy	звёздно-опушённый
Stem	стебель
Steppe	степь
Stiff	жёсткий
Stigma	рыльце (цветка)
Stinging hairs	жгучие волоски
Stipulate	снабжённый прилистниками
Stipule	прилистник
Stocky	коренастый
Stolon	столон
Stony	каменистый
Straight	прямой
Stream bed	русло ручья, сая, маленькой речки
Stream, brook	ручей, сай, маленькая речка
Striated	бороздчатый
String-like	шнуровидный
Style	столбик (цветка)
Suberect	приподнятый
Subshrub	полукустарник
Succulent	сочный
Syncarp	синкарпий
Tail-like	хвостоподобный
Talus	осыпь
Tangled	спутанный
Tapering	суженный
Taproot	стержневой корень
Tendril	усик (растения)
Tepal	листочек околоцветника
Terminal	верхушечный
Ternate	тройчатый (лист)
Thick	толстый
Thin	тонкий
Thorn, spine	колючка
Thread-like	нитевидный
Three follicles	трёхлистовка (плод)
Tip	носик (тонкий конец)
Tomentose	опушённый
Trailing	стелющийся (растение)
Triangular	треугольный
Trichome	трихома
Trifoliate	тройчатый
Trifurcated	трёхветвистый
Trigonous, triquetrous	трёхгранный
Tripartite	трёхраздельный (лист)
Tripartite-pinnatisect	тройчато-перисторассечённый
Tripinnate	трижды перистый
Tripinnatisect	трижды перисторассечённый (лист)
Triquetrous, trigonous	трёхгранный
Trisulcate	трёхборзчатый
Truncated	усечённый
Trunk	ствол (растения)
Tube	трубка
Tuber	клубень
Tubercle	бугорок
Tuberculate	бугорчатый
Tuberiform	клубневидный
Tubular	трубчатый
Tugai	тугай
Twig-like	прутьевидный
Twining	вьющийся
Two-horned	двурогий
Two-lipped (2-lipped)	двугубый
Two-valved (2-valved)	двустворчатый
Umbel	зонтик (соцветие)
Umbellet	вторичный зонтик
Umbelliform	зонтиковидный
Understory	подлесок
Undulate	волнистый
Unequal	неравный
Unirrigated	богара (не орошаемая зона)
Unisexual	однополый
Upper	верхний
Utricle	мешочек (плод)
Valley	долина
Valve	створка
Vegetable garden	огород
Vegetation	растительность
Vegetatively	вегетативный
Vein	жилка
Velutinous, velvety	бархатистый
Velvety, velutinous	бархатистый
Vertical	вертикальный
Verticillaster	полумутовка (соцветие)
Vigorous	мощный
Village	посёлок
Villous, tomentose, hairy	опушённый
Vine	цепляющиеся растение
Walnut forests	ореховый лес
Waste place	места с мусором
Water-eroded	смытый
Wavy	выемчатый
Wedge-shaped	клиновидный
Weed	сорняк
Well	колодец
Wet	сырой
Wheat field	посевы пшеницы
Whorl	мутовка
Wide	широкий (в ширину)
Winding	извилистый
Wing	крыло
Winged petiole	крылатый черешок

Wingless	бескрылый
Withering	отмирают
Woody	одеревенелый
Woolly-hairy	шерстисто опушённый
Wrinkled	морщинистый
Zygomorphic	зигоморфный

Appendix 2

English-Russian Translations of Chemical Terms

25-d-spirosta-3,5-diene	25-d-спирост-3,5-диен
2-methoxy-1,4 naphthoquinone	2-метокси-1,4-нафтохинон
Absinthin	абсинтин
Acanthophylloside	аконтофиллазид
Acetic acid	уксусная кислота
Acetylcholine	ацетилхолин
Acetylnapelline	ацетилнапеллин
Acevaltrate	ацевалтрат
Aconitic acid	акотиновая кислота
Acsinatine	аксинатин
Aesculin	эскулин
Aglycone	агликон
Aksine	аксин
Akuamidine	акуаммидин
Akuamine	акуамин
Alantolactone	алантолактон
Alantone	алантон
Alantopicrine	алантопикрин
Alcohol	спирт
Aldehyde	альдегид
Alginidine	алгинидин
Alginine	алгинин
Alhagidin	алхагидин
Alhagitin	алхагитин
Aliphatic alcohol	алифатический спирт
Alkaloid	алкалоид
Alkamide	алкамид
Alkanin	алканин
Allocryptopine	аллокриптопин
Allyl-isothiocyanate	аллилизотиоцианат
Aloemodin	алоээмодин
Aloperine	алоперин
Alpha-amyrin	а-амирин
Alteramine	альтерамин
Aluminium	алюминий
Amaranthin	амарантин
Amide	амид
Amino acid	аминокисота
Aminoalcohol	аминоспирт
Amygdalin	амигдалин
Amyl alcohol	амиловый спирт
Amyrin	амирин
Anabasine	анабазин
Anabsinthin	анабсинтин
Anagirine	анагирин
Anagyrine	анагирин
Anchusa acid	анхузовая кислота
Anchusin	анхизин
Anemonin	анемонин
Anethole	анетол
Angelic acid	анисовая кислота
Anhydroaustricine	ангидроаустрицин
Anhydroperforine	ангидроперфорин
Anisic acid	ангеликовая кислота
Anonaine	анонаин
Anthocyan	антоциан
Anthocyanidin	антоцианидин
Anthocyanin	антоцианин
Anthracene	антрацен
Anthraglycoside	антрагликозид
Anthranoyllycoctonine	антраноилликоктонин
Anthraquinone	антрахинон
Apigenin	апигенин
Arabinose	арабиноза
Arachic acid	арахиновая кислота
Arachidic acid	арахидиновая кислота
Arctigenin	арктигенин
Arctiin	арктиин
Arctiopicrin	аркциопикрин
Argemonine	аргемонин
Argentine	аргентин
Arnidiol	арнидиол
Aromatic acid	ароматическая кислота
Aromatic aldehyde	ароматический альдигид
Artabasin	артабсин
Arteannuin	артеаннуин
Artelein	артелеин

S. Eisenman et al. (eds.), *Medicinal Plants of Central Asia: Uzbekistan and Kyrgyzstan*,
DOI 10.1007/978-1-4614-3912-7, © The Editor 2012

Artelin	артелин
Artemetin	артеметин
Artemisinin	артемизинин
Asaresin	асарезен
Asaresinol	асарезинол
Asaresinotannol	асарезинотанол
Asarone	азарон
Ascorbic acid	аскорбиновая кислота
Ash	зола
Asparagine	аспарагин
Asperuloside	асперулозид
Astringent substances	вяжущие вещества
Atropine	атропин
Aucubin	аукубин
Austricine	аустрицин
Avenasterol	авенастерол
Avicularin	авикуларин
Bactericidal	бактерицидный
Baicalein	байкалеин
Bakuchiol	бакучиол
Barium	барий
Behenic acid	бегеновая кислота
Benzaldehyde	бензойный альдегид
Benzoic acid	бензойная кислота
Benzyl isothiocyanate	бензилизотиоцианат
Berbamine	бербамин
Berbamunine	бербамунин
Berberine	берберин
Bergamotene	бергамотен
Betaine	бетаин
Bicyclic	бициклический
Bicyclogermacrene	бициклогермакрен
Biogenic amine	биогенный амин
Bitter substances	горькие вещества
Bitters	горечи
Borneol	борнеол
Bornyl acetate	борнил ацетат
Bornyl-isovalerianate	борнил-изовалерианат
Britanin	британин
Bufadienolide	буфадиенолид
Bursic acid	бурсовая кислота
Cadinene	кадинен
Caffeic acid	кофейная кислота
Caffeine	кофеин
Caffeoylquinic acid	кофеоилхинная кислота
Calcium oxalate	оксалат кальция
Campesterin	кампестерин
Campesterol	кампестерол
Camphene	камфена
Camphene	камфена
Camphorol	кемпферол
Canavanine	канаванин
Caoutchouc	каучук
Capronic acid	капроновая кислота
Caprylic acid	каприловая кислота

Carabron	караброн
Carbohydrates	углеводы
Carbolic acid	n-оксибензойная кислота
Cardiac glycoside	сердечный гликозид
Cardinolide	карденолид
Carene	карен
Carotene	каротин
Carotenoid	каротиноид
Carvacrol	карвакрол
Carvone	карвон
Caryophylladienol	кариофилла-диен-ол
Caryophyllene	кариофиллин
Caryophyllene oxide	окись кариофиллина
Catechin	катехин
Catecholamine	катехоламин
Caulosapogenin glycoside	каулосапогенин-гликозид
Cedrene	цедрен
Cedrol	цедрол
Cerotinic acid	церотиновая кислота
Ceryl-alcohol	цериловый спирт
Chatinene	хатинин
Chemical compound	химичесчкое соединение
Chicoric acid	*цикоревая кислота*
Chlorogenic acid	хлорогеновая кислота
Chlorophyll	хлорофилл
Cholesterol	холестерин
Choline	холин
Chromium	хром
Chromone	хромон
Chrysoeriol	хризоэриол
Chrysophanic acid	хризофановая кислота
Cichoriin	цикорин
Cicutine, coniine	коницин
Cinaroside	цинаросин
Cincholic acid	хинолиновая кислота
Cineol	цинеола
Cinnamaldehyde	коричный альдегид
Cinnamamide	циннамамид
Citral	цитраль
Citric acid	лимонная кислота
Citrin	цитрин
Citronellol	цитронеллол
Citronellyl acetate	цитронеллиацетат
Clematine	клематин
Cnicin	кницин
Cobalt	кобальт
Codonopsin	кодонопсин
Codonopsinin	кодонопсинин
Columbamine	колумбамин
Compound	соединение
Condelphine	кондельфин
Conhydrine	конгидрином
Convolamine	конволамин
Convolidine	конволидин
Convolvine	конвольвин

Convolvuline	конвольвулин
Copper	медь
Corydine	коридин
Corytuberine	коритуберин
Coumaric acid	кумаровая кислота
Coumarin	кумарин
Coumarinic acid	кумариновая кислота
Cryptopine	криптопин
Crystalline	кристаллический
Crystals	кристаллы
Cuminaldehyde	куминовый альдегид
Cuminyl alcohol	куминиловый спирт
Cuscohygrine	кускогигрин
Cyanidin	цианидин
Cyanidin-3-glucoside	цианидин-3-глюкозид
Cyanin	цианин
Cyanogenic compound	цианогенныое соединение
Cyanogenic glycoside	цианогенный гликозид
Cyasterone	циастерон
Cyclamine	цикламин
Cyclic alcohol	циклический спирт
Cyclic peptide	циклиновый пептид
Cyclitols	циклитолы
Cyclolignan	циклолигнан
Cyclopropenoid fatty acid	циклопропаноидная жырная кислота
Cymene	цимен
Cymol	цимол
Cytisine	цитизин
Cytosterin	ситостерин
Daucane esters	дауциновые эфиры
Daucane-type sesquiterpene	сесквитерпен, типа дауцин
Daucene	дауцин
Daucosterol	даукостерол
Dehydroabietol	дигидроабиетол
Dehydrothalicmine	дегидроталикмин
Delatine	делатин
Delphiline	дельфелин
Delphinindin	дельфинидин
Delphirine	дельпирин
Delsemine	дельсемин
Delsine	дельсин
Delsoline	дельсолин
Delsosine	делькозин
Deoxypeganine	дезоксипеганин
Derivatives	производные
Diacylglyceride	диацилглицерид
Dictamine	диктамин
Didrovaltrate	дидровалтрат
Dihydroalantolactone	дигидроалантолактон
Dihydroxyacids	диоксикислота
Dillapiole	диллапиол
Dimethamine	диметамин
Diosgenin	диосгенина
Diosmetin	диосметин
Diosmine	диосмин
Dipegene	дипегин
Disaccharide	дисахаридов
Disulfide	дисульфид
Dopamine	допамин
Doremol	доремол
Doremon	доремон
Drupacine	друпацин
Drupanol	друпанол
Drying fatty oil	жырное высыхающее масло
Dubamine	дубамин
Dubinidine	дубинидин
Ecdysone	экдизон
Ecdysterone	экдистерон
Eicosenoic acid	гадолеиновая кислота
Elemol	элемол
Ellagic acid	эллаговая кислота
Emodin	эмодин
Enzyme	фермент
Enzymic hydrolysis	ферментативный гидролиз
Ephedrine	эфедрин
Epi-13-manool	13-эпиманоол
Epicatechin	эпикатехин
Epigallocatechin	эпигалокатехин
Epi-rhododendrin	эпирододендрин
Epoxyacylglyceride	эпоксиацилглицерид
Equisetin	эквизетин
Equisetonin	эквизетонин
Equisetrine	эквизетрин
Eremuran	эремуран
Eremursine	эремурсин
Ergolide	эрголид
Eriodictyol	эриодиктиол
Erucic acid	эруковая кислота
Ervamine	эрвамин
Ervine	эрвин
Ervinine	эрвинин
Erysimine	эризимин
Erysimoside	эризимозид
Esculetin	эскулетин
Esculin	эскулин
Essential oil	эфирное масло
Ester	сложный эфир
Ether	эфир
Ethyl	этил
Ethyl ester	этиловый эфир
Eucalyptol	эвкалиптол
Eudesmine	эудесмин
Eugenol	евгенол
Euphorbin	эуфорбин
Evodine	эводин
Evoxin (haploperin)	эвоксин (хаплоперин)
Evoxoidine	эвоксоидин
Excelsine	эксцельзин
Extriol	экстриол

Faradiol	фарадиол
Farnesferol	фарнезиферол
Fat-like substances	жироподобные вещества
Fatty acid	жирная кислота
Fatty oil	жирное масло
Fenchone	фенхон
Fermononetin	формононетин
Ferruginol	ферругинол
Ferulic acid	феруловая кислота
Fiber	клетчатка
Flavanone glycoside	флаванонгликозид
Flavolignan	флаволигнан
Flavone	флавон
Flavonoid	флавоноид
Flavonol glycoside	флавоновый гликозид
Flavoxanthin	флавоксантин
Flindersine	флиндерсин
Foetidine	фетидин
Folic acid	фоливая кислота
Foliosidine	фолиозидин
Formic acid	муравьиная кислота
Fractions	фракции
Frangula-emodin	франгулаэмодин
Frangula-emodin anthronol	франгулаэмодинантранол
Frangulin	франгулин
Free fatty acid	свободная жирная кислота
Fructose	фруктоза
Fumaric acid	фумаровая кислота
Fumaridine	фумаридин
Fumvailline	фумвайлин
Furocoumarin	фурокумарин
Fustin	фустин
Galanthamine	галантамин
Galiosin	галиозин
Gallic acid	галловая кислота
Gallocatechin	галокатехин
Gallotannin	галлотанид
Genistein	генистеин
Genistin	генистин
Gentianadine	генцианадин
Gentiananine	генциананин
Gentianine	генцианин
Gentioflavine	генциофлавин
Gentiotibetine	генциотибетин
Gentisinic acid	гептизиновая кислота
Gentianaine	генцианаин
Geraniol	гераниол
Germacrene	гермакрин
Gitogenin	гитогенина
Glabric acid	глабровая кислота
Glaucine	глауцин
Glaunidine	глаунидин
Glaunine	глаунин
Glauvine	глауфин
Glucobarbarin	глюкобарбарин
Glucocapparin	глюкокаппарин
Glucofrangulin	глукофрангулин
Glucofructose	глюкофруктоза
Glucose	глюкоза
Glucoside	глюкозид
Glycerin	глицерина
Glycoalkaloid	гликоалкалоид
Glycone	гликон
Glycoperine	гликоперин
Glycoside	гликозид
Glycyrramarin	глицирамарин
Glycyrrhetinic acid	глицирретиниковая кислота
Glycyrrhizic acid	глицирризиновая кислота
Glycyrrhizin	глицирризин
Gossypol	госсипол
Granilin	гранилин
Guaiol	гвайол
Gum	камедь
Gypsogenin	гипсогенин
Haplofidine	хаплофидин
Haplofilidine	хаплофилидин
Haplophytin	хаплофитин
Haplopine	хаплопин
Harmaline	гармалин
Harmalol	гармалол
Harman	гарман
Harmine	гармин
Harpagide	гарпагид
Harpagoside	гарпагосид
Hederagenin derivatives	производные хедерагенина
Hemicellulose	гемицеллюлоза
Hemolytic index	гемолитический индекс
Herniarine	герниарин
Hesperidin	гесперидин
Heterocyclic	гетероциклический
Hexadecanoic acid (palmitic acid)	пальмитиновая кислота (гексадекановая кислота)
Hexynyl disulfide	гексенилдисульфид
Hippeastrine	гиппеастрин
Histamine	гистамин
Homothermopsine	гомотермопсин
Hordenine	горденин
Humulene	хумулен
Hydrocarbons	углеводороды
Hydrocyanic acid	цианистоводородная кислота
Hydroxycinnamic acid	гидроксикоричная кислота
Hydroxytryptamine	гидрокситриптамин
Hygrine	гигрин
Hyoscyamine	гиосциамин
Hypericin	гиперицин
Hyperoside	гиперозид
Hyssopin	гиссопин
Imperatorin	императорин
Incanine	инканин
Inorganic	неорганический

Inosine	инозин
Interoside	интерозид
Intibin	интибин
Inulin	инулин
Iodine	йод
Iridoid glucoside	иридоидный гликозид
Iridoids	иридоиды
Iron	железо
Isoalantolactone	изоалантолактон
Isobaldine	изоболдин
Isobetanin	изобетанин
Isocorydine	изокоридин
Isoflavan	изофлавон
Isoleontine	изолеонтин
Isoliquiritigenin	изоликвиритигенин
Isomenthone	изоментон
Isopsoralen	изопсорален
Isoquercitrin	изокверцитрин
Isoremerin	изорёмерин
Isoreserpiline	изорезерпилин
Isorhamnetin	изорамнетин
Isorubijervine	изорубийервин
Isosalipurposide	изосалипурпозид
Isotalatizine	изоталатизин
Isotanshinone	изотаншинон
Isotetrandrine	изотетрандрин
Isotrifolin	изотрифолин
Isovalerianic acid, isovaleric acid	изовалериановая кислота
Isovaleric acid, isovalerianic acid	изовалериановая кислота
Isovaltrate	изовалтрат
Jalapine	ялапин
Jatrorrhizine	ятроррицин
Jervine	йервин
Juglone	юглон
Juniperin	юниперин
Kaempferol	кампферол
Karakoline	караколин
Karasamine	карасамин
Ketone	кетон
Ketose	кетосахар
Koproporphyrin	копропорфирин
Korseveramine	корсеверамин
Korseveridine	корсеверидин
Korseverinine	корсеверинин
Kusunokinin	кусунокинин
Lactic acid	молочная кислота
Lactone	лактон
Lactose	лактоза
Lactucin	лактуцын
Lactucopicrin	лактупекрин
Lagochilin	лагохиллин
Lappaconidine	лаппаконидин
Lappaconitine	лаппаконитин
Lead	свинец
Leontamine	леонтамин
Leontidine	леонтидин
Leontine	леонтин
Lepidoside	лепитоцид
Leucoanthocyanide	лейкоантоцианид
Leucoanthocyanidin	лейкоантоцианидин
Leucodelphinidin	лукоделфинидин
Leucomisine	леукомизин
Levorotatory	левовращающейся
Licoctonine	ликоктонин
Licorine	ликорин
Lignan	лигнан
Lignin	лигнин
Lignoceric acid	лигиоцериновая кислота
Limonene	лимонен
Limonoid	лимоноид
Linalool	линалоол
Linalyl acetate	линалилацетат
Lindleyin	линдлеин
Linoleic acid	линолевая кислота
Linolenic acid	линоленовая кислота
Lipid	липид
Liquirazide	ликвиритозид
Liquiritin	ликвиритин
Liquitigenin	ликвитигенин
Liriodenine	лириоденин
Lithospermic acid	литосперомвая ислота
Longifolin	лонгифолен
Loroglossine	лороглоссин
Lotaustralin	лотаустралин
Lucidin	луцидин
Lupane	лупан
Lupanine	лупанин
Lupeol	лупеол
Lutein	лютеин
Luteolin	лютеолин
Luteolin 7-glucoside	лютиолин-7-глюкозид
Luteolin 7-rutinoside	лютиолин-7-рутинозид
Lycopene	ликопин
Macroelement	макроэлемент
Magnesium	магний
Magnoflorine	магнофлорин
Maleic acid	малеиновая кислота
Malic acid	яблочная кислота
Manganese	марганец
Manool	маноол
Marubiin	марубиин
Matricarin	матрикарин
Matrine	матрин
Mecambroline	мекамбролин
Melilotic acid	мелилотиновая кислота
Melilotin	мелилотин
Melilotocide	мелилотозид
Melissic acid	мелиссовая кислота
Menthol	ментол

Menthone	ментон
Mesaconitine	мезаконитин
Methoxy-cinnaroic aldehyde	метоксикоричный альдегид
Methyl gallate	метилгаллат
Methyl lachnophyllate	метил лакнофиллат
Methyl-chavicol	метилхавикол
Methyl-coniine	метилкониин
Methylcytidine	метилцитизин
Methylcytisine	метилцитизин
Methyl-evoxin	метилэвоксин
Methyllycaconitine	метилликаконитин
Methylquercetin	метилкверцетин
Microelements	микроэлементы
Mineral salts	минеральные соли
Mollugin	моллугин
Molybdenum	молибден
Monoacylglyceride	моноацилглицерид
Monocaffeyltartaric acid	монокофеил-винная кислота
Monohydroxyacid	монооксикислота
Monosaccharide	моносахарид
Monoterpene	монотерпен
Monticamine	монтикамин
Morphine	морфин
Mucilage	слизи
Mussaenoside	муссаенозид
Mustard essential oil	горчичное эфирное масло
Myrcene	мирцен
Myricitrin	мирицитрин
Myricyl alcohol	мирициловый спирт
Myristic acid	миристиновая кислота
Myristicin	миристицин
Myrtenol	миртенол
Myrtenyl isovalerianate	миртенил изовалерианат
Napelline	напеллин
Naphthoquinone	нафтохинон
Naringenin chalcone	нарингенин халькон
Narwedine	нарведин
n-dimethyl colletine	n-диметилколлетин
Neoline	неолин
Neosophoramine	неософорамин
Neoxanthine	неоксантин
Nepetalactone	непеталактон
Nepodin	неподин
Nerolidol	неролидол
n-heptacosane	n-гептакозан
Nickel	никель
Nicotine	никотин
Nicotinic acid	никотиновая кислота
Nitrogen	азот
Nitrogenous compounds	азотсодержащие соединения
Noradrenaline	норадреналин
Norcorydine	норкоридин
Nortropine	нортропин
n-oxy-benzoic acid	n-оксибензойная кислота
Oblongine	облонгин
Ocimene	оцимен
Octylene	октилен
Oil	масло
Oleanane	*олеанан*
Oleanolic acid	олеаноловая кислота
Oleic acid	олеиновая кислота
Oleoresin	олеорезин
Oligosaccharide	олигосахарид
Oliveramine	оливерамин
Oliveridine	оливеридин
Oliverine	оливерин
Olmelin	олмелин
Omega-3 fatty acid	омега-3 жырная кислота
Onopordopicrin	онопордопикрин
Organic	органический
Organic acid	органическая кислота
Osthol	остхол
Oxalic acid	щавелевая кислота
Oxyacanthine	оксиакантин
Oxymatrin	оксиматрин
Oxypeucedanin	оксипейцеданин
Oxysophocarpine	окись софокарпина
Oxysteroid	оксистероид
Oxytanshinone	окситаншинон
Pachycarpine	пахикарпин
Palmatine	пальматин
Palmitic acid (hexadecanoic acid)	палмитиновая кислота
Pancratine	панкратин
Pantotenic acid	пантотеновая кислота
Paraffin	парафин
Paraoxycoumarin	параоксикумарин
Parfumine	парфумин
Parinaric acid	паринариновая кислота
Parishin	паришин
Patchouli alcohol	пачулиевый спирт
Patrinoside	патринозид
Pectic substances	пектиновые вещества
Pectins	пектины
Pegamine	пегамин
Peganidine	пеганидин
Peganine	пеганин
Peganol	пеганол
Pelargonin	пелларгонин
Pentosan	пентозаны
Peonidin	пеонидин
Perfamine	перфамин
Perforine	перфорин
Phenol	фенол
Phenolcarbonic acid	фенолкарбоновая кислота
Phenolic acid	феноловая кислота
Phenolic glucoside	феноловый гликозид
Phenyl glycoside	фенил гликозид
Phenylbutanoid	фенилбутаноид
Phenylpropanoid	фенилпропаноид

Phenyl-β-naphthylamine	фенил-b-нафтиламин
Phloridzin	флоридзин
Phloroglucinol	флороглюцин
Phospholipid	фосфолипид
Phosphoric acid	фосфорная кислота
Phyllalbine	филлальбин
Phytoecdysone	фитоэкдизон
Phytoecdysteroid	фитоекдистероид
Phytoestrogen	фитоэстрогеном
Phytol	фитол
Phytoncid	фитонцид
Phytosterin	фитостерин
Pigments, dyeing substances	красящие вещества
Pilocarpine	пилокарпин
Pinene	пинен
Pinocamphone	пинокамфон
Piperitone	пиперитон
Plantagonine	плантагонин
Podophyllotoxin	подофилотоксин
Poisonous	ядовитый
Polyenes	полиены
Polyphenol	полифенол
Polysaccharide	полисахарид
Potassium	калий
Prangenidin	прангенидин
Prangenin	прангенин
Prangosine	прангосин
Primveraza	примвераза
Proanthocyanidin	проантоцианидин
Proazulen	проазулен
Propenyl isothiocyanate	пропенил изотиоцианат
Propionic acid	пропионовая кислота
Protein	белок
Protocatechin	протокатехин
Protopine	протопин
Protoporphyrin	протопорфирин
Protopseudohypercin	протопсевдогиперицин
Prulaurasin	прулауразин
Prunasin	пруназин
Prussic acid	синильная кислота
Pseudoconhydrine	псевдоконгидрином
Pseudoephedrine	псевдоэфедрин
Pseudohypercin	псевдогиперицин
Pseudojervine	псевдойервин
Pseudotaraxasterol	псевдотараксастерол
Pseudotropine	псевдотропин
Psoralen	псорален
Pulegone	пулегон
Purine derivatives	производные пурина
Purpurin	пурпурин
Pyrocatechin	пирокатехин
Pyrogallol	пирогаллол
Pyrrolidine	пирролидин
Pyrrolidine alkaloid	пирролидиновый алкалоид
Quercetin	кверцетин
Quercetin arabinoside	арабинозид кверцетина
Quercetin galactoside	галактозид кверцетина
Quercetin triglycoside	тригликозид кверцетина
Quercetin-3-arabinoside	3-арабинозид квецетина
Quercetin-3-galactoside	кверцетин-3-галактозид
Quercetrin	кверцитрин
Quinidine	хинидин
Quinone	хинон
Quinovic acid	хинновая кислота
Ranunculin	ранункулин
Remrefidine	ремрефидин
Remrefine	ремрефин
Reserpine	резерпин
Reserpinine	резерпинин
Reticuline	ретикулин
Rhamnoglucoside	рамноглюкозид
Rhamnoglycoside	рамногликозид
Rhododendrol	рододендрол
Roemeridine	ремеридин
Roemerine	рёмерин
Rosmarinic acid	розмариновая кислота
Royleanone	ройлеанон
Ruberythric acid	руберитриновая кислота
Rubiadin	рубиадин
Rubijervine	рубийервин
Rubiadine	рубиодин
Ruscogenin	рускогенин
Rutin	рутин
Rutinoside	рутиназид
Sabinene	сабинена
Salicylic acid	салициловая кислота
Salidroside	салидросид
Salvicin	сальвицин
Salvicinin	сальвицинин
Salvicinolide	сальвицинолид
Salvicinolin	салвицинолин
Salvifolin	сальвифолин
Salvin	сальвин
Sambulene	самбулен
Sanguinarine	сангвинарин
Santolina alcohol	сантолиновый спирт
Sapogenin	сапогенин
Saponarin	сапонарин
Saponin	сапонин
Saporubin	сапорубин
Saporubinic acid	сапорубиновая кислота
Sclareol	склареол
Scopolamine	скополамин
Scopoletin	скополетин
Scutellarin	скутелляреин
Selenium	селен
Semi-drying oil	полувысыхающее масло
Sesquiterpene	сесквитерпен
Sesquiterpene alcohol	сесквитерпеновый спирт
Sesquiterpene lactone	сесквитерпеновый лактон

Shepherin	шеперин
Silicic acid	кремневая кислата
Silicon	кремний
Silicristin	силикристин
Silimarin	силимарин
Silybin	силибин
Sinapic acid	синапиновая кислота
Sinigrin	синигрин
Skimmianine	скиммианин
Smirnovine	смирновин
Sogdisterone	согдистерон
Songorine	сонгорин
Sophocarpine	софокарпин
Sophoramine	софорамин
Sophoridine	софоридин
Spartein	спартеин
Sparteine	спартеин
Spathulenol	спатуленол
Spherophysine	сферофизин
Spherosine	сферозин
Spinasterol	спинастерол
Stachydrine	стахидрин
Stachyose	стахиоза
Starch	крахмала
Stearic acid	стеариновая кислота
Stearidonic acid	стеаридоновая кислота
Sterin	стерин
Steroid	стероид
Steroidal saponin	стероидный сапонин
Sterol	стерол
Stigmasterin	стигмастерин
Stigmasterol	стигмастерол
Stilbene derivatives	производные стильбена
Strychnine	стрихнин
Substance	вещество
Succinic acid	янтарная кислота
Sucrose	сахароза
Sugar	сахар
Sulfur	сера
Talatizamine	талатизамин
Talatizidine	талатизидин
Talatizine	талатизин
Talicmidine	таликмидин
Tannins	дубильные вещества
Tanshinone	таншинон
Taraxanthin	тараксантин
Taraxasterol	тараксастерол
Taraxerol	тараксерол
Taraxol	тараксол
Tartaric acid	винная кислота
Taspine	таспин
Tatsetine	тацеттин
Taxifolin	таксифолин
Taxodione	таксодион
Terpene	терпен
Terpinen-4-ol	терпинен-4-ол
Trpinene	терпенен
Terpineol	терпенеол
Terpinolene	терпинолен
Thalfine	тальфин
Thalfinine	тальфинин
Thalicmidine	таликмидин
Thalicmine	таликмин
Thalicminine	таликминин
Thalicmitrine	таликмитрин
Thalisopidine	тализопидин
Thalisopine	тализопин
Thalmine	тальмин
Thalminine	тальминин
Thermopsine	термопсин
Thermopsocide	термопсозид
Thioglycoside	тиогликозид
Thiramine	тирамин
Thujone	туйон
Thymol	тимол
Thymoquinone	тимохинон
Titratable organic acids	титруемые органические кислоты
Tocopherol	токоферол
Tolmetin	тальметин
Torachrysone	торахризон
Toxic	токсичный
Trace	следы
trans-carveol	транс-карвеол
Triacanthine	триакантин
Triacylglyceride	триацилглицерид
Triacylglycerol	триацилглицерол
Trichodesmine	триходесмин
Tricyclic	трициклический
Trifolin	трифолин
Trifoside	трифозид
Triglyceride	триглицерид
Triglycoside isorhamnetin	тригликозид изорамнетина
Trihydroxychalcone	тригидрооксихалкон
Trimethoxyl-cinnamic acid	триметоксил коричная кислота
Triterpene	тритерпен
Triterpene alcohol	тритерпеновый спирт
Triterpene glycoside	тритерпеновый гликозид
Triterpene saponin	тритерпеновый сапонин
Triterpenoid saponin	тритерпеноидный сапонин
Triterpenol	тритерпенол
Tropane alkaloid	тропановый алкалоид
Tropine	тропин
Tropinone	тропинон
Tropolone	трополон
Tungsten	вольфрам
Turkesterone	туркестерон
Tussilagin	туссилягин
Tyramine	тирамин
Tyrosine	тирозин

Tyrosol	тиросол
Umbelliferol	умбелиферол
Umbelliferone	умбеллиферон
Undecanoic acid	ундекановая кислота
Unsaturated fatty acid	ненасыщенная жирная кислота
Uronic acid	уроновая кислота
Ursane	урсан
Ursolic acid	урзоловая кислота
Urticin	уртицин
Vaillantine	вайлантин
Valerianic acid	валериановая кислота
Valeric acid	валериановая кислота
Valeride	валерид
Valerine	валерин
Valtrate	валтрат
Vanillic acid	ванилиновая кислота
Vasicinone	вазицинон
Verbascoside	вербаскозид
Verbenone	вербенон
Vincanidine	винканидин
Vincanine	винканин
Vinervine	винервин
Vinervinine	винервинин
Vin amine	винкамин
Vitamin	витамин
Vitexin	витексин
Vitexin-ramnoside	витексинрамнозид
Viticosterone	витикостерон
Volatile	летучий
Vulgarol	вулгарол
Waxes	воск
Xanthamine	ксантумин
Xanthanine	ксантоксин
Xanthanol	ксантанол
Xanthinine	ксантинин
Xanthinosin	ксантинозин
Xanthone	ксантон
Xanthophyll	ксантофилл
Xanthosine	ксантоксин
Xanthostrumarin	ксантострумарин
Xanthumanol	ксантуманол
Xyloglucoside	ксилоглюкозид
Zinc	цинк
Zygofabagine	зигофабагин
α-humulene	а-хумулен
α-hydrojuglone	а-гидроюглон
α-linolenic acid	а-линолиновая кислота
α-terpenyl-acetate	а-терпенилацетат
β-bisabolene	b-бизаболен
β-carboline	b-карболин
β-hydrojuglone	b-гидроюглон
β-sitosterin	b-ситостерин
γ-coniceine	g-конмцеином
γ-terpinene	g-терпинен

Appendix 3

English-Russian Translations of Medical Terms

Abdomen	живот
Abdominal cavity	брюшная полость
Abortifacient	абортирующее (средство)
Abrasion	ссадина
Abscess	нарыв, гнойник
Acetylcholine	ацетилхолин
Acetylcholinesterase	ацетилхолинэстераза
Ache	боль, угрь
Acute tests on animals	острые опыты на животных
Adaptagen	адаптоген
Adrenal gland	надпочечник
Adrenaline	адреналин
Adrenergic	адренергический
Aggregative properties	агрегационные свойства
Aimaline	аймалин
Air sickness	воздушная болезнь
Aldose reductase	альдоз редуктаза
Alimentary	алиментарный
Allergen	аллерген
Allergy	аллергия
Altitude sickness	горная болезнь
Amenorrhea	аменоррея
Anabolic activity	анаболическая активность
Anacidic gastritis	анацыдный гастрит
Anacidity	анацидный
Analeptic	аналептик
Analgesic	болеутоляющее, анальгезирующее (средство)
Anaphylactic shock	анафилактический шок
Anasarca	анасарка
Androgenic action	андрогенное действие
Anemia	малокровие, анемия
Anemonin	анемонин
Anesthesia	анестезия
Anesthetic	анестезирующий
Angiocholitis	ангиохолит
Angioprotector	ангиопротектор
Anorexia	анорексия
Antagonist	антагонист
Anthrax	сибирская язва
Anti-aggregant	антиагрегант
Anti-amnesic	антиамнезийный
Antiarrhythmic	антиаритмический
Anti-asthmatic	противоастматический
Antibacterial	антибактериальный, противобактериальный
Antibiotic	антибиотик
Anticarcinogenic	антикарциногенный
Anticonvulsive	противосудорожный
Anti-cough	противокашлевой
Antidiarrheal	противопоносный
Antiedemic	противоотёчный
Antiemetic	противорвотный
Anti-fibrillant	антифибрилят
Antifungal	антимикотический, противогрибковый
Antigonadotropic	антигонадотропный
Antihelminthic	антигельминтный
Antihistamine	антигистамин
Anti-HIV	анти-ВИЧ
Antihypertensive	противогипертонический
Antihypoxic	антигипоксический
Antihysteric	противоистерический
Anti-inflammatory	противовоспалительный
Antimicrobial	антимикробный
Antioxidant	антиоксидант
Antiparasitic	противопаразитарный
Antiproliferative	антипролиферативный
Antiprotist	протистоцидный
Antipyretic	жаропонижающий
Antisclerotic	антисклеротический
Antiseptic	противосептический
Antispasmodic	противоспазматический
Antithyroid	антитиреоидный
Antitumor	противоопухолевый
Antiulcerogenic	противоязвенный
Antivenom	противоядие
Aorta	аорта

S. Eisenman et al. (eds.), *Medicinal Plants of Central Asia: Uzbekistan and Kyrgyzstan*,
DOI 10.1007/978-1-4614-3912-7, © The Editor 2012

Apoptosis	апоптоз
Apoptotic effect	апоптический эффект
Appetite	аппетит
Arachno-encephalitis	арахноэнцефалит
Arachnoiditis	арахноидит
Arrhythmia	аритмия
Arterial pressure	артериальное давление
Arteriosclerosis	атеросклероз
Arthralgia	артралгия
Arthritis	артрит
Arthrosis	артроз
Ascariasis	аскаридоз
Asphyxia	асфиксия
Asthenia	астения
Asthma	астма
Astringent	вяжущее (средство)
Atonia	атония
Autoimmune	аутоиммунный
Autonomic ganglia	автономные ганглии
Avitaminosis	авитаминоз
Back pain	боли в пояснице
Bacteria	бактерии
Bactericidal	бактерицидный
Bacteriostatic	бактериостатический
Bed bug	клоп
Bedsores	пролежни
Bile	желчь
Bile duct	желчный путь
Bile-stimulant	желчестимулирующий
Bilirubin	билирубин
Bleeding, hemorrhaging	кровотечение
Bloated	вздутие
Blood	кровь
Blood circulation	кровообращение
Blood cleanser	кровоочистительный
Blood coagulation	свёртывание крови
Blood sugar	сахар крови
Bone	кость
Bone fracture	перелом кости
Bone marrow	костный мозг
Bradycardia	брадикардия
Bronchiectasis	бронхоэктаз
Bronchitis	бронхит
Bronchospasm	бронхоспазм
Bronchus	бронх
Brucellosis	бруцеллёз
Bruise	ушиб
Burn	ожог
Butyrylcholinesterase	бутирилхолинэстераза
Ca++ channel	кальциевые каналы
Ballus	мозоль
Calm	успокоение
Cancer	рак (болезнь)
Capillary	капилляр
Capillary strengthening	капилляроукрепляющее
Carbuncle	карбункул

Carcinogenesis	канцерогенез
Cardiosclerosis	кардиосклероз
Cardiotonic	кардиотонический
Caries	кариес
Carminative	ветрогонное (средство)
Carotid nerve	каротидный нерв
Catalepsy	каталепсия
Cataract	катаракта
Cell	клетка
Central nervous system	центральная нервная система
Cerebral cortex	головной мозг
Cervical erosion	эррозия шейки матки
Cervix	шейка матки
Chest	грудь
Cholagogue, choleretic	желчегонное (средство)
Cholecystitis	холецистит
Cholera	холера
Choleresis	желчеотделение
Choleretic action	желчегонное действие
Choleretic, cholagogue	желчегонное (средство)
Cholesterol	холестерин
Cholinergic	холинергический
Cholinesterase	холинэстеразный
Chorea	хорея
Chronic	хронический
Chronotropic	хронотропный
Coagulation	свёртывание
Coating (remedy)	обволакивающее (средство)
Colitis	колит
Common cold	простуда
Compress	компресс
Congenital defect	врождённый дефект
Congestion	закупорка
Conjunctivitis	конъюнктивит
Constipation	запор
Contraceptive	контрацептивный, противозачаточный
Contractility	сократимость
Contraction	сокращение
Convulsions	судорги
Coronary	коронарный
Cough	кашель
Croupous pneumonia	крупозная пневмония
Curare	кураре
Cuts	порезы
Cyclooxygenase (OX) enzyme	циклооксигеназный фермент (ЦОГ)
Cystitis	цистит
Cytotoxicity	цитотоксичность
Deafness	глухота
Decoction	отвар
Decompensation	декомпенсация
Demulcent	мягчительное (средство)
Depression (emotional)	депрессия (эмоциональная)
Depression (physical, physiological)	угнетение

Dermatitis	дерматит
Dermatosis	дерматоз
Desensitization	десенсибилизация
Detoxicant	детоксикант
Detoxify	детоксицировать
Diabetes	диабет
Diaphoretic, sudorific	потогонное (средство)
Diarrhea	понос, диарея
Diathesis	диатез
Digestion	пищеварение
Digestive organs	пищеварительные органы
Digestive system	пищеварительная система
Digestive tract	пищеварительный тракт
Diphtheria	дифтерия
Disease	заболевание
Disinfectant	обеззараживающий
Disinfection	дезинфекция
Dissolve	растворение
Diuresis	диурез
Diuretic	мочегонное (средство)
Dizziness	головокружение
DNA	ДНК
Duodenum	двенанадцатиперстная кишка
Dysentery	дизентерия
Dysmenorrhea	дисменорея
Dyspepsia	диспепсия, плохое пищеварение
Dyspnea	удушье
Dysuria	дизурия
Ear	ухо
Eczema	экзема
Edema	водянка, отёк
Elephantitis	слоновая болезнь
Emetic	рвотное (средство)
Emollient	мягчительное (средство)
Endarteritis	эндартериит
Endocytotic activity	эндоцитозная активаность
Endogenous	эндогенный
Endomyocarditis	эндомиокардит
Enterocolitis	энтероколит
Enzyme	фермент
Enzyme-stimulating	ферментостимулирование
Epilepsy	эпилепсия
Epithelization	эпителизация
Ergotism	эрготизм
Erysipelas	рожа (болезнь)
Erysipelatous inflammation	рожистое воспаление
Erythrocyte	эритроцит
Erythrodermia	эритродермия
Esophagus	пищевод
Estrogen	эстроген
Exophthalmic goiter	базедовая болезнь
Expectorant	отхаркивающее (средство)
External bleeding	наружное кровотечение
Exudative diathesis	экссудативный диатез
Eye	глаз
Facial paralysis	паралич лицевого нерва

Faint	обморок
Fatigue	упадок сил
Febrifuge	жаропонижающее (средство)
Festering wounds	гнойные раны
Fetal hypoxic hypotropia	гипоксическая гипотрофия плода
Fever	лихорадка
Fibrinolytic	фибринолитический
Fibroblast	фибробласт
Fibrosis	фиброз
Flea	блоха
Food poisoning	пищевое отравление
Fragmentation	фрагментация
Free radicals	свободные радикалы
Frostbite	обморожение
Fungal skin disease	лишай
Fungicide	фунгицид
Fungus	грибок
Furuncle	фурункул
Furunculosis	фурункулёз
Galenic preparation	галеновый препарат
Gallbladder	желчный пузырь
Gallstone	желчный камень
Ganglion-blocking	ганглиоблокирование
Gangrene	гангрена
Gargle	полоскание горла
Gastralgia	гастралгия
Gastric disease	желудочное заболевание
Gastric fluid	желудочный сок
Gastric pneumatosis	пневматоз желудка
Gastritis	гастрит
Gastroenteritis	гастроэнтерит
Gastrointestinal tract	желудочнокишечный тракт
Gene expression	экспрессиия ген
General tonic	общеукрепляющее (средство)
Genotoxic	генотоксический
Gingivitis	гингивит
Gland	железа
Goiter	зоб
Gonorrhea	гонорея
Gout	подагра
Gram-negative bacteria	грамотрицательные бактерии
Gram-positive bacteria	грамположительные бактерии
Granulation	грануляция
Guinea pig	морская свинка
Guinea worm	ришта, гвинейский червь
Gum	десна
Gynecological disorders	женские заболевания
Headache	головная боль
Heart	сердце
Heart failure	порок сердца
Helminthosis	гельментоз
Hematuria	гематурия
Hemodynamics	гемодинамика
Hemoglobin	гемоглобин
Hemolytic index	гемолитический индекс
Hemoptysis	кровохаркание

Hemorrhagic shock	геморрагический шок
Hemorrhaging, bleeding	кровотечение
Hemorrhoidal hemorrhage	геморроидальное кровотечение
Hemorrhoids	геморрой
Hemostatic	гемостатический, кровоостанавливающий
Hepatitis	гепатит
Hepatocholecystitis	гепатохолецистит
Hepatoprotector	гепатопротектор
Hepatotoxic	гепатотоксичный
Hernia	грыжа
Herpes	герпес
Herpes simplex type 1	герпис симплекс типа 1
Highmoritis	гайморит
Hippocampus	гиппокампа
Histamine	гистамин
Hoarseness	охриплость
Homeopathy	гомеопатия
Hydrophobia	водобоязнь
Hyperglycemia	гипергликемия
Hyperhydrosis	гипергидроз
Hyperpituitarism	гиперпитуитаризм
Hypertension	гипертония
Hypertensive	гипертензивный
Hyperthyroidism	гипертиреоз
Hypertrophy	гипертрофия
Hypoacidic gastritis	гипоацидный гастрит
Hypochondria	ипохондрия
Hypogastritis	гипогастрит
Hypoglycemic	гипогликемический
Hypolipidemic	гиполипидемический
Hypotension	гипотензия
Hypotensive	гипотензивный
Hypothermic	гипотермический
Hypotonia	гипотония
Hypoxia	гипоксия
Hysteria	истерия
Idiopathic skin atrophy	идиопатическая атрофия кожи
Immunological	иммунологический
Immunosuppression	иммуносупрессия
Immunotoxicity	иммунная токсикация
Implantation	имплантация
Impotence	импотенция
Infected	инфецированный
Infection	инфекция
Infectious diseases	инфекционные заболевания
Inflammation	воспалительный процесс
Influenza	грипп
Infusion	настой
Inhibition	ингибирование, задерживание, затормаживание
Inotropic action	инотропное действие
Insecticide	инсектицид
Insomnia	бессонница
Insulin	инсулин
Internal	внутренний
Intestinal colic	кишечные колики
Intestinal disorder	растройство кишечника
Intestines	кишки, кишечник,
Intoxication	интоксикация
Intracellular	внутриклеточный
Intracranial	внутричерепной
Intravenous injection	внутривенное вливание
Itch	зуд
Jaundice	желтуха
Joint	сустав
Kidney	почка
Kidney stone	почечный камень
Lactation	лактация
Lactogenic	лактогенный
Laryngitis	ларингит
Larynx	гортань
Laxative	слабительное (средство)
Leishmaniasis	лейшманиоз
Lethargic encephalitis	летаргический энцефалит
Leucocytes	лейкоциты
Leucomisine	леукомизин
Leucopenia	лейкопения
Leukemia	лейкемия
Leukorrhea	бели (болезнь)
Libido	либидо
Liver	печень
Local anesthesia	местная анестезия
Lotion	примочка
Low stomach acidity	пониженная кислотность желудка
Lungs	лёгкие
Lupus	волчанка (болезнь)
Lymph nodes	лимфатические узлы
Lymphadenectomy	лимфаденектомия
Lymphadenitis	лимфаденит
Lymphedema	лимфодемия
Lymphoblastoid	лимфобластоид
Macrophage	макрофаг
Malaria	малярия
Malignant	злокачественный
Mammalian	относящийся к млекопитающим
Mannose	манноза
Mastitis	мастит
Measles	корь
Mediator	медиатор
Medulla oblongata	продолговатый мозг
Melancholy	меланхолия
Menopause	климакс
Menorrhagia	меноррагия
Menstruation	менструация
Menstruation cycle	менструальный цикл
Metabolism	обмен веществ
Methicillin-resistance	метицилин-резистентный
Methicillin-sensitive	метицилин-чувствительный
Metropathy	метропатия

Metrorrhagia	метроррагия
Midbrain	средний мозг
Migraine	мигрень
Mitral failure	митральный порок
Molluscicidal	моллюскицидный
Mouse	мышь
Mouth wash	полоскание рта
Mucous membrane	слизистая оболочка
Mucus	слизи (носоглодки)
Multiple sclerosis	рассеянноый склероз
Muscle	мускул
Musculoskeletal	скелетно-мускульный
Myasthenia	миастения
Mydriasis	мидриатический
Mydriatic	мидриатик
Myocardial infarction	инфаркт миокарда
Myocarditis	миокардит
Myodystrophy	миодистрофия
Myometrium	миометрия
Myopathy	миопатия
Narcosis	наркоз
Narcotic	наркотик
Narcotized	наркотизированный
Nausea	тошнота
Necrosis	некроз
Nephritis	нефрит
Neural	невральный
Neuralgia	невралгиия
Neurasthenia	неврастения
Neuritis	неврит
Neurodermatitis	нейродермотит
Neuromuscular	нервно-мышечный
Neuron	нейрон
Neuroprotective	нейрозащитный
Neurosis	невроз
Noradrenaline	норадреналин
Nose	нос
Obesity	ожирение
Obstetric-gynecological	акушерско-гинекологичкский
Ointment	мазь
Osteitis	остит
Osteoarthritis	остеоартрит
Osteomyelitis	остеомиелит
Otitis	отит
Otolaryngology	отоларингология
Pancreas	поджелудочная железа
Papilloma	папиллома
Paradontosis	парадонтоз
Paralysis	паралич
Paralytic	паралитик
Parasite	паразит
Parasympathetic ganglions	парасимпатические ганглии
Paratyphoid	паратиф
Parenteral	парэнтеральный
Parkinson's disease	болезнь Паркинсона
Paronychia	панариций
Parturifacient	родовспомогательный
Paste	паста
Pathogenic	потогенность
Pediculosis (lice infestation)	педикулёз
Pellagra	пеллагра
Penicillin	пеницелин
Pepsin	пепсин
Pepsinogen	пепсиноген
Periodontal disease	пародантоз
Periodontitis	периодонтит
Periostitis	периостит
Peripheral	периферический
Peristalsis	перистальтика
Peritoneal	перитонеальный
Permeability	проницаемость
Pertussis	коклюш
Phagocyte	фагоцит
Phagocytic	фагоцитарный
Pharmacological	фармакологический
Pharyngitis	фарингит
Phthisis	чахотка
Phytoestrogen	фитоэстроген
Pimple	прыщик
Pinworm	острица
Plague, pestilence	чума
Plasma	плазма
Platelet activating factor (PAF)	фактор активации тромбоцитов (ФАТ)
Pleurisy	плеврит
Pneumonia	пневмония, воспаленние лёгких
Poison	яд, отрава
Poliomyelitis	полиомиелит
Poliovirus	полиовирус
Polyarthritis	полиартрит
Polyp	полип
Polyvitamin	поливитамин
Postencephalitic	постэнцефалитический
Poultice	припарка
Powder	порошок
Pressor action	прессорное действие
Proliferation	пролиферация
Prophylactic	профилактика
Prostaglandin	простагландин
Prostate	предстательная железа (простата)
Prostitis	простатит
Prothrombin	протромбин
Psoriasis	псориаз
Psychiatry	психиатрия
Psychomimetic	психомиметический
Psychoneurological diseases	психоневрологические заболевания
Pulpitis	пульпит
Pupil	зрачок
Purgative	рвотный
Pus	гной

Pyelitis	пиелит
Pyoderma	пиодермия
Quench thirst	жаждоутоляющее
Rabies	бешенство
Radiculitis	радикулит
Rash	крапивница, сыпь
Rat	крыса
Recalcification	рекальцификация
Receptor	рецептор
Rectal prolapse	выпадение прямой кишки
Reduced hair growth	плохой рост волос
Reflex excitability	рефлекторная возбудимость
Relax	расслабляться
Remedy	средство
Renal pelvis	почечная лоханка
Reoxygenation	реокисление
Residual	остаточный
Resistance	резистентность, устойчивость
Respiration	дыхание
Respiratory	респераторный
Respiratory disease	респираторное заболевание, болезнь органов дыхания
Restorative	общеукрепляющее
Reticular	ретикулярный
Rhabdomyolysis	рабдомиолиз
Rheumacarditis	ревмакардит
Rheumatic pain	ломота
Rheumatism	ревматизм
Rhinitis	ренит
Rickets	рахит
Rubella	краснуха
Salivation	слюнотечение
Scabies	чесотка
Scarlet fever	скарлатина
Sciatic nerve	седалищный нерв
Scleroderma	склеродермия
Scratch	царапина
Scrofula	скрофулёз, золотуха
Scurvy	цынга
Sea sickness	морская болезнь
Secretion	секреция
Secretory activity	секреторная деятельность
Secretory function	выделительная функуия
Sedative	седативный, успокаивающее (средство)
Seizure	припадок
Sensory	сенсорный
Septicemia	септицемия
Serotonin	серотонин
Shin	голень
Shortness of breath	одышка
Sialorrhea	слюнотечение
Sinus	синус (пазуха)
Sinus cold	насморк
Skin	кожа
Skin diseases	заболевания кожи
Skin ulcer	язва кожи
Sleep	сон
Smallpox	оспа
Smooth muscles	гладкая мускулатура
Snake bites	укус змеи
Snake venom	змеинный яд
Soporific	снотворное средство
Sore throat	ангина
Spasm	спазм
Spastic	спастический
Spastic paresis	спастический парез
Spazmophilia	спазмофилия
Spermatocidic	спермацидный
Spleen	селезёнка
Sprain	вывих
Staphylococcus	стафилокок
Stenocardia	стенокардия
Stimulate	стимулировать, возбуждать
Stomach	желудок
Stomach ache	боли желудка
Stomach catarrh	катар желудка
Stomatitis	стоматит
Streptococcus	стрептокок
Streptozotocin-induced	стрептозотоцин-индуцированный
Stress	стресс
Stress factor	стресфактор
Sudorific, diaphoretic	потогонное (средство)
Sunstroke	солнечный удар
Surgery	хирургия
Swelling	опухоль, оттёк
Sympathetic nervous system	симпатическая нервная система
Sympathomimetic	симпатомиметик
Synapse	синапс
Synergistic	синергичный
Syphilis	сифилис
Systolic	систолический
Tachycardia	тахикардия
Tachyphylaxis	тахифилаксия
Tapeworm	солитёр
Tetanus	столбняк (болезнь)
Throat	горло
Thrombocytes	тромбоциты
Thrombophlebitis	тромбофлебит
Thromboplastic activity	тромбопластическая активность
Thymus	тимус
Thyroid gland	щитовидная железа
Thyroid stimulating hormone (TSH)	тиреотропный гормон (ТТГ)
Thyroidtoxicosis	тиреотоксикоз
Thyroxine (T_4)	тироксин (Т4)
Tick	клещь
Tincture	тинктура, настоика
Tissue	ткань (клетки)
Tongue	язык
Tonic	тонизирующее, укрепляющее (средство)

Tonic action	тонизирующее действие
Tonsillitis	тонзиллит
Tonus	тонус
Tooth	зуб
Toothache	зубная боль
Toxic	токсичный
Tracheitis	трахеит
Tranquilizing effect	транквилизирующий эффект
Trichomoniasis	трихомониаз
Triiodothyronine (T_3)	трийодтиронин (Т3)
Trophic ulcer	трофическая язва
Trypanocidal	трипаноцидный
Tuberculosis	туберкулёз
Tubular necrosis	некроз трубчатых клеток почек
Tumor	опухоль
Tympanites	метеоризм
Typhoid fever	брюшной тиф
Ulcer	язва
Ulcer disease	язвенная болезнь
Upper respiratory	верхние дыхательные пути
Ureter	мочеточник
Urethra	уретра
Urinary incontinence	недержание мочи
Urinary tract	мочевыводящий путь
Urination disorders	расстройство мочеотделения
Urogenital	мочеполовой
Uterine atonia	антония матки
Uterine fibroids	фибриома матки
Uterine hemorrhages	маточное кровотечение
Uterine horn	рог матки
Uterine ulcers	язва матки
Uterus	матка
Vaccine	вакцина
Vagus nerve	блуждающий нерв
Vasoconstrictor	сосудосуживающий
Vasomotor center	сосудодвигательный центр
Vegetative neurosis	вегетативный невроз
Vein	вена
Venereal diseases	венерические болезни
Vermifuge	глистогонное (средство)
Vessel	сосуд
Veterinary medicine	ветеринария (ветеринарная медицина)
Virus	вирус
Vitiligo	витилиго
Vomit	рвота
Wart	бородавка
Weakened	ослабленный
Weakness	слабость
Weariness	усталость
Weight deficiency	низкий вес
Wound	рана
Wound healing	ранозаживляющее (средство)

References

Abdalla S, Al-Khalil S, Afifi F (1991) Observations on the pharmacology of thalicminine, an oxyaporphine alkaloid from *Thalictrum isopyroides*. Gen Pharmacol 22(2):253–257

Abdallah OM (1991) Phenolic glucoside and other constituents of *Dipsacus laciniatus*. Phytochemistry 30(8):2805–2806

Abdel Gawad MM, El Amin SM, Ohigashi H, Watanabe Y, Takeda N, Sugiyama H, Kawanaka M (2000) Molluscicidal saponins from *Anagallis arvensis* against schistosome intermediate hosts. Jpn J Infect Dis 53(1):17–19

Abdukadirov IT, Yakubova MR, Nuriddinov KhR, Mamatkhanov AU, Turakhozhaev MT (2005) Ecdysterone and turkesterone in *Ajuga turkestanica* determined by HPLC. Chem Nat Compd 41(4):475–476

Abdulladzhanova NG, Mavlyanov SM, Dalimov DN (2003) Polyphenols of certain plants of the Euphorbiaceae family. Chem Nat Compd 39(4):399–400

Abdullaeva DU, Shakirov R (2006) Alkaloids from *Korolkowia sewertzovii*. Sevedinine N-oxide. Chem Nat Compd 42(5):623–624

Abdunazarov SKh (2000) Plants of the Fergana valley – *Pulicaria* Gaertn. Bioecological peculiarities. Synopsis of dissertation for the degree of Ph.D. in biology, Kokand (in Uzbek)

Abdurakhmanov TR (1962) About the pharmacology of some species of *Lagochilus*. Synopsis of dissertation for the degree of Ph.D. in medicine, Karaganda (in Russian)

Abduzhabbarova S, Maekh SKh, Yunusov SYu, Yagudaev MR, Kurbakov D (1978) Alkaloids of *Thalictrum isopyroides*. Chem Nat Compd 14(4):400–402

Abourashed E, El-Alfy A, Khan I, Walker L (2003) Ephedra in perspective–a current review. Phytother Res 17(7):703–712

Abuharfeil NM, Maraqa A, Von Kleist S (2000) Augmentation of natural killer cell activity in vitro against tumor cells by wild plants from Jordan. J Ethnopharmacol 71(1–2):55–63

Aida K, Tawata M, Shindo H, Onaya T, Sasaki H, Yamaguchi T, Chin M, Mitsuhashi H (1990) Isoliquiritigenin: a new aldose reductase inhibitor from glycyrrhizae radix. Planta Med 56(3):254–258

Aitbaeva RKh (1972) Antibacterial activities of a preparation of Tian Shan mountain ash. Zdravookhraneniye Kazakhstana (1):76–77 (in Russian)

Aizikov MI, Kurmukov AG, Syrov VN (1978) Physiological activity and correlative changes in protein, carbohydrate digestions and lipid exchange under influence of ecdizones and nerobols. In: Sadritdinov BS (ed) Pharmacology of natural substances. FAN, Tashkent, pp 107–125 (in Russian)

Aizikov MI, Rasulova SA, Kurmukov AG (1984) Hypocoagulatory action of amorphin in experimental atherosclerosis. Dokl Akad Nauk UzSSR (10):49–50

Aizikov MI, Syrov VN, Kurmukov AG, Khushbaktova ZA, Rasulova SA, Dzhukharova MKh, Malikov VM, Batyrov EKh (1986) Effect of polyphenols on lipid metabolism in hyperlipemia and atherosclerosis. Meditsinskii Zh Uzbekistana (12):45–48 (in Russian)

Aizikov MI, Kurmukov AG, Rasulova SA, Sidiakin GP, Sham'ianov ID, Malikov VM (1991) Angioprotecting and hypolipidemic activity of leucomisine on experimental atherosclerosis. Farmakol Toksikol (3):35–37 (in Russian)

Aizikov MI, Kurmukov AG, Rasulova SA (1993a) Influence of leucomisine on the regression of experimental atherosclerosis. Meditsinskii Zh Uzbekistana (4):72–75 (in Russian)

Aizikov MI, Prohorova IR, Kurmukov AG (1993b) Angioprotecting, hypolipidemic and atherosclerotic activity of austricine. In: Tezisui dokladov – 1 konferentsiyakardiologov Moldovui (Proceedings of the 1st conference of cardiologists of Moldova), Kishinev, pp 150–151

Akbarov ZS, Aliev KhU, Sultanov MB (1972) Antiarrhythmic action of thalisopine, novocainamide, and quinidine on experimental arrhythmia models. Farmakol. In: Sultanov MB (ed) Farmakologiia Alkaloidov i Ikh Proizvodnikh (Pharmacology of alkaloids and their derivatives). FAN, Tashkent, pp 129–133 (in Russian)

Akbarov ZS, Aliev KhU, Sultanov MB (1978) Antiarrhythmic action of alkaloids of bisbenzyl, isoquinoline and criptopine series. In: Sadritdinov BS (ed) Farmakologiya prirodnikh veshchestv (Pharmacology of natural substances). FAN, Tashkent, pp 11–29 (in Russian)

Akhmedhodzhaeva KhS (1978) The influence of the alkaloids of *Haplophyllum acutifolium* on sex cycle and reproductive function of rats. In: Sadritdinov BS (ed) Farmakologiya prirodnikh veshchestv (Pharmacology of natural substances). FAN, Tashkent, pp 51–56 (in Russian)

Akhmedhodzhaeva KhS, Kurmukov AG (1975) Haplamine pharmacology. Dokl Akad Nauk UzSSR (8):36 (in Russian)

Akhmedhodzhaeva KhS, Polievtsev NP (1963) Sedative properties of the alkaloid perforine and its synergy with soporifics and narcotics. In: Voprosy biologii i krayevoi meditsinui (Issues of Biology and Regional Medicine), no 4. Izdatelstvo Akademii Nauk UzSSR, Tashkent (in Russian)

Akhmedzhanova VI, Bessonova IA, Yunusov SYu (1974) Haplamine – a new alkaloid from *Haplophyllum perforatum*. Chem Nat Prod 10(1):121–122

Akhondzadeh S, Noroozian M, Mohammadi M, Ohadinia S, Jamshidi AH, Khani M (2003) *Melissa officinalis* extract in the treatment of patients with mild to moderate Alzheimer's disease: a double blind, randomised, placebo controlled trial. J Neurol Neurosurg Psychiatry 74(7):863–866

S. Eisenman et al. (eds.), *Medicinal Plants of Central Asia: Uzbekistan and Kyrgyzstan*,
DOI 10.1007/978-1-4614-3912-7, © The Editor 2012

Akimaliev AA, Putieva ZhM, Alimbaeva PK, Abubakirov NK (1989) Triterpene glucosides of *Dipsacus azureus*. III. Structure of dipsacobioside. Chem Nat Compd 25(2):174–176

Akopov IE (1981) Hemostatic plants. Meditsina, Tashkent (in Russian)

Akopov IE (1990) The most important native medicinal plants and their uses. Meditsina, Tashkent (in Russian)

Akopov IE, Konovalova VA, Mansurov MM (1958). Pharmacology of stachydrine hydrochloride. Farmakol Toksikol 21:44–46

Aksinenko SG, Klimentova DA, Pashinsky VG (2003) Cytoprotective effects of alcohol extract from above-ground part of *Fragaria vesca* L. under the conditions of intoxication by cyclophosphane. Rastitel'nye Resursy 39(4):130–134 (in Russian)

Al-Awadi F, Shoukry M (1988) The lipid lowering effect of an anti-diabetic plant extract. Acta Diabetol 25(1):1–5

Aldashev AA (1979) *Aconitum*: pharmacology, toxicology and uses. Ilim, Frunze (in Russian)

Al-douri NA (2000) A survey of medicinal plants and their traditional uses in Iraq. Pharm Biol 38(1):74–79

Al-Ghamdi MS (2001) The anti-inflammatory, analgesic and antipyretic activity of *Nigella sativa*. J Ethnopharmacol 76(1):45–48

Ali BH, Blunden G (2003) Pharmacological and toxicological properties of *Nigella sativa*. Phytother Res 17(4):299–305

Ali MS, Pervez MK, Saleem M, Tareen RB (2001) Haplophytin-A and B: the alkaloidal constituents of *Haplophyllum acutifolium*. Phytochemistry 57(8):1277–1280

Aliev RK (1945) A wound-healing preparation of the leaves of broad-leaf plantain. Farmatsiya (2):28–29, Meditsina, Moscow (in Russian)

Aliev KhU, Kamilov IK (1967) Pharmacology of protopine and its derivatives. In: Kamilov IK (ed) Farmakologiia Alkaloidov i Glikozidov (Pharmacology of alkaloids and glycosides). FAN, Tashkent, pp 176–178 (in Russian)

Aliev RK, Aliev ND, Rakhimova AK (1961) Materials about the study of the rhizomes of *Geum rivale*. Dokl Akad Nauk AzSSR 17(6):519–524 (in Russian)

Aliev KhU, Zakirov UB, Kamilov IK (1967) Pharmacology of the alkaloid hordenine and its derivatives. In: Kamilov IK (ed) Farmakologiia Alkaloidov i Glikozidov (Pharmacology of alkaloids and glycosides). FAN, Tashkent, pp 114–117

Aliev RK, Prilipko LI, Damirov IA, Islamova NA, Guseinov DYa (1972) Lekarstvenniye rasteniya Azerbaidzhana: spravochnik (Medicinal plants of Azerbaijan: manual). Guseinov DYa (ed). Azerbaidzhanskoye Knizhnoye Izdatelstvo, Baku

Alimbaeva PK (1961) About the pharmacology of *Lagochilus* growing in Kyrgyzstan, vol 15. Trudi Kyrgyzskogo Meditsinskogo Instituta, Frunze (in Russian)

Alimbaeva PK, Akimaliev AA (1975) Phytochemical study of *Scabiosa songarica* Schrenk. Farmatsiya 24(3):33–36, Meditsina, Moscow (in Russian)

Alimbaeva PK, Goncharova AV (1971) Dikorastushchiye lekarstvenniye rasteniya v Kirgizii (farmakolognosticheskiye issledovaniye) [Wild-growing medicinal plants of Kyrgyzstan (pharmagonostic studies)]. Kyrgyzstan, Frunze (in Russian)

Alimbaeva PK, Shambetov TM (1988) Current status of harvesting medicinal plant raw material in the Kyrgyz SSR. Rastitel'nye Resursy 24(2) (in Russian)

Alimbaeva PK, Mukhamedziev MM, Akimaliev AA (1986) Lekarstvenniye rasteniya semeistva vorcyankovuikh florui Kirgizii (Medicinal plants of the family Dipsacaceae from the flora of Kyrgyzstan). Ilim, Frunze (in Russian)

Ali-Shtayeh MS, Abu Ghdeib SI (1999) Antifungal activity of plant extracts against dermatophytes. Mycoses 42(11–12):665–672

Al-Sultan SI, Hussein YA, Hegazy A (2003) Toxicity of *Anagallis arvensis* plant. Pak J Nutr 2(3):116–122

Altimishev AA (1991) Prirodniye tselebnuiye sredstva, izd. 2 (Natural medicinal remedies), 2nd edn. Kyrgyzstan, Bishkek (in Russian)

Amirov MN (1974) A study of the connection between anti-tumor activity and the chemical structures of plant tannins. Zdravookhraneniye Kazakhstana (8):36–38 (in Russian)

Amoros M, Fauconnier B, Girre RL (1987) In vitro antiviral activity of a saponin from *Anagallis arvensis*, Primulaceae, against herpes simplex virus and poliovirus. Antiviral Res 8(1):13–25

Anderson KJ, Teuber SS, Gobeille A, Cremin P, Waterhouse AL, Steinberg FM (2001) Walnut polyphenolics inhibit in vitro human plasma and LDL oxidation. J Nutr 131(11):2837–2842

Aoyagi Y, Takahashi Y, Satake Y, Takeya K, Aiyama R, Matsuzaki T, Hashimotom S, Kurihara T (2006) Cytotoxicity of abietane diterpenoids from *Perovskia abrotanoides* and of their semisynthetic analogues. Bioorg Med Chem 14(15):5285–5291

Appolonova IB (1956) Influences of *Petrosimonia*, *Berteroa incana* and *Epilobium hirsutum* on the activity of the heart. Synopsis of dissertation for the degree of Ph.D. in medicine, Ivanovo (in Russian)

Arif R, Küpeli E, Ergun F (2004) The biological activity of *Centaurea* L. species. GU J Sci 17(4):149–164

Arifkhodzhaev AO, Rakhimov DA (1986) Polysaccharides of saponin-bearing plants. III. Polysaccharides of the epigeal organs of *Biebersteinia multifida*. Chem Nat Compd 22(6):719–720

Arifkhodzhaev AO, Rakhimov DA (1993) Polysaccharides of saponin-bearing plants. IV. Structure of glucans A, B, and C of *Biebersteinia multifida*. Chem Nat Compd 29(2):151–153

Arifkhodzhaev AO, Kondratenko ES (1983) Polysaccharides of saponin-bearing plants. I. Characteristics of the polysaccharides of Allochrusa gypsophiloides. Chem Nat Compd 19:222–223

Arifkhodzhaev AO, Arifkhodzhaev KhA, Kondratenko ES (1985) Polysaccharides of saponin-bearing plants. II. Isolation and characterization of the polysaccharides of *Biebersteinia multifida*. Chem Nat Compd 21(6):714–716

Atta AH, Mouneir SM (2004) Antidiarrhoeal activity of some Egyptian medicinal plant extracts. J Ethnopharmacol 92(2–3):303–309

Atta-ur-Rahman, Atia-tul-Wahab, Sultankhodzhaev MN, Teshebaeva UT, Choudhary MI (2005) Norditerpenoid alkaloids from *Aconitum karakolicum* Rapaics. Nat Prod Res 19(7):713–718

Awaad Amani AS, Maitland DJ, Soliman GA (2006) Antiulcerogenic activity of *Alhagi maurorum*. Pharm Biol 44(4):292–296

Azimov MA, Nazirov ZN (1969) A study of the triterpenes of *Euphorbia jaxartica*. Chem Nat Compd 5(6):517

Azimov MA, Nazirov ZN (1970) Flavonoids of *Euphorbia jaxartica* and *E. lamprocarpa*. Chem Nat Compd 6(2):268

Azizov KN, Aizikov MI, Akhmatova NF et al (1992) The influence of leucomisine on thrombocyte aggregation in healthy volunteers and in CHD patients. Meditsinskii Zh Uzbekistana (8):24–27 (in Russian)

Bagirov VY, Sheichenko VI, Gasanova RY, Pimenov MG (1978) The lactones of *Ferula malacophylla*. Chem Nat Compd 14(4):377–380

Baimukhamedov MA, Komissarenko NF (1990) Coumarins of *Helichrysum maracandicum*. Chem Nat Compd 25(5):617

Bakiev A, Makhkamov S (1987) Zelyonie i Malo Rasprostpanenniye Ovoshchniye Kulturui (Green and not widely used vegetable crops). Mekhnat, Tashkent (in Russian)

Balick MJ, Cox PA (1996) Plants, people, and culture: the science of ethnobotany. W.H. Freeman/Scientific American Library, New York

Bandyukova VA, Khalmatov KhKh, Alimov KhI (1969) Flavonoids of *Centaurea depressa*. Chem Nat Compd 5(4):274–275

Banthorpe DV, White JJ (1995) Novel anthraquinones from undifferentiated cell cultures of *Galium verum*. Phytochemistry 38(1):107–111

Barakat HH, Hussein SAM, Marzouk MS, Merfort I, Linscheid M, Nawwar MAM (1997) Polyphenolic metabolites of *Epilobium hirsutum*. Phytochemistry 46(5):935–941

Barnes J, Anderson LA, Phillipson DJ (2001) St John's wort (*Hypericum perforatum* L.): a review of its chemistry, pharmacology and clinical properties. J Pharm Pharmacol 53(5):583–600

Baser KHC, Nuriddinov KR, Nigmatullaev AM, Aripov KN (1997a) Essential oil of *Mentha asiatica* Boriss. from Uzbekistan. Plant Physiol Biochem 9(4):453–454

Baser KHC, Ozek T, Nuriddinov KhR, Nigmatullaev AM, Khadzimatov KKh, Aripov KhN (1997b) Essential oils of *Mediasia macrophylla* (Regel et Schmalh.) Pimen. and *Foeniculum vulgare* Mill. from Uzbekistan. J Essent Oil Res 9(2):249–250

Basher KHC, Ozek T, Demirchakmak B, Abduganiev BYo, Nuriddinov KhR, Aripov KhN, Doriev AS, Karataeva ChSh (1997) Essential oil of *Perovskia angustifolia* from Kyrgyzystan. Chem Nat Compd 33(3):296–298

Bedir E, Khan IA, Walker LA (2002) Biologically active steroidal glycosides from *Tribulus terrestris*. Pharmazie 57(7):491–493

Bekker NP, Ul'chenko NT, Glushenkova AI (2003) Lipids from the aerial part of *Crambe kotschyana*. Chem Nat Compd 39(2):218

Bekker NP, Ul'chenko NT, Glushenkova AI (2005) Lipids from *Descurainia sophia* seeds. Chem Nat Compd 41(3):346–347

Belodubrovskaya GA et al (2002) Encyclopedic dictionary of medicinal plants. http://www.spcpa.ru/slr/preview/articles/1.html. Accessed 22 Feb 2007 (in Russian)

Berezhinskaya VV, Trutneva EA (1959) Pharmacology of evoxin. Farmakol Toksikol 22(2):117–122 (in Russian)

Berezhinskaya VV, Aleshinskaya EE, Aleskina YaA (1968) Comparative pharmacological study of some alkaloids of aporphine group. Farmakol Toksikol 31(1):44–46 (in Russian)

Bishayee A, Sarkar A, Chatterjee M (1995) Hepatoprotective activity of carrot (*Daucus carota* L.) against carbon tetrachloride intoxication in mouse liver. J Ethnopharmacol 47(2):69–74

Blagojevic P, Radulovic N, Palic R, Stojanovic G (2006) Chemical composition of the essential oils of Serbian wild-growing *Artemisia absinthium* and *Artemisia vulgaris*. J Agric Food Chem 54(13):4780–4789

Blinova KF (1957) *Geum* species as tannid producing plants. Trudy Leningradskogo Khimiko-farmatsevticheskogo Instituta (2):80–90 (in Russian)

Blumenthal M (Senior ed) (1998) The complete German commission E monographs – therapeutic guide to herbal medicine. American Botanical Council, Austin

Bobokhanova LT, Bekturgunava SD (1996) Uzbekistan i Uzbeki (Uzbekistan and Uzbeks). Uzbekiston, Tashkent (in Russian)

Boguslavskaya LI, Dem'yanenko SI, Salam DK (1983) Flavonoids of some species of the genus *Dianthus*. Chem Nat Compd 19(3):366

Bohm BA, Stuessy TF (2001) Flavonoids of the sunflower family (Asteraceae). Springer, Vienna

Bondarenko NV (1972) Alkaloids of *Veratrum lobelianum*. IV. Chem Nat Compd 8(6):800–801

Bos R, Woerdenbag HJ, van Putten FM, Hendriks H, Scheffer JJ (1998) Seasonal variation of the essential oil, valerenic acid and derivatives, and velopotriates in *Valeriana officinalis* roots and rhizomes, and the selection of plants suitable for phytomedicines. Planta Med 64(2):143–147

Botirov EKh, Drenin AA, Makarova AV (2006) A chemical study of flavanoids from medicinal and edible plants. Khimiya rastitelnogo syrya (1):45–48 (in Russian)

Budincevic M, Vrbaski Z, Turkulov J, Dimíc E (1995) Antioxidant activity of *Oenothera biennis* L. Fett 97(7–8):277–280

Bykov BA (1950) Elovyye lesa Tyan'-Shanya ikh istoriya, osobennosti i tekhnologiya (Spruce forests of the Tian-Shan: their history, uniqueness and technology). Isdatelstvo Akademii Nauk, Alma-Ata, pp 21–27 (in Russian)

Cantrell CL, Abate L, Fronczek FR, Franzblau SG, Quijano L, Fischer NH (1999) Antimycobacterial eudesmanolides from *Inula helenium* and *Rudbeckia subtomentosa*. Planta Med 65(4):351–355

Capek P, Rosik J, Kardosova A, Toman R (1987) Polysaccharides from the roots of the marsh mallow (*Althaea officinalis* L., var. *Rhobusta*): structural features of an acidic polysaccharide. Carbohydr Res 164:443–452

Carnat AP, Carnat A, Fraisse D, Lamaison JL (1998) The aromatic and polyphenolic composition of lemon balm (*Melissa officinalis* L. subsp. *officinalis*) tea. Pharm Acta Helv 72(5):301–305

Carvalho JCT, Vignoli VV, de Souza GHB, Ujikawa K, Neto JJ (1999) Antimicrobial activity of essential oils from plants used in Brazilian popular medicine. Acta Hortic 501:77–81

Cervato G, Carabelli M, Gervasio S, Cittera A, Cazzola R, Cestaro B (2000) Anti-oxidant properties of oregano (*Origanum vulgare*) leaf extracts. J Food Biochem 24(6):453–456

Cession-Fossion A (1967) Action of sympathicomimetic amines on the adrenal medulla of the rat perfused in vitro. Arch Int Pharmacodyn Ther 167(2):486–487 (in French)

Cha JD, Jeong MR, Jeong SI, Moon SE, Kim JY, Kil BS, Song YH (2005) Chemical composition and antimicrobial activity of the essential oils of *Artemisia scoparia* and *A. capillaris*. Planta Med 71(2):186–190

Chan K, Islam MW, Kamil M, Radhakrishnan R, Zakaria MNM, Habibullah M, Attas A (2000) The analgesic and anti-inflammatory effects of *Portulaca oleracea* L. subsp. *sativa* (Haw.) Celak. J Ethnopharmacol 73(3):445–451

Chefranova (1954) The genus *Thermopsis* R. Br. and its industrial imporstance. Synopsis of dissertation for the degree of Ph.D. in biology, Leningrad (in Russian)

Chelombito VA, Muravyova DA (1971) About the pharmacology of the alkaloid chelerythrine isolated from bokkonii melkoplodnoi (*Bocconia microcarpa* Maxim). In: Aktualniye Problemui Farmakologii i Farmatsii (Actual problems in pharmacology and pharmacy), Moscow, pp 183–188 (in Russian)

Chemonics International Inc (2001) Biodiversity assessment for Uzbekistan, task order under the biodiversity & sustainable forestry. IQC (BIOFOR) USAID contract number: LAG-I-00-99-00014-00, submitted to: USAID Central Asian Republics Mission, Almaty. http://pdf.usaid.gov/pdf_docs/PNACN475.pdf. Accessed 22 Feb 2007

Chen G, Todorov DK, Zeller WJ (1992) Antitumor activity of thaliblastine (NSC-68075) in experimental ovarian tumor cell lines sensitive and resistant to cisplatin. Cancer Lett 62(2):173–177

Chen B, Teranishi R, Kawazoe K, Takaishi Y, Honda G, Itoh M, Takeda Y, Kodzhimatov OK (2000) Sesquiterpenoids from *Ferula kuhistanica*. Phytochemistry 54(7):717–722

Chen J, Shi Y, Liu J (2003) Determination of noradrenaline and dopamine in Chinese herbal extracts from *Portulaca oleracea* L. by high-performance liquid chromatography. J Chromatogr A 1003(1–2):127–132

Chen Z, Lu Y, Wang Z, Tao X, Wei D (2007) Protective effects of salidroside on hypoxia/reoxygenation injury by sodium hydrosulfite in pc12 cells. Pharm Biol 45(8):604–612

Cheney RH (1963) Therapeutic potential of Eschscholtziae californicae herba. Pharm Biol 3(3):413–416

Cheng G, Bai Y, Zhao Y, Tao J, Liu Y, Tu G, Ma L, Liao N, Xu X (2000) Flavonoids from *Ziziphus jujuba* Mill var. *spinosa*. Tetrahedron 56(45):8915–8920

Chen-Gu Ch (1957) Certain alkaloids from Chinese medicinal plants. Farmakol Toksikol 20(1):49–52 (in Russian)

Chernenko TV, Glushenkova AI, Kotenko LD (2000) Lipids of *Arum korolkowii* tubers. Chem Nat Compd 36(6):565–567

Chernenko TV, Glushenkova AI, Nigmatullaev AM (2002) Lipids from *Mediasia macrophylla* leaves. Chem Nat Compd 38(4):307–309

Chernenko TV, Gazizov FYu, Glushenkova AI, Nigmatullaev AM (2005) Lipids of *Eminium lehmannii* leaves and tubers. Chem Nat Compd 41(2):141–143

Chiang LC, Chiang W, Chang MY, Ng LT, Lin CC (2002) Antiviral activity of *Plantago major* extracts and related compounds in vitro. Antiviral Res 55(1):53–62

Chicco A (1966) *Prangos pabularia*, an Indian medicinal plant. Corriere del Farmacista 21(17):440–441

Chikov PS (1989) Лекарственные растения: справочник, 2-е изд, Lekarstvenniye rasteniya: spravochnik, izd 2 (Medicinal plants: manual, 2nd edn). Agropromizdat, Moscow [in Russian)

Cho J, Kim HM, Ryu JH, Jeong YS, Lee YS, Jin C (2005) Neuroprotective and antioxidant effects of the ethyl acetate fraction prepared from *Tussilago farfara* L. Biol Pharm Bull 28(3):455–460

Cho JY, Yoo ES, Cha BC, Park HJ, Rhee MH, Han YN (2006) The inhibitory effect of triterpenoid glycosides originating from *Sanguisorba officinalis* on tissue factor activity and the production of TNF-alpha. Planta Med 72(14):1279–1284

Chodoeva A, Bosc JJ, Guillon J, Decendit A, Petraud M, Absalon C, Vitry C, Jarry C, Robert J (2005) 8-O-Azeloyl-14-benzoylaconine: a new alkaloid from the roots of *Aconitum karacolicum* Rapcs. and its antiproliferative activities. Bioorg Med Chem 13(23):6493–6501

Chowdhury AR, Sharma S, Mandal S, Goswami A, Mukhopadhyay S, Majumder HK (2002) Luteolin, an emerging anti-cancer flavonoid, poisons eukaryotic DNA topoisomerase I. Biochem J 366(Pt 2):653–661

Chumbalov TK, Bikbulatova TN (1970) Polyphenolic compounds of *Geranium collinum*. IV. Chem Nat Compd 6(4):523–524

Chumbalov TK, Omurkamzinova VB (1968) Flavonoids of the leaves of *Polygonum coriarium*. I. Chem Nat Compd 4(5):272

Chumbalov TK, Bikbulatova TN, Chanysheva IS (1968) Flavone glycosides of *Geranium collinum*. III. Chem Nat Compd 4(6):323

Chumbalov TK, Mukhamed'yarova MM, Chanysheva IS (1970) Flavonoids of *Atraphaxis spinosa*. I. Chem Nat Compd 6(5):639

Chumbalov TK, Mukhamed'yarova MM, Chanysheva IS, Il'yasova MM (1971) Flavonoids of *Atraphaxis pyrifolia* and *spinosa*. II. Chem Nat Compd 7(4):504–505

Coates GH, Cox B (1972) Harmine tremor after brain monoamine oxidase inhibition in the mouse. Eur J Pharmacol 18(2):284–286

Colaric M, Veberic R, Solar A, Hudina M, Stampar F (2005) Phenolic acids, syringaldehyde, and juglone in fruits of different cultivars of *Juglans regia* L. J Agric Food Chem 53(16):6390–6396

Conti A (1962) Study of the antitumor activity of berberine, particularly of the protoberberine alkaloid group. Referativni Zh Biokhimiya (11):1464 (in Russian)

Curini M, Epifano F, Genovese S, Tammaro F, Menghini L (2006) Composition and antimicrobial activity of the essential oil of *Artemisia dracunculus* "Piemontese" from Italy. Chem Nat Compd 42(6):738–739

Danilevskii BL, Tulyaganov NT, Sadritdinov FS (1972) Ataractic actions of alkaloids haplofilidine, perforine, gentianine, and tetrahydroberberine compared to meprotan and aminazin. Dokl Akad Nauk UzSSR 29(3):37–38 (in Russian)

De Combarieu E, Fuzzati N, Lovati M, Mercalli E (2003) Furostanol saponins from *Tribulus terrestris*. Fitoterapia 74(6):583–591

de Meyts P, Cession-Fossion A (1966) Ephedrine as a sympathicomimetic amine with indirect action in the rat. Comptes Rendus des Séances de la Société de Biologie et de Ses Filiales 160(11):2224–2227 (in French)

de Meyts P, Cession-Fossion A (1967) D(−)ephedrine and L(+)psiephedrine as sympathomimetic amines with indirect action in the rat. Comptes Rendus des Séances de la Société de Biologie et de Ses Filiales 161(5):1162–1164 (in French)

de Meyts P, Cession-Fossion A (1968) Vasomotor action of D(−)-ephedprine and L(+) psi-ephedrine in rats. Arch Int Pharmacodyn Ther 174(1):233–242 (in French)

de Sousa AC, Alviano DS, Blank AF, Alves PB, Alviano CS, Gattass CR (2004) *Melissa officinalis* L. essential oil: antitumoral and antioxidant activities. J Pharm Pharmacol 56(5):677–681

Deepak M, Dipankar G, Prashanth D, Asha MK, Amit A, Venkataraman BV (2002) Tribulosin and β-sitosterol-D-glucoside, the anthelmintic principles of *Tribulus terrestris*. Phytomedicine 9(8): 753–756

Deineka VI, Grigor'ev AM, Borzenko ON, Staroverov VM, Trubitsyn MA (2004) HPLC analysis of anthocyanins: cyanidine glycosides from fruits of plants of the *Prunus* genus. Pharm Chem J 38(8):437–440

Delova GB, Guskova IN (1974) Antibacterial and antifungal effects of essential oil from some species of *Ziziphora*. In: Sobolevskaya KA (ed) Complex studies of the useful plants of Siberia. Nauka, Novosibirsk, pp 131–146 (in Russian)

Demirezer LO, Gurbuz F, Guvenalp Z, Stroch K, Zeeck A (2006) Iridoids, flavonoids and monoterpene glycosides from *Galium verum* subsp. *verum*. Turk J Chem 30(4):525–534

Derksen GC, Niederlander HA, van Beek TA (2002) Analysis of anthraquinones in *Rubia tinctorum* L. by liquid chromatography coupled with diode-array UV and mass spectrometric detection. J Chromatogr A 978(1–2):119–127

Desbène S, Hanquet B, Shoyama Y, Wagner H, Lacaille-Dubois M (1999) Biologically active triterpene saponins from callus tissue of *Polygala amarella*. J Nat Prod 62(6):923–926

Dey M, Ripoll C, Pouleva R, Dorn R, Aranovich I, Zaurov D, Kurmukov A, Eliseyeva M, Belolipov I, Akimaliev A, Sodombekov I, Akimaliev D, Lila MA, Raskin I (2008) Plant extracts from central Asia showing antiinflammatory activities in gene expression assays. Phytother Res 22(7):929–934

Dilman VM, Kozhina IC, Klyuchnikova LA, Kibalchich PN, Kusov YuYu (1968) Study of the abilities and consistuents of *Lithospermum officinale*: perspectives on use in oncology. Vopr Onkol 14(7):86–91 (in Russian)

Dobrokhotova K, Chudinov V (1966) Medicinal plants of Kazakhstan. Kazakhstan, Alma-Ata (in Russian)

Dolya VS, Fursa NS, Litvinenko VI (1973a) Flavonoids of the seeds of some species of the family Cruciferae. Chem Nat Compd 9(6):766

Dolya VS, Shkurupii EN, Fursa NS (1973b) The oils of four plants. Chem Nat Compd 9(5):578–579

Dozortseva PM (1958) Mellictine – a new preparation for treatment of spastic paralysis. Meditsinskaya promuishlennost SSSR (11):54 (in Russian)

Dozortseva PM (1959) The pharmacology of the alkaloid methyllycaconitine (mellictine). Farmakol Toksikol 22(1):34–38 (in Russian)

Du J, He ZD, Jiang RW, Ye WC, Xu HX, But PPH (2003) Antiviral flavonoids from the root bark of *Morus alba* L. Phytochemistry 62(8):1235–1238

Dvornyk AS, Pererva TP, Kukakh VA (2002) Screening substances derived from cultures of medicinal plants for antimutagenic activity in the *Escherichia coli*-bacteriophage lambda system. Tsitologii i Genetika 36(2):3–10 (in Ukrainian)

Dzhakupova NU (1968) The medicinal plant *Delphinium confusum* in the western Tian Shan. Vestnik Akademii Nauk KazSSR (11):66–68 (in Russian)

Dzhangaliev AD, Salova TN, Turekhanova PM (2003) The wild fruit and nut plants of Kazakhstan. Hortic Rev 29:305–371

Dzhumabaev TZ (1972) The pharmacology of *Berberis oblonga* and *Berberis nummularia*. Synopsis of dissertation for the degree of Ph.D., Ryazan

Dzhumabaev TZ, Ibragimov II, Lutfullin KL (1970) Effect of the sum of alkaloids in the root and bark of the stem of the oblong barberry on the blood coagulation system. Meditsinskii Zh Uzbekistana (8):22–25

Dzhumaev K (1980) The most prospective wild-growing aromatic plants of the Surkhan-Sherabad valley and their economic value. In: Aktualnuiye voprosy izucheniya I ispolzobaniya efirnikh masel (Proceedings of 3rd symposium "Important questions in the study of using essential oil bearing plants and essential oil"). Simferopol, 24–26 Sept 1980, pp 96–97 (in Russian)

Eliseeva M (2005) Internation cooperative biodiversity groups (ICBG) Central Asia 2005 progress report (unpublished report)

El-Mallah MH, El-Shami SM, Hassanein MM (2003) Detailed studies on some lipids of *Silybum marianum* (L.) seed oil. Grasas y Aceites 54(4):397–402

El-Razek MHA, Ohta S, Ahmed AA, Hirata T (2001) Monoterpene coumarins from *Ferula ferulago*. Phytochemistry 57(8):1201–1203

Emami SA, Amin-Ar-Ramimeh E, Ahi A, Kashy MRB, Schneider B, Iranshahi M (2007) Prenylated flavonoids and flavonostilbenes from *Sophora pachycarpa* roots. Pharm Biol 45(6):453–457

Enneking D (1995) The toxicity of *Vicia* species and their utilisation as grain legumes, vol 6, Centre for legumes in mediterranean agriculture (CLIMA) occasional publication. University of Western Australia, Nedlands

Erdogrul OT (2002) Antibacterial activities of some plant extracts used in folk medicine. Pharm Biol 40(4):269–273

Ermishina OA, Kurmukov AG, Salikhova RE, Aizikov MI (1982) Effect of ecdysterone on enzyme levels during experimental myocardial infarction. Meditsinskii Zh Uzbekistana (11):90–91 (in Russian)

Ersoz T, Calis I (1991) C-glucosylflavones from *Gentiana olivieri*. Hacettepe Universitesi Eczacilik Fakultesi Dergisi 11:29–36

Eshbakova KA, Saidkhodzhaev AI (2001) Triterpenoids and sterols from three species of *Pulicaria*. Chem Nat Compd 37(2):196–197

Eshbakova KA, Sagitdinova GV, Levkovich MG, Rasulev BF, Abdullaev ND, Malikov VM (1997) Diterpenoids of *Pulicaria salviifolia* IV. Structures of salvicinolide and salvicinolin. Chem Nat Compd 33(4):458–461

Esmaeili A, Rustaiyan A, Nadimi M, Masoudi S, Tadavon F, Sedaghat S, Ebrahimpur N, Haiyzadeh E (2005) Volatile constituents of *Centaurea depressa* M.B. and *Carduus pycnocephalus* L. two compositae herbs growing wild in Iran. J Essent Oil Res 17(5):539–541

Estrada A, Katselis GS, Laarveld B, Barl B (2000) Isolation and evaluation of immunological adjuvant activities of saponins from *Polygala senega* L. Comp Immunol Microbiol Infect Dis 23(1):27–43

Evans WC (2002) Production of crude drugs. In: Evans WC (ed) Trease and Evans pharmacognosy, 15th edn. W.B. Saunders, Edinburgh, pp 61–66

Evdokimova NI, Kurmukov AG (1972) Effect of skimmianine in comparison with meprotan on a change in the EEG (electroencephalogram) of rabbits. Meditsinskii Zh Uzbekistana (5):61 (in Russian)

Fakhrutdinov SF (1971) Comparative pharmacological study of the methiodides of corytuberine alkaloids and iodides of magnoflorine and O, O-diacetylmagnoflorine. In: Sultanov MB (ed) Farmakologiia Alkaloidov i Serdechnykh Glikozidov. FAN, Tashkent, pp 155–158 (in Russian)

Fakhrutdinov SF, Sultanov MB (1972) Pharmacology of liriodenine, thalicminine, corydine and isocorydine hydrochlorides. In: Sultanov MB (ed) Farmakologiia Alkaloidov Serdechnykh Glikozidov. FAN, Tashkent (in Russian)

Faramarzi MA, Moghimi M, Monsef-Esfahani HR, Shahverdi AR, Khodaee S (2008) Chemical composition and antimicrobial activity of essential oils from *Geum kokanicum*. Chem Nat Compd 44(6):811–813

Farsam H, Amanlou M, Dehpour AR, Jahaniani F (2000) Anti-inflammatory and analgesic activity of *Biebersteinia multifida* DC. root extract. J Ethnopharmacol 71(3):443–447

Farzami B, Ahmadvand D, Vardasbi S, Majin FJ, Khaghani Sh (2003) Induction of insulin secretion by a component of *Urtica dioica* leave extract in perifused Islets of Langerhans and its in vivo effects in normal and streptozotocin diabetic rats. J Ethnopharmacol 89(1):47–53

Fazly Bazzaz BS, Haririzadeh G (2003) Screening of Iranian plants for antimicrobial activity. Pharm Biol 41(8):573–583

Fecka I, Cisowski W (2005) Tannins and flavonoids from the *Erodium cicutarium* herb. Z Naturforsch B: Chem Sci 60(5):555–560

Fecka I, Gasiorowski K, Brokos B (1997) The phytochemical analysis and immunotropic activity evaluation of polyphenolic fraction from the common storks bill herb (*Erodium cicutarium* (L.) L'Herit.). Herba Pol 43(3):214–221

Feng YL, Li HR, Xu LZ, Yang SL (2007) 27-Nor-triterpenoid glycosides from the barks of *Zygophyllum fabago* L. J Asian Nat Prod Res 9(6):505–510

Fontana E, Hoeberechts J, Nicola S, Cros V, Battista Palmegiano G, Peiretti PG (2006) Nitrogen concentration and nitrate/ammonium ratio affect yield and change the oxalic acid concentration and fatty acid profile of purslane (*Portulaca oleracea* L.) grown in a soilless culture system. J Sci Food Agric 86(14):2417–2424

Frank BS, Michelson WB, Panter KE, Gardner DR (1995) Ingestion of poison hemlock (*Conium maculatum*). West J Med 163(6):573–574

Fruentov NK (1972) Lekarstvennuiye rasteniya Dalnego Vostoka (Medicinal plants of the Far East). Khabarovskoye knizhhoye izdatelstvo, Khabarovsk (in Russian)

Fu B, Su BN, Takaishi Y, Honda G, Ito M, Takeda Y, Kodzhimatov OK, Ashurmetov O (2001) A bis-sesquiterpene and sesquiterpenolides from *Inula macrophylla*. Phytochemistry 58(7):1121–1128

Fu Y, Hsieh TC, Guo J, Kunicki J, Lee MY, Darzynkiewicz Z, Wu JM (2004) Licochalcone-A, a novel flavonoid isolated from licorice root (*Glycyrrhiza glabra*), causes G2 and late-G1 arrests in androgen-independent PC-3 prostate cancer cells. Biochem Biophys Res Commun 322(1):263–270

Fukudu T, Ito H, Yoshida T (2003) Antioxidative polyphenols from walnuts (*Juglans regia* L.). Phytochemistry 63(7):795–801

Fursa NS, Litvinenko VI (1970) Lepidoside from *Lepidium perfoliatum*. Chem Nat Compd 6(1):119

Fushiya S, Kabe Y, Ikegaya Y, Takano F (1998) (+)-rhododendrol and epi-rhododendrin suppress the NO production by activated macrophages in vivo. Planta Med 64(7):598–602

Gadgoli C, Mishra SH (1999) Antihepatotoxic activity of p-methoxy benzoic acid from *Capparis spinosa*. J Ethnopharmacol 66(2): 187–192

Gammerman AF (1960) Kurs Farmakognozii (Pharmacognosy course), Medgiz (in Russian)

Gammerman AF (1967) Kurs Farmakognozii, izd. 6 (Pharmacognosy course), 6th edn. Meditsina, Leningrad (in Russian)

Gammerman AF, Kadaev GN, Yatsenko-Hmelevsky AA (1990) Lekarstvennuiye Rasteniya (Medicinal plants). Vuishaya shkola, Moscow (in Russian)

Gan PA (1970) Ekologicheskiye osnovui introduktsii I lesorazvedeniya v poyasye yelovuikh lesov Tyan-Shan (Basic ecology of introduction and reforestation in the Spruce Belt of the Tian-Shan). Ilim, Frunze (in Russian)

Gan LX, Han XB, Chen YQ (1986a) Astrasieversianins IX, XI and XV, cycloartane derived saponins from *Astragalus sieversianus*. Phytochemistry 25(6):1437–1441

Gan LX, Han XB, Chen YQ (1986b) The structures of thirteen astrasieversianins from *Astragalus sieversianus*. Phytochemistry 25(10):2389–2393

Ganenko TV, Isaev MI, Gromova AS, Abdullaev ND, Lutskii VI, Larin MF, Semenov AA, Gorovits MB, Abubakirov NK (1986) Triterpene glycosides and their genins from *Thalictrum foetidum*. IV. Structure of cyclofoetigenin B. Chem Nat Compd 22(3):315–319

Gapparov AM, Razzakov NA, Aripova SF (2007) Alkaloids of *Convolvulus subhirsutus* from Uzbekistan. Chem Nat Compd 43(3):291–292

Garbacki N, Tits M, Angenot L, Damas J (2004) Inhibitory effects of proanthocyanidins from *Ribes nigrum* leaves on carrageenin acute inflammatory reactions induced in rats. BMC Pharmacol 4(1):25

Gauthaman K, Ganesan AP, Prasad RN (2003) Sexual effects of puncturevine (*Tribulus terrestris*) extract (protodioscin): an evaluation using a rat model. J Altern Complement Med 9(2):257–265

Gazizov FYu, Glushenkova AI (1997) Phospholipids of the seeds of *Fumaria vaillantii*. Chem Nat Compd 33(6):627–629

Geetha S, Sai Ram M, Singh V, Ilavazhagan G, Sawhney RC (2002) Anti-oxidant and immunomodulatory properties of seabuckthorn (Hippophae rhamnoides)–an in vitro study. J Ethnopharmacol 79(3):373–378

Georgadze VH (1938) Pharmacology of the alkaloids from *Sophora*. Fiziologicheskii Zh SSSR 25(1–2):179–195 (in Russian)

Germano MP, De Pasquale R, D'Angelo V, Catania S, Silvari V, Costa C (2002) Evaluation of extracts and isolated fraction from *Capparis spinosa* L. buds as an antioxidant source. J Agric Food Chem 50(5):1168–1171

Ghafari H, Yasa N, Mohammadirad A, Dehghan G, Zamani MJ, Nikfar S, Khorasani R, Minaie B, Abdollahi M (2006) Protection by *Ziziphora clinopoides* of acetic acid-induced toxic bowel inflammation through reduction of cellular lipid peroxidation and myeloperoxidase activity. Hum Exp Toxicol 25(6): 325–332

Ghannadi A, Zolfaghari B (2003) Compositional analysis of the essential oil of *Lallemantia royleana* (Benth. in Wall.) Benth. from Iran. Flavour Frag J 18(3):237–239

Gibbons S, Oluwatuyi M, Veitch NC, Gray AI (2003) Bacterial resistance modifying agents from *Lycopus europaeus*. Phytochemistry 62(1):83–87

Gilani AH, Janbaz KH, Lateef A, Zaman M (1994) Ca^{++} channel blocking activity of *Artemisia scoparia* extract. Phytother Res 8(3):161–165

Glasl S, Presser A, Gunbilig D, Werner I, Narantuya S, Haslinger E, Jurenitsch J, Kubelka W (2001) Highly hydroxylated guaianolides of *Achillea asiatica* and Middle European *Achillea* species. Phytochemistry 58(8):1189–1194

Goel HC, Prasad J, Singh S, Sagar RK, Prem Kumar I, Sinha AK (2002) Radioprotection by a herbal preparation of *Hippophae rhamnoides*, RH-3, against whole body lethal irradiation in mice. Phytomedicine 9(1):15–25

Golcu A, Dolaz M, Dığrak M, Serin S (2002) The biological activity of dyer's madder (*Rubia tinctorium* L.). In: Yayli N, Kucuk M (eds) Proceedings of the 1st international congress on the chemistry of natural products. Karadeniz University, Trabzon, pp 255–258

Golovina LA, Nikonov GK (1973) The structure of drupanol – a new phenol from *Psoralea drupacea*. Chem Nat Compd 9(1):7–9

Golovkova AG (1990) Rastitelnost Kirgizii (ratsionalnoye ispolzovaniye i okhrana) [Vegetation of Kyrgyzstan (rational use and protection)]. Ilim, Frunze (in Russian)

Gorkin VZ (1964) Current progress in the study of the nature and physiological role of mitochondrial monoamine oxidase. Voprosy meditsinskoi khimii (10):115–134 (in Russian)

Gozler B, Gozler T, Freer AJ, Shamma M (1988) Some unusual pavine and isopavine alkaloids from *Roemeria refracta*. J Nat Prod 51(4):760–764

Gozler B, Kivçak B, Gozler T, Shamma M (1990a) Two new benzyltetrahydroisoquinoline alkaloids from *Roemeria refracta*. J Nat Prod 53(3):666–668

Gozler B, Ozic P, Freyer AJ, Shamma M (1990b) Morphinandienone alkaloids from *Roemeria refracta*. J Nat Prod 53(4):986–988

Greger H, Hofer O (1980) New unsymmetrically substituted tetrahydrofurofuran lignans from *Artemisia absinthium*. Assignment of the relative stereochemistry by lanthanide induced chemical shifts. Tetrahedron 36(24):3551–3558

Grinkevich NE (Senior ed) (1991) Lekarstvennuiye Rasteniya: Sprav. Posobiye (Medicinal plants: infomative manual). Vuishaya shkola, Moscow (in Russian)

Gudej J (1991) Flavonoids, phenolic acids and coumarins from the roots of *Althaea officinalis*. Planta Med 57(3):284–285

Gudej J, Tomczyk M (2004) Determination of flavonoids, tannins and ellagic acid in leaves from *Rubus* L. species. Arch Pharm Res 27(11):1114–1119

Guil-Guerrero JL, Rodríguez-García I (1999) Lipids classes, fatty acids and carotenes of the leaves of six edible wild plants. Eur Food Res Technol 209(5):313–316

Guil-Guerrero JL, García Maroto FF, Giménez A (2001) Fatty acid profiles from forty-nine plant species that are potential new sources of γ-linolenic acid. J Am Oil Chem Soc 78(7):677–684

Gulcin I, Kufrevioglu OI, Oktay M, Buyukokuroglu ME (2004) Antioxidant, antimicrobial, antiulcer and analgesic activities of nettle (*Urtica dioica* L.). J Ethnopharmacol 90(2–3):205–215

Gulluce M, Sahin F, Sokmen M, Ozer H, Daferera D, Sokmen A, Polissiou M, Adiguzel A, Ozkan H (2007) Antimicrobial and antioxidant properties of the essential oils and methanol extract from *Mentha longifolia* L. ssp. *Longifolia*. Food Chem 103(4): 1449–1456

Gupta SPK (2002) Rustom Jal Vakil (1911–1974) father of modern cardiology – a profile. J Indian Acad Clin Med 3(1):100–104

Gurbuz I, Ustun O, Yesilada E, Sezik E, Kutsal O (2003) Anti-ulcerogenic activity of some plants used as folk remedy in Turkey. J Ethnopharmacol 88(1):93–97

Gusakova SD, Khomova TV (1997) Lipids of *Ziziphora pedicellata*. Chem Nat Compd 33(6):633–635

Gusakova SD, Kholmatov KhK, Umarov AU (1976) Composition of the seed oils of *Origanum tyttanthum* and *Mentha asiatica*. Chem Nat Compd 12(2):134–139

Gusinin IA (1962) Toksikologiya Yadovituikh Rasteniya (Toxicology of Poisonous Plants). Izdaniye selkhozliteraturi zhurnalov i plakatov, Moscow (in Russian)

Haddad M, Larchevêque M (2003) Diastereocontrolled synthesis of (−)–codonopsinine. Synlett 2:274–276

Hamilton AC (2004) Medicinal plants, conservation and livelihoods. Biodivers Conserv 13:1477–1517

Han JW, Lee BG, Kim YK, Yoon JW, Jin HK, Hong S, Lee HY, Lee KR, Lee HW (2001) Ergolide, sesquiterpene lactone from *Inula britannica*, inhibits inducible nitric oxide synthase and cyclo-oxygenase-2 expression in RAW 264.7 macrophages through the inactivation of $NF\text{-}_{K}B$. Br J Pharmacol 133(4):503–512

Harrison DM (1990) Steroidal alkaloids. Nat Prod Rep 7(2):139–147

Hartleb I, Seifert K (1995) Triterpenoid saponins from *Verbascum songaricum*. Phytochemistry 38(1):221–224

Hofer O, Szabó G, Greger H (1986) 2-Hydroxy-4-methoxy-trans-cinnamic acid as a precursor of herniarin in *Artemisia dracunculus*. Org Chem Biochem 117(10):1219–1222

Hosseinimehr SJ, Pourmorad F, Shahabimajd N, Shahrbandy K, Hosseinzadeh R (2007) In vitro antioxidant activity of *Polygonium hyrcanicum*, *Centaurea depressa*, *Sambucus ebulus*, *Mentha spicata* and *Phytolacca americana*. Pak J Biol Sci 10(4):637–640

Hou BL, Li ZL, Li X (2005) A new flavonoid glycoside from the roots and stems of *Sphaerophysa salsula*. Yao Xue Xue Bao 40(6):533–535 (in Chinese)

Hsu FL, Chen YC, Cheng JT (2000) Caffeic acid as active principle from the fruit of *Xanthium strumarium* to lower plasma glucose in diabetic rats. Planta Med 66(3):228–230

Hu C, Kitts DD (2004) Luteolin and luteolin-7-O-glucoside from dandelion flower suppress iNOS and COX-2 in RAW264.7 cells. Mol Cell Biochem 265(1–2):107–113

Hu KF, Hou XY, Gong YT, Abbas A (2006) Analysis of physical-chemical properties and fatty acid compositions of *Hibiscus trionum* L. seed oil. Food Sci 27(11):455–457 (in Chinese)

Hvattum E (2002) Determination of phenolic compounds in rose hip (*Rosa canina*) using liquid chromatography coupled to electrospray ionisation tandem mass spectrometry and diode-array detection. Rapid Commun Mass Spectrom 16(7):655–662

Hwang SB, Chang MN, Garcia ML, Han QQ, Huang L, King VF, Kaczorowski GJ, Winquist RJ (1987) L-652,469 – a dual receptor antagonist of platelet activating factor and dihydropyridines from *Tussilago farfara* L. Eur J Pharmacol 141(2):269–281

Iacobellis NS, Cantore PL, Capasso F, Senatore F (2005) Antibacterial activity of *Cuminum cyminum* L. and *Carum carvi* L. essential oils. J Agric Food Chem 53(1):57–61

Iauk L, Lo Bue AM, Milazzo I, Rapisarda A, Blandino G (2003) Antibacterial activity of medicinal plant extracts against periodontopathic bacteria. Phytother Res 17(6):599–604

Ibragimov II, Dzhumabaev TZ (1971) Influence of *Berberis oblonga* on blood coagulation. In: Problemui Svyortuivaniya Krovi i Gemostaza (Problems of blood coagulation and hemostasis), vol 33. Nauchni Trudy Kubanskogo Med Instituta (Scientific Works of Kuban Medical Institute), Krasnodar. pp 110–114 (in Russian)

Ibragimov FI, Ibragimova VS (1960) Osnovnuiye Lekarstvennuiye Sredstva Kitaiskoi Meditsinui (The basic medicinal remedies of Chinese medicine). Medgiz, Moscow (in Russian)

Ibragimov AYa, Khazanovich RL (1972) Ursolic acid from *Agrimonia asiatica*. Chem Nat Compd 8(3):390–391

Ibragimova MU, Israilov IA, Yunusov MS, Yunusov SYu (1974) Alkaloids of *Fumaria vaillantii* structure of vaillantine. Chem Nat Compd 10(4):481–482

Idzumi I, Conti A (1962) Study of the activity of the alkaloid berberine on the intestines and uterus in mice. Referativnii Zh Biokhimii (12):1504 (in Russian)

Iida A, Kiuchi F, Ito M, Honda G, Mizushina Y, Yoshida H, Sarsenbaev K (2003) Phenylethanoid glycosides targeting mammalian DNA polymerases. Nat Med 57(4):146–149

Iinuma M, Tanaka T, Mizuno M, Shirataki Y, Yokoe I, Komatsu M, Lang FA (1990) Two flavanones in *Sophora leachiano* and some related structures. Phytochemistry 29(8):2667–2669

Iinuma M, Ohyama M, Tanaka T (1995) Six flavonostilbenes and a flavanone in roots of *Sophora alopecuroides*. Phytochemistry 38(2):519–525

Ikan R, Ravid U (1971) The isolation and identification of ecdysterone from *Ajuga iva*. Planta Med 20(1):33–35

Ilarionova M, Maneva K, Todorov D (1980) Pathomorphological studies on animals bearing experimental tumors treated with thaliblastine. Dokl Bolgarskoi Akad Nauk 33(5):719–722

Iordanov D, Nikolov P, Boichinov A (1970) Fitoterapiya. Lecheniye Lekarstvennuimi Travami. Vtoroye dopolnennoye Russkoye Izdaniye. (Phytotherapy: treatment with medicinal herbs, 2nd Russian ed). Meditsina i Fizkultura, Sofia (in Russian)

Isakov ShI (1969) Study of medicinal plant material used by Kyrgyz people. Synopsis of dissertation for the degree of Ph.D. in medicine, Ashkabad (in Russian)

Ishmukhamedov NA, Sultanov MB (1965) Pharmacology of the new alkaloid alginidine. In: Farmakologiia Alkaloidov (Pharmacology of alkaloids). Nauka, Tashkent, pp 342–346 (in Russian)

Iskanderova SD, Sharipova UC (1992) Influence of *Ecdysten* on physical working capacity of patients, after myocardial infarction. Meditsinskii Zh Uzbekistana (3):14–16 (in Russian)

Ivancheva S, Stantcheva B (2000) Ethnobotanical inventory of medicinal plants in Bulgaria. J Ethnopharmacol 69(2):165–172

Ivancheva S, Manolova N, Serkedjieva J, Dimov V, Ivanovska N (1992) Polyphenols from Bulgarian medicinal plants with anti-infectious activity. Basic Life Sci 59:717–728

Ivanova VM (1963) Study of the roots and rhizomes of *Patrinia intermedia* as a source for new medicinal preparations. Synopsis of dissertation for the degree of Ph.D. in Pharmacology, Moscow (in Russian)

Jafarian-Dehkordi A, Emami SA, Saeidi M, Sadeghi H (2004) Cytotoxicologic studies of the extracts of Iranian *Juniperus sabina* and *Platycladus orientalis* on cancer cells. J Res Med Sci 9(5):7–11

Jamshidzadeh A, Niknahad H (2006) Hepatoprotective activity of *Berberis integerrima* Bge extract in rats treated with CCl4: in vitro and in vivo studies. Toxicol Lett 164(Suppl 1):S310

Jamwal KS, Anand KK, Chopra IC (1962) Pharmacological properties of a crystalline substance (osthol) isolated from *Prangos pabularia*. Arch Int Pharmacodyn Ther 138:400–411

Jassbi AR, Mehrdad M, Soleimani M, Mirzaeian M, Sonboli A (2005) Chemical composition of the essential oils of *Bunium elegans* and *Bunium caroides*. Chem Nat Compd 41(4):415–417

Ji L, Xu Z, Pan G, Yang G (1997) GC-MS analysis of constituents of essential oils from stems of *Ephedra sinica* Stapf, *E. intermedia* Schrenk et C.A. Mey. and *E. equisetina* Bge. Zhongguo Zhong Yao Za Zhi 22(8):489–492 (in Chinese)

Jo EH, Hong HD, Ahn NC, Jung JW, Yang SR, Park JS, Kim SH, Lee YS, Kang KS (2004) Modulations of the Bcl-2/Bax family were involved in the chemopreventive effects of licorice root (*Glycyrrhiza uralensis* Fisch.) in MCF-7 human breast cancer cell. J Agric Food Chem 52(6):1715–1719

Johansson S, Göransson U, Luijendijk T, Backlund A, Claeson P, Bohlin L (2002) A neutrophil multitarget functional bioassay to detect anti-inflammatory natural products. J Nat Prod 65(1):32–41

Jung JI, Chung E, Seon MR, Shin HK, Kim EJ, Lim SS, Chung WY, Park KK, Park JH (2006a) Isoliquiritigenin (ISL) inhibits ErbB3 signaling in prostate cancer cells. Biofactors 28(3–4):159–168

Jung JI, Lim SS, Choi HJ, Cho HJ, Shin HK, Kim EJ, Chung WY, Park KK, Park JH (2006b) Isoliquiritigenin induces apoptosis by depolarizing mitochondrial membranes in prostate cancer cells. J Nutr Biochem 17(10):689–696

Kahkonen MP, Hopia AI, Vuorela HJ, Rauha J, Pihlaja K, Kujala TS, Heinonen M (1999) Antioxidant activity of plant extracts containing phenolic compounds. J Agric Food Chem 47(10):3954–3962

Kanazawa M, Satomi Y, Mizutani Y, Ukimura O, Kawauchi A, Sakai T, Baba M, Okuyama T, Nishino H, Miki T (2003) Isoliquiritigenin inhibits the growth of prostate cancer. Eur Urol 43(5):580–586

Kanokmedhakul K, Kanokmedhakul S, Phatchana R (2005) Biological activity of anthraquinones and triterpenoids from *Prismatomeris fragrans*. J Ethnopharmacol 100(3):284–288

Karamenderes C, Khan S, Tekwani BL, Jacob MR, Ikhlas IA (2006) Antiprotozoal and antimicrobial activities of *Centaurea* species growing in Turkey. Pharm Biol 44(7):534–539

Karimov VA, Shomakhmudov A (1993) Khalk tabobati va zamonavii imli tibda kullaniladigan shifobakhsh usimliklar. Avicenna, Toshkent (in Uzbek)

Karimov A, Telezhenetskaya MV, Lutfullin KL, Yunusov SYu (1975) Alkaloids of *Echium vulgare* and *Berberis oblonga*. Chem Nat Compd 11(3):452

Karimov A, Abdullaev ND, Telezhenetskaya MV, Lutfullin K, Yunusov SYu (1976) Oblongine – a new quaternary alkaloid from *Berberis oblonga*. Chem Nat Compd 12(1):111–112

Karimov A, Telezhenetskaya MV, Lutfullin KL, Yunusov SYu (1977) Berberis alkaloids. The new alkaloid oblongamine. Chem Nat Compd 13(1):68–70

Karimov A, Tashkhodzhaev B, Rashkes YV, Makhmudov MK, Mil'grom EG (1993a) *Berberis* alkaloids. XXI. Intebrine – a new N-benzylisoquinoline alkaloid from *Berberis integerrima*. Chem Nat Compd 29(1):53–57

Karimov A, Vinogradova VI, Shakirov R (1993b) *Berberis* alkaloids. XXII. Interbrinine and intebrimine – new alkaloids from *Berberis integerrima*. Chem Nat Compd 29(1):57–60

Karimova SU, Israilov IA, Yunusov MS, Yunusov SYu (1980) Alkaloids of *Glaucium fimbrilligerum*. Chem Nat Compd 16(2):177–180

Karimova SU, Israilov IA, Vezhnik F, Yunusov MS, Slavik Yu, Yunusov SYu (1983) Alkaloids of *Glaucium fimbrilligerum*. II. Chem Nat Compd 19(4):464–467

Kasimu R, Tanaka K, Tezuka Y, Gong ZN, Li JX, Basnet P, Namba T, Kadota S (1998) Comparative study of seventeen *Salvia* plants: aldose reductase inhibitory activity of water and MeOH extracts and liquid chromatography-mass spectrometry (LC-MS) analysis of water extracts. Chem Pharm Bull 46(3):500–504

Katsube T, Tabata H, Ohta Y, Yamasaki Y, Anuurad E, Shiwaku K, Yamane Y (2004) Screening for antioxidant activity in edible plant products: comparison of low-density lipoprotein oxidation assay, DPPH radical scavenging assay, and Folin-Ciocalteu assay. J Agric Food Chem 52(8):2391–2396

Katsube T, Imawaka N, Kawano Y, Yamazaki Y, Shiwaku K, Yamane Y (2006) Antioxidant flavonol glycosides in mulberry (*Morus alba* L.) leaves isolated based on LDL antioxidant activity. Food Chem 97(1):25–31

Kaurichev IC (Senior ed) (1989) Pochvovedeniye, 4 izd (Soil science), 4th edn. Agropromizdat, Moscow (in Russian)

Kawasaki Y, Goda Y, Yoshihira K (1992) The mutagenic constituents of *Rubia tinctorum*. Chem Pharm Bull 40(6):1504–1509

Keneshov BM, Kuliev ZA, Vdovin AD, Abdullaev ND, Makhmatkulov AB, Nishanov AA (1997a) Proanthocyanidins of *Polygonum coriarium*. III. Structures of proanthocyanidins T1 and T2. Chem Nat Compd 33(4):453–457

Keneshov BM, Kuliev ZA, Vdovin AD, Abdullaev ND, Nishanov AA (1997b) Proanthocyanidins of *Polygonum coriarium*. IV. Structures of proanthocyanidins T3 and T4. Chem Nat Compd 33(5): 548–553

Kennedy DO, Little W, Scholey AB (2004) Attenuation of laboratory-induced stress in humans after acute administration of *Melissa officinalis* (Lemon Balm). Psychosom Med 66:607–613

Khaidav Ts (1965) Medicinal plants used in Mongolian Folk medicine. In: Tez. Doklad. 41-I Itog, Nauch. Konf. Chenovits. Med. Inst. (Proceedings of the 41st scientific conference, Chernovitsa Medical Institute), pp 41–45 (in Russian)

Khakimov MKh, Tsukervanik IP (1948) Essential oil of *Achillea filipendulina* Lam., vol 1. Trudy Instituta Khimii, Tashkent (in Russian)

Khalilova AZ, Litvinov IA, Beskrovnyi DV, Gubaidullin AT, Shakurova ER, Nuriev IR, Khalilov LM, Dzhemilev UM (2004) Isolation and crystal structure of taraxasteryl acetate from *Onopordum acanthium*. Chem Nat Compd 40(3):254–257

Khalmatov KhKh (1964) Dikorastushchiye Lekarstvenniye Rasteniya Uzbekistana (Wild-growing medicinal plants of Uzbekistan). Meditsina, Tashkent (in Russian)

Khalmatov KhKh, Khabibov ZKh (1976) Kovrak – Ferula vonyuchaya – Ferula assafoetida L.Sp Pl. In: Uzbekistonning shifobakhsh usimliklar. Meditsina, Tashkent, pp 49–50 (in Uzbek)

Khalmatov KhKh, Kosimov AI (1992) Russian-Latin-Uzbek dictionary of medicinal plants (herbal plants). Ibn Sino, Tashkent (in Russian)

Khalmatov KhKh, Kosimov AI (1994) Dorivor Usimliklar (Malummotnoma) [Medicinal plants (references)]. Ibn Sino, Tashkent (in Uzbek)

Khalmatov KhKh, Kharlamov IA, Alimbayeva PK, Karriev MO, Khaetov IH (1984) Osnovnuiye Lekarstvennuiye Rasteniya Srednei Azii (The main medicinal plants of Central Asia). Meditsina, Tashkent (in Russian)

Khamdamov I (1972) Farmakologiia Alkaloidov i Ikh Proizvodnikh (Phamacology of alkaloids and their derivatives). Sultanov MB (ed). FAN, Tashkent (in Russian)

Khamidov II, Tashkhodzhaev B, Aripova SF, Telezhenetskaya MV, Karimov AK (1996) *Berberis* alkaloids. XXXVII. Investigation of the alkaloids of *B. oblongata* and *B. integerrima*. Crystal structure of 8-trichloromethyldihydroberberine. Chem Nat Compd 32(6):876–879

Khamidov II, Aripova SF, Karimov AK (2003) *Berberis* alkaloids. XLI. Alkaloids from leaves of cultivated *Berberis oblonga*. Chem Nat Compd 39(4):407

Khamidullina EA, Gromova AS, Lutsky VI, Owen NL (2006) Natural products from medicinal plants: non-alkaloidal natural constituents of the *Thalictrum* species. Nat Prod Rep 23(1):117–129

Khanov M, Kurmukov AG, Sultanov MB, Akhmedkhodzhaeva KhS (1968) Influence of vincarin on central nervous system. Farmakol Toksikol 31(5):538–541 (in Russian)

Khanov M, Sultanov MB, Egorova TA (1971) Pharmacological study of the alkaloid codonopsine extracted from *Codonopsis clematidea*. In Sultanov MB (ed) Farmakologiia Alkaloidov i Serdechnikh Glikozidov. Akademii nauk UzSSR, Tashkent. pp 210–212 (in Russian)

Khanov M, Kurmukov AG, Akhmedkhodzhaeva KhS, Sultanov MB (1972a) Effect of vincarine on experimental heart arrhythmias. In: Sultanov MB (ed) Farmakologiia Alkaloidov i Ikh Proizvodnikh (Pharmacology of alkaloids and their derivatives). FAN, Tashkent, pp 22–25 (in Russian)

Khanov M, Kurmukov AG, Akhmedkhodzhaeva KhS (1972b) Influence of vincarin on experimental cardiac arrhythmia. In: Pharmacology of alkaloids and their derivatives. FAN, Tashkent (in Russian)

Kharlamova TV (2007) Reaction of frangula-emodin with α -bromoalkylmethylketones. Chem Nat Compd 43(4):391–394

Khashimov AM, Shakirov R, Yunusov SYu (1970) A study of the alkaloids of *Veratrum lobelianum*. Chem Nat Compd 6(3):338–340

Khodzhaeva MA (1994) Carbohydrates of *Allium*. XIII. Glucofructans of *Allium suvorovii*. Chem Nat Compd 30(2):162–165

Khodzhaeva MA, Kondratenko ES (1984) Carbohydrates of *Allium*. VI. Stachyose from the seeds of *Allium suvorovii*. Chem Nat Compd 20(6):745–746

Khodzhaeva MA, Turakhozhaev MT (1992) Carbohydrates of *Allium* IX. Polysaccharides of *A. suvorovii*. Chem Nat Compd 28(6):540–541

Khodzhimatov M (1989) Dikorastushchiye Lekarstvennuiye Rasteniya Tadzhikistana (Wild-growing medicinal plants of Tadjikistan). Glavnaya nauchnaya redaktsiya Tadzhikskoi Sovetckoi entsiklopedii, Dushanbe (in Russian)

Kim SY, Gao JJ, Kang HK (2000) Two flavonoids from the leaves of *Morus alba* induce differentiation of the human promyelocytic leukemia (HL-60) cell line. Biol Pharm Bull 23(4):451–455

Kim SR, Park MJ, Lee MK, Sung SH, Park EJ, Kim J (2002) Flavonoids of *Inula britannica* protect cultured cortical cells from necrotic cell death induced by glutamate. Free Radic Biol Med 32(7):596–604

Kim HK, Choi YH, Erkelens C, Lefeber AW, Verpoorte R (2005) Metabolic fingerprinting of *Ephedra* species using ^{1}H-NMR spectroscopy and principal component analysis. Chem Pharm Bull 53(1):105–109

Kim MR, Lee YJ, Lee HH, Aryal DK, Kim YG, Kim SK, Woo ER, Kang KW (2006) Antioxidative effects of quercetin-glycosides isolated from the flower buds of *Tussilago farfara* L. Food Chem Toxicol 44(8):1299–1307

Kimura T, Jyo M, Nakamura N, Komatsu K, Hattori M, Shimotohno K, Shimotohno K, Kakiuchi N (2003) Inhibitory effect of Tibetan medicinal plants on viral polymerases. J Tradit Med 20(6):243–250

Kisiel W, Barszcz B (2000) Further sesquiterpenoids and phenolics from *Taraxacum officinale*. Fitoterapia 71(3):269–273

Klejdus B, Vitamvasova-Sterbova D, Kuban V (2001) Identification of isoflavone conjugates in red clover (*Trifolium pratense*) by liquid chromatography–mass spectrometry after two-dimensional solid-phase extraction. Anal Chim Acta 450(1–2):81–97

Klimentova DA, Aksinenko SG, Gorbacheva AV, Nesterova YuV, Pashinskiy VG (2005) Antiulcerous properties of *Fragaria vesca* (Rosaceae) aboveground part infusion. Rastiteljrnye Resursy 41(2):129–133 (in Russian)

Knox YM, Hayashi K, Suzutani T, Ogasawara M, Yoshida I, Shiina R, Tsukui A, Terahara N, Azuma M (2001) Activity of anthocyanins from fruit extract of *Ribes nigrum* L. against influenza A and B viruses. Acta Virol 45(4):209–215

Knox YM, Suzutani T, Yoshida I, Azuma M (2003) Anti-influenza virus activity of crude extract of *Ribes nigrum* L. Phytother Res 17(2):120–122

Kobaisy M, Tellez MR, Dayan FE, Mamonov LK, Mukanova GS, Sitpaeva GT, Gemejieva NG (2005) Composition and phytotoxic activity of *Nepeta pannonica* L essential oil. J Essent Oil Res 17(6):704–707

Kobayashi A, Hachiya A, Ohuchi A, Kitahara T, Takema Y (2002a) Inhibitory mechanism of an extract of *Althaea officinalis* l. on endothelin-1-induced melanocyte activation. Biol Pharm Bull 25(2):229

Kobayashi T, Song QH, Hong T, Kitamura H, Cyong JC (2002b) Preventative effects of the flowers of *Inula britannica* on autoimmune diabetes in C57BL/KsJ mice induced by multiple low doses of streptozotocin. Phytother Res 16(4):377–382

Kocsis A, Szabo LF, Podanyi B (1993) New bis-iridoids from *Dipsacus laciniatus*. J Nat Prod 56(9):1486–1499

Kogure K, Yamauchi I, Tokumura A, Kondou K, Tanaka N, Takaishi Y, Fukuzawa K (2004) Novel antioxidants isolated from plants of the genera *Ferula*, *Inula*, *Prangos* and *Rheum* collected in Uzbekistan. Phytomedicine 11(7–8):645–651

Kokoska L, Polesny Z, Rada V, Nepovim A, Vanek T (2002) Screening of some Siberian medicinal plants for antimicrobial activity. J Ethnopharmacol 82(1):51–53

Kolesnikov MP, Gins VK (2001) Phenolic substances in medicinal plants. Appl Biochem Microbiol 37(4):392–399

Komarov VL (Senior ed) (1934) Flora of the USSR, vols 1–30. Izdatelstvo Academii Nauk SSSR, Leningrad (in Russian)

Komissarenko NF, Levashova IG (1988) Phototoxic compounds of the genus *Dictamnus*. In: Vtoraya Resp. Konf. Po Med. Bot. – Kiev (Proceedings of the 2nd Repulican conference on medicinal botany, Kiev), Nauka, Kiev, pp 258–259 (in Russian)

Komissarenko NF, Levashova IG (1989) Flavonoids and coumarins of some plant from Siberia and the far East. In: Novuiye Lekarstvennuiye Preparatui iz Rastenii Sibiri I Dalnego Vostoka (New medical preparations from plants of Siberia and the far East), vol 2, Tomsk, pp 81–82 (in Russian)

Konishi T, Shimada Y, Nagao T, Okabe H, Konoshima T (2002) Antiproliferative sesquiterpene lactones from the roots of *Inula helenium*. Biol Pharm Bull 25(10):1370–1372

Konovalov DA, Nesterova TA (2003). Phytochemical study of some Caucasian species of Asteraceae and the possibility of their allergenic properties. Vestnik VGU. Seriya: Khimiya. Biologiya. Farmatsiya (2):205–207 (in Russian)

Konrad L, Müller HH, Lenz C, Laubinger H, Aumüller G, Lichius JJ (2000) Antiproliferative effect on human prostate cancer cells by a stinging nettle root (*Urtica dioica*) extract. Planta Med 66(1):44–47

Koo HN, Hong SH, Song BK, Kim CH, Yoo YH, Kim HM (2004) *Taraxacum officinale* induces cytotoxicity through TNF-alpha and IL-1alpha secretion in Hep G2 cells. Life Sci 74(9):1149–1157

Korolyuk EA, Konig V, Tkachev AV (2002) Compounds in the essential oil of *Ziziphora clinopodioides* Lam. from the Altai region and Altai Republic. Khim Rastitelnogo Sirya (1):49–52 (in Russian)

Kotenko LD, Yakubova MYa, Tselishcheva NA, Turakhozhaev MT, Badalbaeva TA (1994) Quantitative determination of the total iridoids in plants of the genus *Lagochilus*. Chem Nat Compd 30(6):669–672

Kovalenko VN (1946) About the pharmacochemical and pharmacological characteristics of the alkaloid dictamnine from *Dictamnus albus turkestanicus*. Farmatsiya 9(5):20

Kovaleva NG (1971) Treatment with plants. Meditsina, Moscow (in Russian)

Krasnov EA, Demidenko LA (1981) Herbacetin glycosides in plants of the family Crassulaceae and their biological activity. In: Materialui 4-go Vseros. Sezda Farmatsevtov (Proceedings of the 4th all-union conference in pharmacology), Voronezh, USSR, pp 439–440

Krasnov EA, Saratikov AS, Surov YuP (1979) Semeistva Tolstyankovuikh. Plants of the family Crassulaceae. Tomskii gosudarstvenui universitet, Tomsk (in Russian)

Krenn L, Wiedenfeld H, Roeder E (1994) Pyrrolizidine alkaloids from *Lithospermum officinale*. Phytochemistry 37(1):275–277

Krilov GV (1972) Travui Zhizni i Ikh Iskateli (Herbs for life and their seekers). Zapad-Sibirskoye Knezhnoye Izdatelstvo, Novosibirsk (in Russian)

Kruglikova-Livova RP (1952) About the pharmacology of sophocarpine. Farmakol Toksikol 15(6):51–55 (in Russian)

Kuchin NN (1955) About using an infusion of the flowers of *Crataegus songarica* for the treatment of patients with rheumatism. Synopsis of dissertation for the degree of Ph.D. in Medical Sciences, Alma-Ata (in Russian)

Kudryashev S (1932) Essential oils from wild plants of the central region of the Gissar mountains. Parfums de France 12:98–101

Kulikov VV (1975) Lekarstvennuiye Rasteniya Altaiskogo Kraya, 2-I izd (Medicinal plants of the Altai territory), 2nd edn. Altaickoye khizhnoye isdatelstvo, Barnaul (in Russian)

Kumarasamy Y, Cox PJ, Jaspars M, Nahar L, Sarker SD (2002) Screening seeds of Scottish plants for antibacterial activity. J Ethnopharmacol 83(1–2):73–77

Kumarasamy Y, Byres M, Cox PJ, Jaspars M, Nahar L, Sarker SD (2007) Screening seeds of some Scottish plants for free radical scavenging activity. Phytother Res 21(7):615–621

Kurbatova NV, Muzychkina RA, Mukhitdinov NM, Parshina GN (2003) Comparative phytochemical investigation of the composition and content of biologically active substances in *Marrubium vulgare* and *M. alternidens*. Chem Nat Compd 39(5):501–502

Kurlin VA, Zapesochnaya GG (1986) Terpenoids of the rhizomes of *Rhodiola linearifolia*. Chem Nat Compd 22(5):607

Kurmukov AG (1956) Pharmacological investigations of some Cruciferous plants of Uzbekistan for the presence of cardiac glycosides. In: Nauchnuiye rabotui studentov Tash. Gos. Med. Inst. (Scientific works by students of Tahskent State Medical Institute). Izdatelstvo Akademii Nauk UzSSR, Tashkent, pp 59–65 (in Russian)

Kurmukov AG (1967a) Electroencephalogram analysis of influences of reserpine, vincamine, ervamine and ervamidine on CNS. In: Kamilov IK (ed) Farmakologiia Alkaloidov i Glikozidov (Pharmacology of alkaloids and glycosides). FAN, Tashkent, pp 74–79 (in Russian)

Kurmukov AG (1967b) Pharmacology of the alkaloid vinervinine. In: Kamilov IK (ed) Farmakologiia Alkaloidov i Glikozidov (Pharmacology of alkaloids and glycosides). FAN, Tashkent, pp 79–82 (in Russian)

Kurmukov AG (1968a) Alkaloids as stimulants of uterine activity. Dokl Akad Nauk UzSSR 25(12):23–24 (in Russian)

Kurmukov AG (1968b) On the pharmacology of the alkaloid ervamidine (akuamidine). Farmakol Toksikol 31(1):47 (in Russian)

Kurmukov AG (1970) Pharmacological investigations of some indole alkaloids. Synopsis of dissertation for the degree of doctor of medicine, Tashkent (in Russian)

Kurmukov AG (1975) The pharmacology of the alkaloid ervine. Farmakol Toksikol 38(1):82–84 (in Russian)

Kurmukov AG (1976) Pharmacology of licorice and its preparations (Influences on the hypothalamo-hypophyseal-adrenal systems). In: Sultanov MB (ed) Farmakologiia Rastitelnuikh Veshchestv (Pharmacology of plant substances). FAN, Tashkent, pp 174–187 (in Russian)

Kurmukov AG (1978) The pharmacology of the indole alkaloids akuammine, aimaline, vinervine and the derivatives ervamine and pseudocopsinine. In: Sadritdinov BS (ed) Farmakologiia Prirodnikh Veshchestv (Pharmacology of natural substances). FAN, Tashkent, pp 32–51 (in Russian)

Kurmukov AG (1987) About the anti-inflammatory action of the lactone leucomisine. Meditsinskii Zh Uzbekistana (9):72–75 (in Russian)

Kurmukov AG, Akhmedkhodzhaeva KhS (1975) The experimental influence of psoralen and infekundin on the sexual cycle and embryogenesis. Meditsinskii Zh Uzbekistana (8):35 (in Russian)

Kurmukov AG, Akhmedkhodzhaeva KhS (1994) Estrogennuiye lekarstvennuiye preparatui iz rastenii roda Ferula (Estrogenic medicinal preparations of plants in the Genus *Ferula*). Ibn Sino, Tashkent (in Russian)

Kurmukov AG, Ermishina OA (1986) Influence of ecdysterone on the size of the necrotic zone after experimental myocardial infarction. In: Otsenka Razmera i Taktika Lecheniya Infarkta Miokarda (Estimation of the size, and methods for treating, myocardial infarction), Tomsk, pp 62–63 (in Russian)

Kurmukov AG, Ermishina OA (1991) The effect of ecdysterone on experimental arrhythmias and changes in the hemodynamics and myocardial contractility induced by coronary artery occlusion. Farmakol Toksikol 54(1):27–29 (in Russian)

Kurmukov AG, Kurmukova NA (1992) Influence of ecdysterone on prenatal hypertrophy, caused by infringement of placental-uterine circulation. In: Eksperimentalnaya farmalogiya klinishi: "Tezisui dokladov". Vinnytsia, Kiev, pp 42–43

Kurmukov AG, Rasulova SA (1983) About anti-sclerotic and cardio-protecting actions of leucomisine. In: Trudy Tash. Med. Inst. "Ishemicheskaya Blezn Serdtsa" (Scientific works of Tashkent State Medical Institute "Ischemic Heart-disease"). Tashkent, pp 55–58 (in Russian)

Kurmukov AG, Saidkasymov TK (1968) Mechanism of the effect of ervinine on the central nervous system. Meditsinskii Zh Uzbekistana 9:6–9

Kurmukov AG, Saidkasymov TK (1969) About the main mechanism of action of ervinine. Meditsinskii Zh Uzbekistana (3):6–9 (in Russian)

Kurmukov AG, Sultanov MB (1965) Pharmacology of *Vinsumine*. In: Farmakologiia Alkaloidov (Pharmacology of alkaloids). Nauka, Tashkent, pp 143–157 (in Russian)

Kurmukov AG, Sultanov MB (1971) Vinkametrin – a new stimulator of child delivery. In: Sultanov MB (ed) Farmakologiia Alkaloidov i Serdechnikh Glikozidov (Pharmacology of alkaloids and cardiac glycosides). FAN, Tashkent (in Russian)

Kurmukov AG, Syrov VN (1976) About the anabolic activity of the phytoecdysone ecdysterone isolated from *Rhaponticum carthamoides*. Farmakol Toksikol (6):690–693 (in Russian)

Kurmukov AG, Zakirov UB (1992) Alkaloidyi Preparatui Tselebnuikh Trav dlya Lecheniya Gipertenzivnuikh Sostoyanii (Alkaloids and preparations of medicinal herbs for treating hypertensive conditions). Ibn Sino (Avicenna), Tashkent (in Russian)

Kurmukov AG, Akhmedkhodzhaeva KhS, Sirov VN, Sidyakin VN (1976) Phytoestrogens from Central Asian plants. Rastitel'nye Resursy (4):515–525 (in Russian)

Kurmukov AG, Syrov VN, Akhmedkhodzhaeva KhS, Aizikov MI (1977) Hormonal activity of plant substances. In: Tezisui Dokladov Vsesoyuznoi Konferentsii. Leningrad-Pushkin, 17–19 Sept 1977, p 225 (in Russian)

Kurmukov AG, Syrov VN, Ermishina OA (1980) Effects of Phytoecdysones and sapogenin 6-ketoderivatives on cardiac muscle metabolism. J Mol Cell Cardiol (12):86

Kurmukov AG, Makhmudov O, Aizikov MI (1982) The influence of fruticine on the content of major compounds for lipid metabolism in rabbits with experimental atherosclerosis. In: Materialov yubileinoi nauchno-prakticheskoi konferentsii posvyashchonnoi 50 lyetiyu sanatoriya "Uzbekistan" v gorodye Kislovodskye po problemye "Etapnoye lecheniye IBS", Tashkent

Kurmukov AG, Aizikov MI, Rasulova SA (1984a) Hypolipidemic and antisclerotic activities of fruticine in experiments and clinics. In: Demiologiya, diagnostika, klinika, techeniye I reabilitatsiya serdechno-sosudistuikh zabolevanii (Demology, diagnostics, clinics, progression and rehabilitation of cardiovascular diseases). Kanaus, pp 288–289 (in Russian)

Kurmukov AG, Aizikov MI, Rasulova SA (1984b) Hypocoagulative effects of amorphine for experimental atherosclerosis. Dokl Akad Nauk UzSSR (10):49–50 (in Russian)

Kurmukov AG, Aizikov MI, Rasulova SA, Makhmudov OKh (1986) Hypolipidemic, hypocholesterolemic and antisclerotic effects of the rotenoid glycoside amorphine. Farmakol Toksikol 49(6):54–57

Kurmukov AG, Aizikov MI, Rasulova SA, Sidiakin GP, Sham'ianov ID, Malikov VM (1991a) The angioprotector and hypolipidemic activity of leucomisine on experimental atherosclerosis. Farmakol Toksikol 54(3):35–37 (in Russian)

Kurmukov AG, Nazrullaev SS, Abdullakhodzhaeva DG (1991b) Pharmacological treatment of hypoxic conditions with plant proanthocyanidins. In: Farmakologicheskaya korrektsiya gipoksicheskikh sostoyanii. Chast II. (Pharmacological treatment of hypoxic conditions, part 2) Materials of the 2nd All-Union Conference, Grodno, pp 245–246 (in Russian)

Kurmukova NA (2000a) Fetal hypoxic hypotrophy and its correction by ecdysten introduction. Patologiya (2):24–27 (in Russian)

Kurmukova NA (2000b) Seasonal stability (resistance) of fetus against uteroplacental hemodynamics disturbance and its correction by ecdysten introduction. Chem Nat Compd (special edn):148–150 (in Russian)

Kurmukova ShR, Aizikov MI (1997) Studies of the anti-inflammatory action of a mixture of leucomisine and ascorbic acid. In: Trudy Tash Med Inst "Farmakoterapiya patalogicheskikh sostoyanii" (Scientific works of Tashkent State Medical Institute, "Pharmacotherapy of pathological conditions"), vol 1. Tashkent, pp 65–68 (in Russian)

Kurmukova NA, Kurbanov SD (1999) Evaluation of immune state condition of women with "Inuterine Delayed Embro Development" during the treatment with ecdysterone. Pediatriya (special edn):250 (in Russian)

Kurmukova ShR, Aizikov MI, Zakirov NU (1997a) Study of the effectivity of leucomisine and leucomisine mixed with ascorbic acid on experimental izadrinic myocarditis in rats. In: III Kongress Assotsiatsii Kardiologov Stran Tsentralnoi Azii (III congress of association of cardiologists in Middle Asia), Tashkent, pp. 173 (in Russian)

Kurmukova ShR, Zakirov NU, Aizikov MI (1997b) Cardio-protecting action of leucomisine L-AK on experimental myocarditis. Kimye va farmatsiya (3):37–40 (in Russian)

Kurochkin EI (1998) Lekarstvennuiye Rasteniya (Medicinal plants). Parus, Samara (in Russian)

Kwon DH, Kwon HY, Kim HJ, Chang EJ, Kim MB, Yoon SK, Song EY, Yoon DY, Lee YH, Choi IS, Choi YK (2005) Inhibition of hepatitis B virus by an aqueous extract of *Agrimonia eupatoria* L. Phytother Res 19(4):355–358

Ladigina EA, Morozova RS (1987) Fitoterapiya (Phytotherapy). Meditsina, Leningrad (in Russian)

Lam ANC, Demasi M, James MJ, Husband AJ, Walker C (2004) Effect of red clover isoflavones on cox-2 activity in murine and human monocyte/macrophage cells. Nutr Cancer 49(1):89–93

Larsen E, Kharazmi A, Christensen LP, Christensen SB (2003) An anti-inflammatory galactolipid from rose hip (*Rosa canina*) that inhibits chemotaxis of human peripheral blood neutrophils in vitro. J Nat Prod 66(7):994–995

Lee SJ, Chung HY, Maier CGA, Wood AR, Dixon RA, Mabry TJ (1998) Estrogenic flavonoids from *Artemisia vulgaris* L. J Agric Food Chem 46(8):3325–3329

Lee SM, Min BS, Lee CG, Kim KS, Kho YH (2003) Cytotoxic triterpenoids from the fruits of *Zizyphus jujuba*. Planta Med 69(11): 1051–1054

Li LM, Liao X, Peng SL, Ding LS (2005) Chemical constituents from the seeds of *Ziziphus jujuba* var. *spinosa* (Bunge) Hu. J Integr Plant Biol 47(4):494–498

Li L, Tsao R, Yang R, Kramer JKG, Hernandez M (2007) Fatty acid profiles, tocopherol contents, and antioxidant activities of heartnut (*Juglans ailanthifolia* var. *cordiformis*) and Persian walnut (*Juglans regia* L.). J Agric Food Chem 55(4):1164–1169

Liao WT, Beal JL, Wu WN, Doskotch RW (1978a) Alkaloids of *Thalictrum* XXVI. New hypotensive and other alkaloids from *Thalictrum minus* race B. Lloydia 41(3):257–270

Liao WT, Beal JL, Wu WN, Doskotch RW (1978b) Alkaloids of *Thalictrum* XXVII. New hypotensive aporphine-benzylisoquinoline derived dimeric alkaloids from Thalictrum minus race B. Lloydia 41(3):271–276

Lin LZ, He XG, Lindenmaier M, Yang J, Cleary M, Qiu SX, Cordell GA (2000) LC-ESI-MS study of the flavonoid glycoside malonates of red clover (*Trifolium pratense*). J Agric Food Chem 48(2):354–365

Lipkin A, Anisimova V, Nikonorova A, Babakov A, Krause E, Bienert M, Grishin E, Egorov T (2005) An antimicrobial peptide Ar-AMP from amaranth (*Amaranthus retroflexus* L.) seeds. Phytochemistry 66(20):2426–2431

Liu X, Cui Y, Yu Q, Yu B (2005) Triterpenoids from *Sanguisorba officinalis*. Phytochemistry 66(14):1671–1679

Logendra S, Ribnicky DM, Yang H, Poulev A, Ma J, Kennelly EJ, Raskin I (2006) Bioassay-guided isolation of aldose reductase inhibitors from *Artemisia dracunculus*. Phytochemistry 67(14):1539–1546

Lopez TA, Cid MS, Bianchini ML (1999) Biochemistry of hemlock (*Conium maculatum* L.) alkaloids and their acute and chronic toxicity in livestock. A review. Toxicon 37(6):841–865

Lugmanova MR, Muzafarov AM, Mikhailenko OI, Fedorov NI (2007) Results revealing alkaloid-bearing species in the flora of the mountain-forest-steppe zones in the southern Urals. Vestnik Bashkirskogo Universiteta (Bashkirsk University) (1) (in Russian)

Lutfullin KL, Tadzhibaev MM, Abdullaev UA, Malikov VM, Yunusov SYu (1976) Alkaloids of *Reseda luteola*. Chem Nat Compd 12(5):559–563

Lutfullin KL, Tadzhibaev MM, Malikov VM, Abdullaev UA, Rakhmankulov U (1977) Dynamics of the accumulation of the alkaloids of *Reseda luteola* structure of lutine and lutinine. Chem Nat Compd 13(6):696–699

Ma Yu-H, Yang J-R (2006) Protective effect of stachydrine on myocardial ischemia-reperfusion injury in rats. Zhongguo Shiyan Fangjixue Zazhi 12(5):40–42 (in Chinese)

Ma SC, Du J, But PPH, Deng XL, Zhang YW, Ooi VEC, Xu HX, Lee SHS, Lee SF (2002a) Antiviral Chinese medicinal herbs against respiratory syncytial virus. J Ethnopharmacol 79(2):205–211

Ma ZJ, Li X, Li N, Wang JH (2002b) Stilbenes from *Sphaerophysa salsula*. Fitoterapia 73(4):313–315

Ma ZJ, Tian J, Fujii I, Ebizuka Y, Li X (2003) A new lignan from the seeds of *Sphaerophysa salsula*. Nat Prod Res 17(6):423–425

Ma ZJ, Fujii I, Ebizuka Y, Li X (2004a) A new lignan from the seeds of *Sphaerophysa salsula*. J Asian Nat Prod Res 6(4):265–269

Ma ZJ, Fujii I, Ebizuka Y, Li X, Shimomaki S, Sakano Y (2004b) Flavonoids from the seeds of *Sphaerophysa salsula*. J Asian Nat Prod Res 6(1):69–73

Makhlayuk VP (1967) Lekarstvennuiye Rasteniya v Narodnoi Meditsinye, izd. 2 (Medicinal plants in folk medicine), 2nd edn. Privolskoye Knezhnoye Izdatelstvo, Saratov (in Russian)

Makhlayuk VP (1992) Lekarstvennuiye Rasteniya v Narodnoi Meditsinye (Medicinal plants in folk medicine). Niva Rossii, Moscow (in Russian)

Makhmatkulov AB, Kuliev ZA, Vdovin AD, Yagudaev MR, Malikov VM (1992) Proanthocyanidins of *Polygonum coriarium*. I. Chem Nat Compd 28(1):48–54

Makhmatkulov AB, Kuliev ZA, Vdovin AD, Malikov VM (1994) Proanthocyanidins of *Polygonum coriarum*. II. Chem Nat Compd 30(2):214–222

Makhmudov MM (2001) Country pasture/forage resource profiles: Uzbekistan. http://www.fao.org/ag/AGP/AGPC/doc/Counprof/Uzbekistan/uzbekistan.htm. Accessed 12 Jan 2008

Maksudov NKh (1964) Poluchenie Efirnykh Masel (Terpenov) i Ikh Primenenie pri Mochekamennoi Bolezni [Production of essential oils (Terpenes) and their use for the treatment of Urolithic diseases]. Meditsina, Tashkent (in Russian)

Mallabaev A, Sidyakin GP (1976) Artemidinol – a new isocoumarin from *Artemisia dracunculus*. Chem Nat Compd 12(6):729

Mallabaev A, Yagudaev MR, Saitbaeva IM, Sidyakin GP (1970) The isocoumarin artemidin from *Artemisia dracunculus*. Chem Nat Compd 6(4):479

Mallabaev A, Saitbaeva IM, Sidyakin GP (1971) The isocoumarin artemidinal from *Artemisia dracunculus*. Chem Nat Compd 7(3):248–249

Mamatkhanov AU, Yakubova MR, Syrov VN (1998) Isolation of turkesterone from the epigeal part of *Ajuga turkestanica* and its anabolic activity. Chem Nat Compd 34(2):150–154

Mamedov N (2005) Adaptogenic, geriatric, stimulant and antidepressant plants of Russian Far East. J Cell Mol Biol 4:71–75

Mamedov N, Craker LE (2001) Medicinal plants used for the treatment of bronchial asthma in Russia and Central Asia. J Herbs Spices Med Plant 8(2/3):91–118

Mamedov N, Gardner Z, Craker LE (2004) Medicinal plants used in Russia and Central Asia for the treatment of selected skin conditions. J Herbs Spices Med Plant 11(1/2):191–222

Mamitov AM (1965) Classification, vertical belts and provinces of Kyrgyz soils. In: Kovda VA, Lobova EV (eds) Geography and classification of Asian soils. Nauka, Moscow, pp 50–51 (in Russian)

Mandal SC, Dhara AK, Ashok Kumar CK, Maiti BC (2001) Neuropharmacological activity of *Xanthium strumarium* Linn extract. J Herbs Spices Med plant 8(1):69–77

Mansurov MM (1967) Effect of *Ferula asafetida* on blood coagulability. Meditsinskii Zh Uzbekistana (6):46–49 (in Russian)

Marco JM, Sanz-Cervera JF, Corral J, Carda M, Jakupovic J (1993) Xanthanolides from *Xanthium*: absolute configuration of xanthanol, isoxanthanol and their C-4 epimers. Plant Physiol Biochem 34(6):1569–1576

Martino E, Ramaiola I, Urbano M, Bracco F, Collina S (2006) Microwave-assisted extraction of coumarin and related compounds from *Melilotus officinalis* (L.) Pallas as an alternative to Soxhlet and ultrasound-assisted extraction. J Chromatogr A 1125(2):147–151

Mashkovskii MD (1953) The pharmacology of tropacin. Farmakol Toksikol (5):3–8 (in Russian)

Mashkovskii MD (1984) Lekarstvennuiye Sredstva: Posobiye po Farmakoterapii Dlya Vrachei, 9 izd (Medicinal remedies: manual on pharmacotherapy for doctors), 9th edn. Meditsina, Moscow (in Russian)

Massagetov PC (1932) Tian Shan expedition of NIKhFI. Khim. Farmatsev, Promishlennost (4–5): .134–c.142 (in Russian)

Matsuhisa M, Shikishima Y, Takaishi Y, Honda G, Ito M, Takeda Y, Shibata H, Higuti T, Kodzhimatov OK, Ashurmetov O (2002) Benzoylphloroglucinol derivatives from *Hypericum scabrum*. J Nat Prod 65(3):290–294

Maznev N (2004) Bolshaya Entsiklopediya Narodnoi Meditsinui (Big encyclopedia of traditional medicines). Martin, Moscow (in Russian)

McCutcheon AR, Roberts TE, Gibbons E, Ellis SM, Babiuk LA, Hancock RE, Towers GH (1995) Antiviral screening of British Columbian medicinal plants. J Ethnopharmacol 49(2):101–110

MedicineLib.ru. Medicinal plants. Ochnui tsvet pashennui. http://medicinelib.ru/travi/O/travi158.html. Accessed 12 Jan 2008 (in Russian)

Mikanagi Y, Yokoi M, Ueda Y, Saito N (1995) Flower flavonol and anthocyanin distribution in subgenus *Rosa*. Biochem Syst Ecol 23(2):183–200

Milkowska-Leyck K, Borkowski B, Rozanski M (1999) Selected polyphenolic compounds Part III. Biological properties of iridoids. Herba Pol 45(3):232–241

Mimaki Y, Kuroda M, Fukasawa T, Sashida Y (1999) Steroidal saponins from the bulbs of *Allium karataviense*. Chem Pharm Bull 47:738–743

Mimaki Y, Fukushima M, Yokosuka A, Sashida Y, Furuya S, Sakagami H (2001) Triterpene glycosides from the roots of *Sanguisorba officinalis*. Phytochemistry 57(5):773–779

Minayeva VG (1991) Lekarstvennuiye Rasteniya Sibiri, 5 izd (Medicinal plants of Siberia), 5th edn. Nauka, Novosibirsk (in Russian)

Ministry of Environmental Protection (1998) Kyrgyz Republic biodiversity strategy and action plan. Republic of Kyrgyzstan Ministry of Environmental Protection, Bishkek

Mircheva J, Stoychkov J (1976) Effect of thaliblastine on trans-plantable tumors in mice. Biomed Express 25:280–281

Moallem SA, Niapour M (2008) Study of embryotoxicity of *Perovskia abrotanoides*, an adulterant in folk-medicine, during organogenesis in mice. J Ethnopharmacol 117(1):108–114

Mohammadreza VR (2008) Essential oil composition and biological activity of *Ziziphora clinopodioides* Lam. from Iran. Res J Pharmacol 2(2):17–19

Momin RA, De Witt DL, Nair MG (2003) Inhibition of cyclooxygenase (COX) enzymes by compounds from *Daucus carota* L. seeds. Phytother Res 17(8):976–979

Morita H, Yun YS, Takeya K, Itokawa H, Shirota O (1997a) Thionation of segetalins A and B, cyclic peptides with estrogen-like activity from seeds of *Vaccaria segetalis*. Bioorg Med Chem 5(3):631–636

Morita H, Yun YS, Takeya K, Itokawa H, Yamada K, Shirota O (1997b) Vaccaroid A, a new triterpenoid saponin with contractility of rat uterine from *Vaccaria segetalis*. Bioorg Med Chem Lett 7(8):1095–1096

Mukhamed'yarova MM, Chumbalov TK (1979) Flavonoids of *Rumex confertus*. Chem Nat Compd 15(6):758

Mukhamedziev MM, Alimbaeva PK (1969) Triterpene glycosides of *Dipsacus azureus*. Chem Nat Compd 5(5):376–377

Muminova BA, Batirov EKh, Yuldashev MP, Inamova ZG (2006) Flavonoids from *Sophora griffithii* and *Goebelia pachycarpa*. Chem Nat Compd 42(1):108–109

Murai M, Tamayama Y, Nishibe S (1995) Phenylethanoids in the herb of *Plantago lanceolata* and inhibitory effect on arachidonic acid-induced mouse ear edema. Planta Med 61(5):479–480

Muravyova DA (1978) Farmakognoziya (Pharmacognosy). Meditsina, Moscow (in Russian)

Murillo-Alvarez JI, Encarnación DR, Franzblau SG (2001) Antimicrobial and cytotoxic activity of some medicinal plants from Baja California Sur (Mexico). Pharm Biol 39(6):445–449

Muzychkina RA (2000) Certain Kazakhstan *Galium* species. I. Chem Nat Compd 36(3):321–322

Nachman RJ, Olsen JD (1983) Ranunculin: a toxic constituent of the poisonous range plant bur buttercup (*Ceratocephalus testiculatus*). J Agric Food Chem 31(6):1358–1360

Nahrstedt A, Butterweck V (1997) Biologically active and other chemical constituents of the herb of *Hypericum perforatum* L. Pharmacopsychiatry 30:129–134

Naidovich LP, Trumneva EA, Tolkachev ON, Vasileva VD (1976) Chemical composition of indigenous species of the Berberidaceae family. Interrelation of chemical structure and pharmacological activity. Farmatsiia 25(4):33–38 (in Russian)

Nakanishi T, Inada A, Kambayashi K, Yoneda K (1985) Flavonoid glycosides of the roots of *Glycyrrhiza uralensis*. Phytochemistry 24(2):339–341

Nan JX, Park EJ, Kim HJ, Ko G, Sohn DH (2000) Antifibrotic effects of the methanol extract of *Polygonum aviculare* in fibrotic rats induced by bile duct ligation and scission. Biol Pharm Bull 23(2):240–243

Nanaeva MT (1960) About the pharmacology of *Artemisia viridis*. In: 2-ya Konf. Fiziologov, Biokhimikov I farmakologov Srednei Azii i Kazakhstana (Proceedings of the 2nd conference of physiologists, biochemists and pharmacologists of Central Asia and Kazakhstan), Frunze, pp 239–240 (in Russian)

Narsullaev AS, Yunusov MS, Matveev VM, Sabirov SS (1989) 14-Acetylnudicaulidine and 18-deoxylycoctonine – new diterpene alkaloids from *Delphinium confusum*. Chem Nat Compd 25(1):41–42

Nasrullaev FD, Makhsudova BT (1991) Flavonoids of *Lagochilus platycalyx*. Chem Nat Compd 27(4):511–512

National Biodiversity Strategy Project Steering Committee (1998) Republic of Uzbekistan biodiversity conservation: national strategy and action plan, Tashkent. http://bpsp-neca.brim.ac.cn/books/actpln_uzbek/index.html. Accessed online 22 Feb 2007

Nazrullaev CC, Akhmerov RN, Kurmukov AG (1990) Effect of the antihypoxant catacin on miocardial energy metabolism. Meditsinskii Zh Uzbekistana 11:7–9 (in Russian)

Neef H, Declercq P, Laekeman G (1995) Hypoglycaemic activity of selected European plants. Phytother Res 9(1):45–48

Newton SM, Lau C, Wright CW (2000) A review of antimycobacterial natural products. Phytother Res 14(5):303–322

Nikitina EV (1962) Flora I Rastitelnost Pastbishch i Senokosov Khrebta Kirgizskii Ala-Too (The flora and vegetation of pastures and ranges in the Kyrgyz Ala-Too). Izdatelstvo Akademii Nauk KyrgSSR, Frunze (in Russian)

Nishanov AA, Sultankhodzhaev MN, Yunosov MS, Yusupova IM, Tashkhozhaev B (1991) Alkaloids of *Aconitum talassicum*. Chem Nat Compd 27(1):82–86

Nogaller AM, Nikolaeva VG, Alexeyev VF (1987) Fitoterapiya zabolevanii pishchevareniya (Phytotherapy of digestive system diseases). Pavlov Medical Institute, Ryazan (in Russian)

Norn S, Kruse PR (2004) Cardiac glycosides: from ancient history through Withering's foxglove to endogeneous cardiac glycosides. Dansk Medicinhistorisk Arbog:119–132 (in Danish)

Norouzi-Arasi H, Yavari I, Chalabian F, Kiarostami V, Ghaffarzadeh F, Nasirian A (2006) Chemical constituents and antimicrobial activities of the essential oil of *Acroptilon repens* (L.) DC. Flavour Frag J 21(2):247–249

Nosal MA, Nosal IM (1959) Lekarstvennuiye Rasteniya i Sposobui ikh Primrniya v Narodye (Medicinal plants and the ways that they are used by people). Meditsina, Kiev (in Russian)

Nuraliev Yu (1989) Lekarstvennuiye Rasteniya, 2-ye izd (Medicinal plants), 2nd edn. MAORIF, Dushanbe (in Russian)

Nuraliev YuN, Zubaidova TM (1994) Hypocholesterinemic and hypolipidemic effects of ether oil of *Origanum tyttanthum* Gontsch. Khim Farm Zh 28(10):63–64 (in Russian)

Nurmukhamedova MR, Kasymov ShZ, Abdullaev ND, Sidyakin GP, Yagudaev MR (1985) Diterpenoids of *Pulicaria salviifolia* I. Structures of salvin and salvinin. Chem Nat Compd 21(2):188–191

Nurmukhamedova MR, Abdullaev ND, Sidyakin GP (1986) Diterpenoids of *Pulicaria salviifolia*. II. Structure of salvicin. Chem Nat Compd 22(3):277–279

Ogolevitz GS (1951) Entsiklopedicheskii Slovar Lekarstvennuikh, Efiromaslichnuikh i Yadovituikh Rastenii (Encyclopedic dictionary of medicinal, essential oil and poisonous plants). Gos Izdatelstvo selskokhozyaistvennoi literatury, Moscow (in Russian)

Oh H, Kim DH, Cho JH, Kim YC (2004) Hepatoprotective and free radical scavenging activities of phenolic petrosins and flavonoids isolated from *Equisetum arvense*. J Ethnopharmacol 95(2–3):421–424

Okasaka M, Takaishi Y, Kashiwada Y, Kodzhimatov OK, Ashurmetov O, Lin AJ, Consentino LM, Lee KH (2006) Terpenoids from *Juniperus polycarpus* var. *seravschanica*. Phytochemistry 67(24):2635–2640

Omurkamzinova VB, Maurel' ND, Bikbulatova TN (1991) Flavonoids of *Biebersteinia multifida*. Chem Nat Compd 27(5):636–637

Oomah BD, Ladet S, Godfrey DV, Liang J, Girard B (2000) Characteristics of raspberry (*Rubus idaeus* L.) seed oil. Food Chem 69(2):187–193

Oprean L, Mărculescu A, Cristea V, Gaspar E (2007) The antibacterial activity of the volatile oils extracted from medicinal plants. Acta Aliment Hung 36(3):303–309

Orhan DD, Aslan M, Aktay G, Ergun E, Yesilada E, Ergun F (2003) Evaluation of hepatoprotective effect of *Gentiana olivieri* herbs on subacute administration and isolation of active principle. Life Sci 72(20):2273–2283

Orhan I, Sener B, Choudhary MI, Khalid A (2004) Acetylcholinesterase and butyrylcholinesterase inhibitory activity of some Turkish medicinal plants. J Ethnopharmacol 91(1):57–60

Ortiz JG, Nieves-Natal J, Chavez P (1999) Effects of *Valeriana officinalis* extracts on [3h]flunitrazepam binding, synaptosomal [^{3}H]

GABA uptake, and hippocampal [^{3}H]GABA release. Neurochem Res 24(11):1373–1378
Ostad SN, Montazeran H, Minaee B, Monsef HR (2003) The study of acute toxicity of *Biebersteinia multifida* total extract in mice. Toxicol Lett 144(Suppl 1):s63
Oszmianski J, Wojdylo A, Lamer-Zarawska E, Swiader K (2007) Antioxidant tannins from Rosaceae plant roots. Food Chem 100(2):579–583
Overk CR, Yao P, Chadwick LR, Nikolic D, Sun Y, Cuendet MA, Deng Y, Hedayat AS, Pauli GF, Farnsworth NR, van Breemen RB, Bolton JL (2005) Comparison of the in vitro estrogenic activities of compounds from hops (*Humulus lupulus*) and red clover (*Trifolium pratense*). J Agric Food Chem 53(16):6246–6253
Ozek G, Ozek T, Iscan G, Baser KHC, Hamzaoglu E, Duran A (2007) Comparison of hydrodistillation and microdistillation methods for the analysis of fruit volatiles of *Prangos pabularia* Lindl., and evaluation of its antimicrobial activity. S Afr J Bot 73(4):563–569
Pages N, Fournier G, Baduel C, Tur N, Rusnac M (1996) Sabinyl acetate, the main component of *Juniperus sabina* L'Hérit. essential oil, is responsible for antiimplantation effect. Phytother Res 10(5):438–440
Pak VV (2005) Medicine plants of folk medicine used for treatment of gastro-intestinal problems in Fergana valley. (Food Tech), Korean Food Reseach Institute 18(4):150–157
Pakudina ZP, Sadykov AS, Zuparov A (1970) Kaempferol glycosides from the flowers of *Alcea nudiflora*. Chem Nat Compd 6(5):643–644
Panter KE, Keeler RF, Buck WB (1985a) Induction of cleft palate in newborn pigs by maternal ingestion of poison hemlock. Am J Vet Res 46(6):1368–1371
Panter KE, Keeler RF, Buck WB (1985b) Congenital skeletal malformations induced by maternal ingestion of *Conium maculatum* (poison hemlock) in newborn pigs. Am J Vet Res 46(10): 2064–2066
Park CJ, Park CB, Hong SS, Lee HS, Lee SY, Kim SC (2000) Characterization and cDNA cloning of two glycine- and histidine-rich antimicrobial peptides from the roots of shepherd's purse, *Capsella bursa-pastoris*. Plant Mol Biol 44(2):187–197
Pastura G, Mesiti M, Saitta M, Romeo D, Settineri N, Maisano R, Petix M, Giudice A (1999) Lymphedema of the upper extremity in patients operated for carcinoma of the breast: clinical experience with coumarinic extract from *Melilotus officinalis*. La Clinica Terapeutica 150(6):403–408 (in Italian)
Pavlov NV (1947) Rastitelnoye Suiryo Kazakhstana (Rasteniya; ikh Veshchestva i Ispolzovaniye) [Plant raw material of Kazakhstan (Plants: their substances and uses)]. Izdatelstvo Akademii Nauk SSSR, Moscow (in Russian)
Peng JN, Ma CY, Ge YC (1994) Chemical constituents of *Rhodiola kirilowii* (Regel) Regel. Zhongguo Zhong Yao Za Zhi 19(11):676–677 (in Chinese)
Peng WH, Hsieh MT, Lee YS, Lin YC, Liao J (2000) Anxiolytic effect of seed of *Ziziphus jujuba* in mouse models of anxiety. J Ethnopharmacol 72(3):435–441
Petrova VP (1972) Catechins and leucoanthocyanins from the fruits of wild hawthorn species (*Crataegus* L.). Ukr Botanichnii Zh 29(2):144–147 (in Ukrainian)
Pirniyazov AZ, Abdulladzhanova NG, Mavlyanov SM, Kamaev FG, Dalimov DN (2003) Polyphenols from *Vitis vinifera* seeds. Chem Nat Compd 39(4):349–355
Plant Resources of the USSR (1985) Fam. Magnoliaceae – Limoniaceae, vol 1. Nauka, Leningrad (in Russian)
Plant Resources of the USSR (1986) Fam. Paeoniaceae – Thymelaeaceae, vol 2. Nauka, Leningrad (in Russian)
Plant Resources of the USSR (1987) Fam. Hydrangeaceae – Haloragaceae, vol 3. Nauka, Leningrad (in Russian)
Plant Resources of the USSR (1988) Fam. Rutaceae – Elaeagnaceae, vol 4. Nauka, Leningrad (in Russian)
Plant Resources of the USSR (1990) Fam. Caprifoliaceae – Plantaginaceae, vol 5. Nauka, Leningrad (in Russian)
Plant Resources of the USSR (1991) Fam Hippuridaceae – Lobeliaceae, vol 6. Nauka, Leningrad (in Russian)
Plant Resources of the USSR (1993) Fam. Asteraceae, vol 7. Nauka, Saint Petersburg (in Russian)
Plekhanova NV, Turdumambetov KT, Fedorchenko GP (1985) Lechebnui Sakhar iz Sornuikh Rastenii Kirgizii (Medicinal sugar from weedy plants of Kyrgyzstan). Ilim, Frunze
Plesca-Manea L, Pârvu AE, Pârvu M, Taămaş M, Buia R, Puia M (2002) Effects of *Melilotus officinalis* on acute inflammation. Phytother Res 16(4):316–319
Polievtsev NP (1962a) The synergism of an alkaloid, dubinidine, with hypnotics and narcotics. Dokl Akad Nauk UzSSR 19(5):50–54 (in Russian)
Polievtsev NP (1962b) About some analgesic properties of the alkaloid dubinidine. In: Farmakologiia Alkaloidov (Pharmacology of alkaloids), vol 1. Izdatelstvo Akademii Nauk UzSSR, Tashkent, pp 138–144 (in Russian)
Polievtsev NP (1965) The influence of dubinidine on the central nervous system. In: Farmakologiia Alkaloidov (Pharmacology of alkaloids). Nauka, Tashkent, pp 60–80 (in Russian)
Polievtsev NP, Sultanov MB (1971) Pharmacology of a saponin extracted from the plant *Acanthophyllus gypsophiloides*. In: Sultanov MB (ed) Farmakologiia Alkaloidov i Serdechnuikh Glikozidov (Pharmacology of alkaloids and cardiac glycosides). FAN, Tashkent, pp 240–243 (in Russian)
Polievtsev NP, Kamilov IK, Evdokimova NI (1967) Pharmacology of the alkaloid skimmianine. In: Kamilov IK (ed) Farmakologiia Alkaloidov i Glikozidov (Pharmacology of alkaloids and glycosides). FAN, Tashkent, pp 35–42 (in Russian)
Poludenny LV, Zhuravlev YuP (2000) Zagotovka, Vuirashchivanie I pererabotka lekarstvennuikh rastenii (Preparation, management and processing of medicinal plants). Moscow Agricultural Academy (MCXA), Moscow (in Russian)
Pratov U (ed) (1998) Red data book of the Republic of Uzbekistan: rare and endangered species of plants and animals, vol 1, Plants. Chinor ENK, Tashkent (in Russian)
Prokhorova IR, Kurmukov AG (1997) Pre-clinical research of angioprotecting and hypolipidemic activity of the phytoestrogenic preparation *Tefestrol*. In: III Kongress Kardiologov Stran Tsentralnoi Azii (3rd congress of cardiologists of Central Asian countries), Tashkent, p 175 (in Russian)
Prokhorova IR, Aizikov MI, Kurmukov AG (1992a) Antisclerotic trend of lactone A. In: Eksperimentalnaya Farmakologiya Klinikye (Tezisui Dokladov Simposiuma-soveshchaniya s Mezhdunarodnuim uchastiyem Vinnitskogo Med Inst) [Clinical pharmacology experiments. (Proceedings of international symposium, Vinnitsa Medical Institute)] Vinnitsa Medical Institute, Vinnitsa-Kiev, pp 70–71 (in Russian)
Prokhorova IR, Aizikov MI, Kurmukov AG (1992b) Hypolipidemic activity of ferutinin. Meditsinskii Zh Uzbekistana (6):63 (in Russian)
Prokhorova IR, Aizikov MI, Kurmukov AG (1993) Angioprotective, hypolipidemic and antiatherosclerotic activity of austricine. In: Tezisui Dokladov: 1-aya Konferentsiya Kardiologov Moldovui (Proceedings of the 1st conference of Moldavian cardiologists), Kishinev, pp 150–151 (in Russian)
Pulatova TP (1969) Alkaloid content of some plants of the family Labiatae. Chem Nat Compd 5(1):55
Putieva ZhM, Mukhamedziev MM (1998) Triterpene glycosides of *Dipsacus azureus*. Chem Nat Compd 34(3):341–342
Putieva ZhM, Mzhel'skaya LG, Kondratenko ES, Abubakirov NK (1970) Triterpenoid glycosides of *Acanthophyllum gypsophiloides*. Chem Nat Compd 6(4):506–507
Putieva ZhM, Mzhel'skaya LG, Gorovits TT, Kondratenko ES, Abubakirov NK (1975) Triterpenoid glycosides of *Acanthophyllum*

gypsophiloides. IV. Structure of acanthophyllosides B and C. Chem Nat Compd 11(6):756–761

Putieva ZhM, Gorovits TT, Kondratenko ES, Abubakirov NK (1979) Triterpene glycosides of *Acanthophyllum gypsophiloides*. VI. Structure of acanthophylloside D. Chem Nat Compd 15(2):148–151

Puupponen-Pimia R, Nohynek L, Meier C, Kahkonen M, Heinonen M, Hopia A, Oksman-Caldentey KM (2001) Antimicrobial properties of phenolic compounds from berries. J Appl Microbiol 90(4):494–507

Rabinovich MI, Leskov AI, Gladkikh AS (1966) In: Mater Konf Fiziol, Biokhim, Farmakol. Uchastiem Prakt Vrachei, Ufa, pp 181–184 (in Russian)

Rakhimov DA (1997) Polysaccharides of *Eremurus*. XXX. Method of isolating a polysaccharide and its pharmacological activity. Chem Nat Compd 33(2):207

Rakhimov KD, Vermenichev SM, Lutskii VI, Gromova AS, Ganenko TV, Semenov AA (1987) Triterpene alkaloids of *Thalictrum foetidum* L. and *Thalictrum minus* L. and their antitumorigenic activity. Pharm Chem J 21(12):848–850

Rakhimov DA, Salimova ESh, Zakirova RP (2003) Polysaccharides of plant tissue culture. III. Polysaccharides of *Codonopsis clematidea* callus culture. Chem Nat Compd 39(1):46–47

Rakhimova IV, Pulatova TP (1972) Study of the antimicrobial properties of some plants in the family Lamiaceae. In: Phytoncids: results, prospects and research problems, Kiev, pp 94–96 (in Russian)

Rakhmanberdyeva RK, Talipova M, Gazizov F, Rakhimov DA (2002) Carbohydrates and lipids of *Gleditsia triacanthos* seeds. Chem Nat Compd 38(1):24–26

Rakhmatullaev TU, Yunusov SYu (1972a) Alkaloids of *Dipsacus azureus*. Chem Nat Compd 8(3):401

Rakhmatullaev TU, Yunusov SYu (1972b) Alkaloids of *Gentiana olivieri*. Chem Nat Compd 8(3):344–345

Raldugin VA, Demenkova LI, Pentegova VA (1991) Terpenoids of the oleoresin of the Schrenk spruce. Chem Nat Compd 27(3):279–282

Raskin I, Ripoll C (2004) Can an apple a day keep the doctor away? Curr Pharm Des 10(27):3419–3429

Rauha JP, Remes S, Heinonen M, Hopia A, Kähkönen M, Kujala T, Pihlaja K, Vuorela H, Vuorela P (2000) Antimicrobial effects of Finnish plant extracts containing flavonoids and other phenolic compounds. Int J Food Microbiol 56(1):3–12

Razzakov NA, Aripova SF (2004) Confolidine, a new alkaloid from the aerial part of *Convolvulus subhirsutus*. Chem Nat Compd 40(1):54–55

Razzakova DM, Bessonova IA, Yunusov SYu (1973) Alkaloids of *Haplophyllum acutifolium*. Chem Nat Compd 9(2):199–202

Razzakova DM, Bessonova IA, Yunusov SYu (1986) Components of *Haplophyllum acutifolium*. Chem Nat Compd 22(3):363–364

Red Data Book of Kirghiz SSR (1985) Redkiye i Nakhodyashchiyesya Pod Ugrozoi Ischeznoveniya Vidui Shuvotnikh i Rastenii (The rare and endangered flora and fauna species). Izdatelstvo Akademii Nauk KyrgSSR, Frunze (in Russian)

Revina TA, Shustova TN (1982) Study of *Galium* species in connection with their prospective medicinal use. In: Rastitelnuiye Resursui Yuzhnoi Sibiri, ikh Ratsionalnoye Ispolzovaniye i Okhrana (Plant resources of Southern Siberia, their rational use and conservation), Tomsk, pp 91–94 (in Russian)

Rhiouani H, Settaf A, Lyoussi B, Cherrah Y, Lacaille-Dubois MA, Hassar M (1999) Effects of saponins from *Herniaria glabra* on blood pressure and renal function in spontaneously hypertensive rats. Therapie 54(6):735–739

Rhiouani H, Lyoussi B, Settaf A, Cherrah Y, Hassar M (2001) Antihypertensive effect of *Herniaria glabra* saponins in the spontaneously hypertensive rats. Ann Pharm Fr 59(3):211–214 (in French)

Ribalko KS (1978) Natural sesquiterpene lactones. Meditsina, Moscow (in Russian)

Ribnicky DM, Poulev A, Watford M, Cefalu WT, Raskin I (2006) Antihyperglycemic activity of Tarralin, an ethanolic extract of *Artemisia dracunculus* L. Phytomedicine 13(8):550–557

Ringbom T, Segura L, Noreen Y, Perera P, Bohlin L (1998) Ursolic acid from *Plantago major*, a selective inhibitor of cyclooxygenase-2 catalyzed prostaglandin biosynthesis. J Nat Prod 61(10):1212–1215

Ripoll C, Schmidt B, Ilić N, Poulev A, Dey M, Kurmukov AG, Raskin I (2007) Anti-inflammatory effects of a sesquiterpene lactone extract from chicory (*Cichorium intybus* L.) roots. Nat Prod Commun 2(7):717–722

Rojas-Vera J, Patel AV, Dacke CG (2002) Relaxant activity of raspberry (*Rubus idaeus*) leaf extract in guinea-pig ileum in vitro. Phytother Res 16(7):665–688

Rowe GG, Castillo CA, Crumpton CW, Maxwell GM (1965) Systemic and coronary hemodynamic effects of pseudoephedrine. Proc Soc Exp Biol Med 118:682–684

Rustamov B, Tulyaganov N, Sadritdinov F (1974) Pharmacology of the alkaloid peganol. Meditsinskii Zh Uzbekistana (12):22–23 (in Russian)

Ryazantseva Z (Senior ed) (1965) Klimat Kirgizskoi SSR (Climate of the Kirghiz SSR). Ilim, Frunze (in Russian)

Ryu JH, Jeong YS, Sohn DH (1999) A new bisabolene epoxide from *Tussilago farfara*, and inhibition of nitric oxide synthesis in LPS-activated macrophages. J Nat Prod 62(10):1437–1438

Saadali B, Boriky D, Blaghen M, Vanhaelen M, Talbi M (2001) Alkamides from *Artemisia dracunculus*. Phytochemistry 58(7): 1083–1086

Saatov Z, Usmanov BZ, Abubakirov NK (1977) Phytoecdyzones of *Ajuga turkestanica*. IV. Chem Nat Compd 13(3):359

Saatov Z, Gorovits MB, Abubakirov NK (1990) Phytoecdysteroids of the plant *Dianthus hoeltzeri*. Chem Nat Compd 26(6):717–718

Saatov Z, Syrov VN, Mamatkhanov AU, Abubakirov NK (1994) Phytoecdysteroids of *Ajuga* and their biological activity. 1. Distribution and chemical structures of the compounds isolated. Chem Nat Compd, (30)2:138–145

Saatov Z, Agzamkhodzhaeva DA, Syrov VN (1999) Distribution of phytoecdysteroids in plants of Uzbekistan and the possibility of using drugs based on them in neurological practice. Chem Nat Compd 35(2):186–191

Sadritdinov FS (1968) Effect of the alkaloids, dubinidine and haplophilidine, on higher neural activity. Meditsinskii Zh Uzbekistana (6):71–73 (in Russian)

Sadritdinov FS (1971a) Comparative study of the antiinflammatory properties of alkaloids from *Gentiana* plants. In: Sultanov MB (ed) Farmakologiia Alkaloidov i Serdechnykh Glikozidov (Pharmacology of alkaloids and cardiac glycosides). FAN, Tashkent, pp 146–148 (in Russian)

Sadritdinov FS (1971b) Antiinflammatory activity of bis(benzylisoquinoline) alkaloids from *Thalictrum* plants in relation to their chemical structure. In: Sultanov MB (ed) Farmakologiia Alkaloidov i Serdechnykh Glikozidov (Pharmacology of alkaloids and cardiac glycosides). FAN, Tashkent, pp 122–125 (in Russian)

Sadritdinov FS (1973) Effects of the alkaloids thalicmine and thalicmidine on the central nervous system. Meditsinskii Zh Uzbekistana 12:40–43 (in Russian)

Sadritdinov FS, Khamdamov I (1975) Pharmacological properties of the alkaloid thalicmine. Farmakol Toksikol 38(4):490–491

Sadritdinov FS, Kurmukov AG (1980) Farmakologiya Rastitelnuikh Alkaloidov i Ikh Primeneniye v Meditsinye (Pharmacology of plant alkaloids and their use in medicine). Meditsina, Tashkent (in Russian)

Sadritdinov FS, Sultanov MB (1971) Antiinflammatory activity of the alkaloids thalmine and O-methylthalicberine. In: Sultanov MB (ed) Farmakologiia Alkaloidov i Serdechnykh Glikozidov (Pharmacology of alkaloids and cardiac glycosides). FAN, Tashkent, pp 120–122 (in Russian)

Sadritdinov FS, Tulyaganov N (1967) Pharmacology of the alkaloids gentianine and gentianadine. In: Kamilov IK (ed) Farmakologiia Alkaloidov i Glikozidov (Pharmacology of alkaloids and glycosides). FAN, Tashkent, pp 128–137 (in Russian)

Sadritdinov FS, Tulyaganov N (1972) Pharmacology of the new alkaloid gentianadine. In: Sultanov MB (ed) Farmakologiia Alkaloidov i Ikh Proizvodnikh (Pharmacology of alkaloids and their derivatives). FAN, Tashkent, pp 152–154 (in Russian)

Sadritdinov FS, Sultanov MB, Ismailov ZF (1971) Effect of some aporphine alkaloids on the behavior of animals. In: Sultanov MB (ed) Farmakologiia Alkaloidov i Serdechnykh Glikozidov (Pharmacology of alkaloids and cardiac glycosides). FAN, Tashkent, pp 127–132 (in Russian)

Sadyrbekov DT, Suleimenov EM, Tikhonova EV, Atazhanova GA, Tkachev AV, Adekenov SM (2006a) Component composition of essential oils from four species of the genus *Achillea* growing in Kazakhstan. Chem Nat Compd 42(3):294–297

Sadyrbekov DT, Ryazantsev OG, Kenesov BK, Raldugin VA, Atazhanova GA, Adekenov SM (2006b) Spectral properties of methyl-2-cis-lachnophyllate. Chem Nat Compd 42(4):493–494

Saeed SA, Gilani AH, Majoo RU, Shah BH (1997) Anti-thrombotic and anti-inflammatory activities of protopine. Pharmacol Res 36(1):1–7

Sagdullaev BT, Shakhidoyatov RKh, Khodzhaeva MA, Chernenko TV, Turakhozhaev MT, Abduazimova MA (2001) Lipids and carbohydrates from *Althaea nudiflora* and *A. armeniaca* Roots. Chem Nat Compd 37(3):208–212

Sagitdinova GV, Saidkhodzhaev AI (1977) The structures of juniferin and juniferinin. Chem Nat Compd 13(6):665–670

Sagitdinova GV, Saidkhodzhaev AI, Malikov VM (1978) The structures of juferin and juniferidin. Chem Nat Compd 14(6):693–694

Sagitdinova GV, Eshbakova KA, Khushbaktova ZA, Malikov VM, Olimov V (1992) Flavonoid pulicarpin from *Pulicaria salviifolia* and its hypolipidemic activity. Chem Nat Compd 28(3–4):286–288

Sagitdinova GV, Eshbakova KA, Malikov VM (1994) Diterpenoids of *Pulicaria salviifolia*. III. The structure of salvicinin. Chem Nat Compd 30(2):226–228

Sagitdinova GV, Makhmudov MK, Tashkhodzhaev B, Mal'tsev II (1996) Labdanoids of *Marrubium anisodon*. Chem Nat Compd 32(1):43–46

Sahin F, Gulluce M, Daferera D, Sokmen A, Sokmen M, Polissiou M, Agar G, Ozer H (2004) Biological activities of the essential oils and methanol extract of *Origanum vulgare* ssp. *vulgare* in the Eastern Anatolia region of Turkey. Food Control 15(7):549–557

Saidkasimov TK (1960) About the pharmacology of the alkaloid ervinine. In: Voprosy Biologii I Krayevoi Meditisinui (Issues of biology and regional medicine). Izdatelstvo Akademii Nauk UzSSR, Tashkent, pp 242–247 (in Russian)

Saidkhodzhaev AI, Nikinov GN (1973) The structure of ferutinin. Chem Nat Compd 9(1):25–26

Saidkhodzhaev AI, Nikinov GN (1974) The structure of ferutinol. Chem Nat Compd 10(2):178–181

Sairafianpour M, Christensen J, Staerk D, Budnik BA, Kharazmi A, Bagherzadeh K, Jaroszewski JW (2001) Leishmanicidal, antiplasmodial, and cytotoxic activity of novel diterpenoid 1,2-quinones from *Perovskia abrotanoides*: new source of tanshinones. J Nat Prod 64(11):1398–1403

Sakhobiddinov SS (1948) Dikorastushchiye Lekarstvennuiye Rasteniya Srednei Azii (Wild-growing medicinal plants of Central Asia). Gosudarstvennoye Izdatelstvo UzSSR, Tashkent (in Russian)

Salehi P, Sonboli A, Eftekhar F, Nejad-Ebrahimi S, Yousefzadi M (2005) Essential oil composition, antibacterial and antioxidant activity of the oil and various extracts of *Ziziphora clinopodioides* subsp. *rigida* (BOISS.) RECH. f. from Iran. Biol Pharm Bull 28(10):1892–1896

Salimov BT, Turgunjov KK, Tashkhodzhaev B, Dzhakhangirov FN (2004) Structure and antiarhythmic activity of 12-acetyl-12-epinapelline, a new diterpenoid alkaloid from *Aconitum soongoricum*. Chem Nat Compd 40(2):151–155

Samikov K, Shakirov R, Yunusov SYu (1989) Alkaloids of *Korolkowia sewerzowii*. Structure of korselidine. Chem Nat Compd 25(1):28–30

San Feliciano A, Gordaliza M, Miguel del Corral JM, Castro MA, García-Grávalos MD, Ruiz-Lázaro P (1993) Antineoplastic and antiviral activities of some cyclolignans. Planta Med 59(3):246–249

Sang SM, Lao AN, Leng Y, Gu ZP, Chen ZL, Uzawa J, Fujimoto Y (2000) Segetoside F a new triterpenoid saponin with inhibition of luteal cell from the seeds of *Vaccaria segetalis*. Tetrahedron Lett 41(48):9205–9207

Sang SM, Xia Z, Lao A, Cao L, Chen Z, Uzawa J, Fujimoto Y (2003) Studies on the constituents of the seeds of *Vaccaria segetalis*. Heterocycles 59(2):811–821

Sardari S, Amin G, Micetich RG, Daneshtalab M (1998) Phytopharmaceuticals. Part 1. Antifungal activity of selected Iranian and Canadian plants. Pharm Biol 36(3):180–188

Sarkisyants RG (1969a) Influence of *Ferula* of experimental arterial pressure. Meditsinskii Zh Uzbekistana (9):23–24

Sarkisyants RG (1969b) Influence of *Ferula* on the fuction of the gastro-intestinal tract of experimental animals. In: Tezisui 1 Nauchnoi Konferentsii Tsentralnoi Nauchno-Issledovatelskoi Laboratorii (Proceedings of the 1st conference of the Central Research Scientific Laboratory), Samarkand, pp 19–20

Sarkisyants RG (1972) Pharmacological characterization of *Ferula* (galenic forms, asafetids and sodium salts of galbanic acids). Synopsis of dissertation for the degree of Ph.D. in medicine (in Russian)

Sarkisyants RG, Azizova OM (1971) Influence of *Ferula* on progression and development of the Arthus reaction. Meditsinskii Zh Uzbekistana (9):46–47

Savona G, Bruno M, Rodriguez B, Marco JL (1987) Triterpenoids from *Salvia deserta*. Phytochemistry 26(12):3305–3308

Savoskin IP, Pekker EG, Kadirova RB (1971) Amino acids and vitamin potential of some plant species in the bean family (Leguminosae) found growing the the south-east Altai. In: Rastitelnuiye Bogatstva Sibiri (Plant riches of Siberia). Nauka, Novosibirsk, pp 104–109 (in Russian)

Schinella GR, Giner RM, Recio MC, Mordujovich de Buschiazzo P, Ríos JL, Máñez S (1998) Anti-inflammatory effects of South American *Tanacetum vulgare*. J Pharm Pharmacol 50(9):1069–1074

Schmidt JH, Wells R (1990) Evidence for the presence of gossypol in malvaceous plants other than those in the "cotton tribe". J Agric Food Chem 38(2):505–508

Schmidt BM, Ilic N, Poulev A, Raskin I (2007) Toxicological evaluation of a chicory root extract. Food Chem Toxicol 45(7):1131–1139

Schröder H, Schubert-Zsilavecz M, Reznicek G, Cart J, Jurenitsch J, Haslinger E (1993) A triterpene saponin from *Herniaria glabra*. Phytochemistry 34(6):1609–1613

Schultheiss PC, Knight AP, Traub-Dargatz JL, Todd FG, Stermitz FR (1995) Toxicity of field bindweed (*Convolvulus arvensis*) to mice. Vet Hum Toxicol 37(5):452–454

Seifert K, Preiss A, Johne S, Schmidt J, Lien NT, Lavaud C, Massiot G (1991) Triterpene saponins from *Verbascum songaricum*. Phytochemistry 30(10):3395–3400

Selivanova LN (1954) Phamacological study of the alkaloids allocriptopine, protopine and berberine. Synopsis of dissertation for the degree of Ph.D., Moscow (in Russian)

Semple SJ, Pyke SM, Reynolds GD, Flower RL (2001) In vitro antiviral activity of the anthraquinone chrysophanic acid against poliovirus. Antiviral Res 49(3):169–178

Sener B (1994) Recent results in the search for bioactive compounds from Turkish medicinal plants. Pure Appl Chem 66(10/11): 295–2298

Serbin AG, Borisov MI, Chernobai VT (1972a) Flavonoids of *Bidens tripartita*. I. Chem Nat Compd 8(1):119–120

Serbin AG, Borisov MI, Chernobai VT (1972b) Flavonoids of *Bidens tripartita*. II. Chem Nat Compd 8(4):439–441

Serbin AG, Borisov MI, Chernobai VT (1972c) Coumarins of *Bidens tripartita*. Chem Nat Compd 8(5):652

Serbin AG, Borisov MI, Chernobai VT, Kovalev IP, Gordienko VG (1975) Flavonoids of *Bidens tripartita*. III. Chem Nat Compd 11(2):160–162

Seredin RM, Sokolov SD (1969) Lekarstvennuiye Rasteniya i Ikh Primeneniye, izd 2 (Medicinal plants and their uses), 2nd edn. Stavropolskoye knizhnoye izdatelstvo, Stavropol (in Russian)

Sezik E, Yesilada E, Shadidoyatov H, Kulivey Z, Nigmatullaev AM, Aripov HN, Takaishi Y, Takeda Y, Honda G (2004) Folk medicine in Uzbekistan. I. Toshkent, Djizzax, and Samarqand provinces. J Ethnopharmacol 92(2–3):197–207

Sezik E, Aslan M, Yesilada E, Ito S (2005) Hypoglycaemic activity of *Gentiana olivieri* and isolation of the active constituent through bioassay- directed fractionation techniques. Life Sci 76(11): 1223–1238

Shafiee A, Morteza-Semnani K, Amini M (1998) (+)-Bulbocapnine-beta-N-oxide from *Glaucium fimbrilligerum*. J Nat Prod 61(12):1564–1565

Shahidi Bonjar GH (2004) Evaluation of antibacterial properties of Iranian medicinal-plants against *Micrococcus luteus*, *Serratia marcescens*, *Klebsiella pneumoniae* and *Bordetella bronchoseptica*. Asian J Plant Sci 3(1):82–86

Shakirov R, Kul'kova VV, Nakhatov I (1995) Alkaloids of *Veratrum lobelianum* verdinine and 3,15-DI-O-(2-methylbutyroyl)germine. Chem Nat Compd 31(1):79–82

Shamansurov ShSh, Sultanov MB (1967) Use of vincanine hydrochloride in neurologic practice. Zh Nevropatol Psikh 67(12):1807–1809 (in Russian)

Sharapov IM (1959) Pharmacology of peganine. Farmakol Toksikol 22(1):69–73 (in Russian)

Sharifi AM, Darabi R, Akbarloo N (2003) Study of antihypertensive mechanism of *Tribulus terrestris* in 2K1C hypertensive rats: role of tissue ACE activity. Life Sci 73(23):2963–2971

Sharon N, Lis H (2004) History of lectins: from hemagglutinins to biological recognition molecules. Glycobiology 14(11):53R–62R

Shikishima Y, Takaishi Y, Honda G, Ito M, Takeda Y, Kodzhimatov OK, Ashurmetov O (2001) Phenylbutanoids and stilbene derivatives of *Rheum maximowiczii*. Phytochemistry 56(4):377–381

Shimanov VG (1972) Gormonalnaya Aktivnost Pastbishchnuikh Rastenii I vliyaniye ikh na Plodovitost Karakulskikh Ovets (Hormonal activity of pasture plants and their influence on the fertility of karakul sheep). FAN, Tashkent (in Russian)

Shreter AI (1975) Lekarstvennaya Flora Sovetskogo Dalnego Vostoka (Medicinal flora of the Soviet Far East). Meditsina, Moscow (in Russian)

Shvarev IF, Tsetlin AL (1972) Anti-blastic properties of berberine and its derivatives. Farmakol Toksikol 35:73–75

Sidjimov AK, Tawara JN, Stermitz FR, Rithner CD (1998) An isopavine alkaloid from *Thalictrum minus*. Phytochemistry 48(2):403–405

Simsek I, Aytekin F, Yesilada E, Yildirimli S (2004) An ethnobotanical survey of the Beypazari, Ayas, and Güdül district towns of Ankara province (Turkey). Econ Bot 58(4):705–720

Singh NP, Lai HC (2004) Artemisinin induces apoptosis in human cancer cells. Anticancer Res 24(4):2277–2280

Singh VP, Yadav B, Pandey VB (1999) Flavanone glycosides from *Alhagi pseudalhagi*. Phytochemistry 51(4):587–590

Sinitsin GC (1959) About the use of medicinal plants in traditional and veterinary medicine in the western Tian-Shan. Zhurnal Vestnik Akademii Nauk KazakSSR Alma-Ata (4):92–94 (in Russian)

Sivropoulou A, Papanikolaou E, Nikolaou C, Kokkini S, Lanaras T, Arsenakis M (1996) Antimicrobial and cytotoxic activities of *Origanum* essential oils. J Agr Food Chem 44(5):1202–1205

Sklarovsky LY (1972) Tselebnuiye Svoistva Pishchevuikh Rastenii (Healing properties of edible plants). Rosselkhozuzdat, Moscow (in Russian)

Sokmen A, Jones BM, Erturk M (1999) Antimicrobial activity of extracts from the cell cultures of some Turkish medicinal plants. Phytother Res 13(4):355–357

Sokolov SYa, Zamotaev IP (1985) Spravochnik Po Lekarstvennuim Rasteniyam. Fitoterapiya. (Guide to medicinal plants. Phytotherapy). Meditsina, Moscow (in Russian)

Sokolov SYa, Zamotaev IP (1989) Spravochnik Po Lekarstvennuim Rasteniyam. Fitoterapiya, izd. 2 (Guide to medicinal plants. Phytotherapy), 2nd edn. Nedra, Moscow (in Russian)

Sonboli A, Mirjalili MH, Hadian J, Ebrahimi SN, Yousefzadi M (2006) Antibacterial activity and composition of the essential oil of *Ziziphora clinopodioides* subsp. *bungeana* (Juz.) Rech. f. from Iran. Z Naturforsch 61(9–10):677–680

Sonnenbichler J, Scalera F, Sonnenbichler I, Weyhenmeyer R (1999) Stimulatory effects of silibinin and silicristin from the milk thistle *Silybum marianum* on kidney cells. J Pharmacol Exp Ther 290(3):1375–1383

Spiridonov NA, Konovalov DA, Arkhipov VV (2005) Cytotoxicity of some Russian ethnomedicinal plants and plant compounds. Phytother Res 19(5):428–432

Stahl-Biskup E (2002) Essential oil chemistry of the genus thymus – a global view. In: Stahl-Biskup E, Saez F (eds) Thyme: the genus thymus, vol 24, Medicinal and aromatic plants – industrial profiles. Taylor & Francis, London, pp 75–124

Stepanenko GA, Gusakova SD, Umarov AU (1980) Composition of the coats and kernels of the seeds of *Nepeta pannonica* and *Lavandula vera*. Chem Nat Compd 16(5):434–439

Stevens KL, Riopelle RJ, Wong RY (1990) Repin, a sesquiterpene lactone from *Acroptilon repens* possessing exceptional biological activity. J Nat Prod 53(1):218–221

Stoychkov J, Miloushev AL (1980) Enhancement of antitumor effect of cyclophosphamide by thaliblastine in Lewis lung carcinoma and L1210 leukemia in mice. Biomedicine 33(2):42–44

Su BN, Takaishi Y, Tori M, Takaoka S, Honda G, Ito M, Takeda Y, Kodzhimatov OK, Ashurmetov O (2000) Macrophyllidimers A and B, two novel sesquiterpene dimers from the bark of *Inula macrophylla*. Tetrahedron Lett 41(9):1475–1479

Su BN, Takaishi Y, Yabuuchi T, Kusumi T, Tori M, Takaoka S, Honda G, Ito M, Takeda Y, Kodzhimatov OK, Ashurmetov O (2001) Sesquiterpenes and monoterpenes from the bark of *Inula macrophylla*. J Nat Prod 64(4):466–471

Su BN, Cuendet M, Farnsworth NR, Fong HHS, Pezzuto JM, Kinghorn AD (2002) Activity-guided fractionation of the seeds of *Ziziphus jujuba* using a cyclooxygenase-2 inhibitory assay. Planta Med 68(12):1125–1128

Suladze TSh, Vachnadze VYu, Tsakadze DM, Gedevanishvili MD, Tsutsunava LE, Malazoniya NA (2006) Alkaloid accumulation dynamics in *Veratrum lobelianum* growing in Georgia and biological activity of jervine. Chem Nat Compd 42(1):71–74

Sultankhodzhaev MN (1993) Karakanine – a new alkaloid from the epigeal part of *Aconitum karakolicum*. Chem Nat Compd 29(1):51–52

Sultankhodzhaev MN, Tadzhibaev MM (1976) Phenyl-β-naphthylamine from three species of plants. Chem Nat Compd 12(3):361

Sultankhodzhaev MN, Yunusov MS, Yunusov SYu (1973) Alkaloids of *Aconitum karakolicum*. Chem Nat Compd 9(1):129–130

Sultankhodzhaev MN, Yunusov MS, Yunusov SYu (1986) Alkaloids of the epigeal part of *Aconitum karakolicum* structure of karasamine and of 1-benzoylkarasamine. Chem Nat Compd 22(2):192–195

Sultanov MB (1959a) Pharmacology of the alkaloid of vincanidine. Izvestiya Akademii Nauk UzSSR (Seriya Meditsinskaya) (1):29–37 (in Russian)

Sultanov MB (1959b) The strychnine-like action of the alkaloid vincanine. Izvestiya Akad Nauk UzSSR (Seriya Meditsinskaya) (3):38–40 (in Russian)

Sultanov MB (1960) About the emetic action of the alkaloid vincanidine. Meditsinskii Zh Uzbekistana (6):30 (in Russian)

Sultanov MB, Kurmukov AG (1965) Pharmacology of the alkaloid ervamine. In: Farmakologiia Alkaloidov (Pharmacology of alkaloids). FAN, Tashkent, pp 128–142 (in Russian)

Sultanova BA, Lazkov GA, Lebedeva LP, Ionov RN (1998) Preview list of vascular plants for protection and entering in the Red Data Book. Nauka i Novye Technologii (2):119–127

Sumnevich GP (1942) Wild-growing fruit trees and shrubs. In: Suirevuiye resursii Uzbekistana (Raw material resources of Uzbekistan), vol 2(1). Tashkent, pp 10–26 (in Russian)

Supek Z (1946) Effect of berberine on the uterus in test animals. Farmakol Toksikol 9(6):12–14 (in Russian)

Suzutani T, Ogasawara M, Yoshida I, Azuma M, Knox YM (2003) Anti-herpesvirus activity of an extract of *Ribes nigrum* L. Phytother Res 17(6):609–613

Svechnikova A, Umarova RU, Gorovits MB, Abdullaev ND, Abubakirov NK (1983) Triterpene glycosides of *Astragalus* and their genins. XI. Cyclosiversioside G – a triglycoside of *Astragalus sieversianus*. Chem Nat Compd 19(3):296–299

Syrov VN (1984) Mechanism of the anabolic action of phytoecdysteroids in mammals. Biologicheskiye Nauki (11):16–20 (in Russian)

Syrov VN (1994) Phytoecdysteroids: their biological effects in the body of higher animals and the outlook for their use in medicine. Eksperimentalnaya i Klinicheskaya farmakologiia 57(5):61–66 (in Russian)

Syrov VN, Kurmukov AG (1975) On the anabolic action of viticosterone E. Dokl Akad Nauk UzSSR 32(6):31–32 (in Russian)

Syrov VN, Kurmukov AG (1975a) The anabolic and androgenic properties of ecdysterone and methyl-androstenediol. In: Voprosy Farmakologii I Toksikologii (Issues in pharmacology and toxicology). Meditsina, Tashkent, pp 22–25 (in Russian)

Syrov VN, Kurmukov AG (1975c) Effect of Viticosterone E on the weight of organs and their protein content in rats of different ages. Dokl Akad Nauk UzSSR (9):40–41 (in Russian)

Syrov VN, Kurmukov AG (1976a) Anabolic activity of phytoecdysone-ecdysterone isolated from *Rhaponticum carthamoides* (Willd.) Iljin. Farmakol Toksikol 39(6):690–693 (in Russian)

Syrov VN, Kurmukov AG (1976b) The anabolic properties of turkesterone phytoecdysone and turkesterone tetraacetate in experiments on male rats. Problemy Endokrinologii 22(3):107–112 (in Russian)

Syrov VN, Kurmukov AG (1976c) About the biological activity of cyasterone in tests on male rats. Biologicheskie Nauka (5):72–74 (in Russian)

Syrov VN, Kurmukov AG (1976d) Influence of ecdysterones on haematopoiesis of experimental posthemorrhagic anemia in lab animals. In: Aktualniye Voprosy Akusherstva i Ginekologii (Current issues in obstetrics and gynecology), Tashkent, pp 62–63 (in Russian)

Syrov VN, Kurmukov AG (1977) Tonic properties of ecdysterone isolated from *Rhaponticum carthamoides*. Dokl Akad Nauk UzSSR (12):27–30 (in Russian)

Syrov VN, Kurmukov AG (1980) Influence of phytoecdysones, their 6-ketoanalogs, and nerobol on anabolic processes in heart muscle (experimental data). In: Materialui IV Nauchno-Prakticheskoi Konferentsii IV Glavnogo Upravleniya pri Min. Zdravye UzSSR (Proceedings of the 4th scientific-practioners conference of 4th General Directorate of the Ministry of Health of the UzSSR), pp 119–122 (in Russian)

Syrov VN, Aizikov MI, Kurmukov AG (1975a) Effect of ecdysterone on the content of protein, glycogen, and fat in white rat liver, heart, and muscle. Dokl Akad Nauk UzSSR (8):37–38 (in Russian)

Syrov VN, Kurmukov AG, UsmanovBZ (1975b) About the anabolic action of turkesterone and tetra-acetate of turkesterone. Dokl Akad Nauk UzSSR (2):32–34 (in Russian)

Syrov VN, Kurmukov AG, Akhmedkhodzhaeva KhS (1976) About the anabolic and androgenic actions of drupanol. Dokl Akad Nauk UzSSR (1):34–36 (in Russian)

Syrov VN, Matveev SB, Kurmukov AG, Islambekov US (1986) Effect of ecdysterone and nerobol on the healing of experimental bone fractures. Meditsinskii Zh Uzbekistana (3):67–69 (in Russian)

Szabo I, Vonhaz G, Pallag A, Nemeth S (2006) In vitro specific antimicrobial properties of *Hibiscus trionum* L. extractive solutions. In: Proceedings of 4th conference of the association for medicinal and aromatic plants of South-East European countries. Iaşi, România, Alma Mater, Sibiu, 28–31 May 2006

Tabatabaei-Anaraki M, Chalabian F, Masoudi S, Rustaiyan A (2007) The chemical composition and in vitro antibacterial activities of the oil of *Ziziphora clinopodioides* Lam. from Iran. Planta Med 73(9):94

Taborska E, Dostal J, Bochorakova H, Veznik F (1988) Alkaloids of *Papaver argemone* L. and *Papaver pavoninum* Fisch. et Mey. from the Argemonorhoeades FEDDE section. Collect of Czech Chem Commun 53(8):1845–1850

Tada Y, Shikishima Y, Takaishi Y, Shibata H, Higuti T, Honda G, Ito M, Takeda Y, Kodzhimatov OK, Ashurmetov O, Ohmoto Y (2002) Coumarins and γ-pyrone derivatives from *Prangos pabularia*: antibacterial activity and inhibition of cytokine release. Phytochemistry 59(6):649–654

Tadzhibaev YuA, Mukhamedova KhS, Akramov ST (1977) Phospholipids of the seeds of two species of *Erysimum*. Chem Nat Compd 13(5):517–520

Tairov T (1969) Steroid sapogenins of *Asparagus persicus*. Chem Nat Compd 5(4):277

Tajadod G, Majd A (2007) Antibacterial effects of alcoholic extracts of different parts of *Mentha longifolia* and *Ziziphora tenuior*. Planta Med 73(9):176

Takeda Y, Tomonari M, Arimoto S, Masuda T, Otsuka H, Matsunami K, Honda G, Ito M, Takaishi Y, Kiuchi F, Khodzhimatov OK, Ashurmetov OA (2008) A new phenolic glucoside from an Uzbek medicinal plant, *Origanum tyttanthum*. J Nat Med 62(1):71–74

Talakal TS, Dwivedi SK, Sharma SR (1995) In vitro and in vivo antitrypanosomal activity of *Xanthium strumarium* leaves. J Ethnopharmacol 49(3):141–145

Tamas M, Doina S, Simina T (2006) Comparative phytochemical research of *Galium verum* L. and *G. mollugo* L. Notulae Botanicae Horti Agrobotanici Cluj-Napoca 34(1):18–20

Tamayo C, Richardson MA, Diamond S, Skoda I (2000) The chemistry and biological activity of herbs used in Flor-Essence™ herbal tonic and Essiac. Phytother Res 14(1):1–14

Tamemoto K, Takaishi Y, Chen B, Kawazoe K, Shibata H, Higuti T, Honda G, Ito M, Takeda Y, Kodzhimatov OK, Ashurmetov O (2001) Sesquiterpenoids from the fruits of *Ferula kuhistanica* and antibacterial activity of the constituents of *F. kuhistanica*. Phytochemistry 58(5):763–767

Tanaka N, Takaishi Y, Shikishima Y, Nakanishi Y, Bastow K, Lee KH, Honda G, Ito M, Takeda Y, Kodzhimatov OK, Ashurmetov O (2004) Prenylated benzophenones and xanthones from *Hypericum scabrum*. J Nat Prod 67(11):1870–1875

Tashbaev KhI, Sultanov MB (1962) Anticonvulsive action of the alkaloid talizopine. Farmakologiia Alkaloidov, Akademii Nauk UzSSR, Inst Khim Rast Veshchestv (1):210–219 (in Russian)

Tashbaev KhI, Sultanov MB (1965) Antiepileptic and sedative properties of thalizopine dihydriodide. Farmakologiia Alkaloidov, Akademii Nauk UzSSR, Inst Khim Rast Veshchestv (2):197–212 (in Russian)

Tashkhodzhaev B, Aripova SF, Turgunov KK, Abdilalimov O (2004) Stereochemistry of the alkaloids codonopsine and codonopsinine. Chem Nat Compd 40(6):618–619

Tashmukhamedova MA, Mukhina OA, Syrov VN, Khushbaktova ZA, Katkova SP, Kosovskii MI (1992) Hypoglycemic activity of bicyclic diterpenoids of the clerodane series as compared to adebit and maninil. Problemy Endokrinologii 38(1):48–50 (in Russian)

Tawata M, Aida K, Noguchi T, Ozaki Y, Kume S, Sasaki H, Chin M, Onaya T (1992) Anti-platelet action of isoliquiritigenin, an aldose reductase inhibitor in licorice. Eur J Pharmacol 212(1):87–92

Teslov SV (1960) Pharmacognostic studies of some Central Asian species in the genus *Thermopsis* R.Br. Synopsis of dissertation for the degree of Ph.D. in pharmacology, University of Tartu, Tartu (in Russian)

Tezuka Y, Kasimu R, Li JX, Basnet P, Tanaka K, Namba T, Kadota S (1998) Constituents of roots of *Salvia deserta* Schang. (Xinjiang-Danshen). Chem Pharm Bull 6(1):107–112

Tezuka Y, Fan W, Kasimu R, Kadota S (1999) Screening of crude drug extracts for prolyl endopeptidase inhibitory activity. Phytomedicine 6(3):197–203

Tigno XT, de Guzman F, Flora AM (2000) Phytochemical analysis and hemodynamic actions of *Artemisia vulgaris* L. Clin Hemorheol Microcirc 23(2–4):167–175

Todd FG, Stermitz FR, Schultheis P, Knight AP, Traub-Dargatz J (1995) Tropane alkaloids and toxicity of *Convolvulus arvensis*. Phytochemistry 39(2):301–303

Todorov DK, Zeller WJ (1992) Antiproliferative activity of the non-myelotoxic antitumour agent of plant origin thaliblastine on two human glioma cell lines. J Cancer Res Clin Oncol 118(2):83–86

Tolmachev AI (Senior ed) (1976) Atlas Arealov I Resursov Lekarstvennuikh Rastenii SSSR (Atlas of distributions and resources of medicinal plants of the USSR), Moscow (in Russian)

Tolomeo M, Grimaudo S, Di Cristina A, Roberti M, Pizzirani D, Meli M, Dusonchet L, Gebbia N, Abbadessa V, Crosta L, Barucchello R, Grisolia G, Invidiata F, Simoni D (2005) Pterostilbene and 3′-hydroxypterostilbene are effective apoptosis-inducing agents in MDR and BCR-ABL-expressing leukemia cells. Int J Biochem Cell Biol 37(8):1709–1726

Trutneva EA, Berezhinskaya VV (1960) Pharmacology of Lupinan group of alkaloids. Farmakol Toksikol (5):445–449

Tsevegsuren N, Christie WW, Lösel D (1998) *Tanacetum* (*Chrysanthemum*) *corymbosum* seed oil-a rich source of a novel conjugated acetylenic acid. Lipids 33(7):723–727

Tsitsina NB (Senior ed) (1962) Atlas Lekarstvennuikh Rastenii SSSR (Atlas of medicinal plants of the USSR). Izdatelstvo Meditsinskaya literatura, Moscow (in Russian)

Tsukurvanik IP, Simkhaev NG (1948) Study of *Ferula sumbul* Hooker f. roots. In: Trudy Instituta Khimii Akademii Nauk UzSSR: Khimicheskiye Issledovaniya Dikikh Rastenii Srednei Azii, vuip. 1 (Works of the institute of chemistry of the academy of science of Uzbekistan: chemical studies of the wild plants of Central Asia), vol 1. Izdatelstvo Akademii Nauk UzSSR, Tashkent, pp 24–26 (in Russian)

Tsukervanik IP, Bersutskii VP, Burtseva CV, Aizikovich BI (1935) Composition of (B) roots of *Ferula pseudoreosolinum* Lipsky. Bull Univ Asie Centrale (21):55–68 (in Russian)

Tulyaganov NT (1976) Cholinergic properties of quinazoline alkaloids from *Peganum harmala*. In: Sultanov MB (ed) Farmakologiia Rastitelnuikh Veshestv (Pharmacology of plant substances). FAN, Tashkent, pp 101–110 (in Russian)

Tulyaganov NT, Sadritdinov FS (1968) Effects of gentianine on the central nervous system. Meditsinskii Zh Uzbekistana (11):51–53 (in Russian)

Tulyaganov NT, Danilevskii BL, Sadritdinov FS (1971) Comparative sedative activity of gentianine, gentianadine, meprotan, and aminazine. In: Sultanov MB (ed) Farmakologiia Alkaloidov i Serdechnykh Glikozidov (Pharmacology of alkaloids and cardiac glycosides). FAN, Tashkent, pp 148–151 (in Russian)

Tulyaganov NT, Dzhakhangirov FN, Sadritdinov FS, Khamdamov I (1976) Pharmacology of aconite alkaloids. In: Sultanov MB (ed) Farmakologiia Rastitelnuikh Veshestv (Pharmacology of plant substances). FAN, Tashkent, pp 76–91 (in Russian)

Tulyaganov N, Sadritdinov FS, Suleimanova GA (1986) Pharmacological characteristics of desoxypeganine hydrochloride. Farmakol Toksikol 49(3):37–40 (in Russian)

Tunon H, Olavsdotter C, Bohlin L (1995) Evaluation of anti-inflammatory activity of some Swedish medicinal plants. Inhibition of prostaglandin biosynthesis and PAF-induced exocytosis. J Ethnopharmacol 48(2):61–76

Turdumambetov K (1995) Carbohydrates from plants in the genus *Cousinia*. Synopsis of dissertation for the degree of Ph.D. in chemistry, Bishkek (in Russian)

Turdumambetov K, Rakhimov DA, Malikova MKh (2007) Oligo- and polysaccharides from *Cousinia umbrosa*. Chem Nat Compd 43(3):308–309

Turker AU, Camper ND (2002) Biological activity of common mullein, a medicinal plant. J Ethnopharmacol 82(2–3):117–125

Turker AU, Gurel E (2005) Common mullein (*Verbascum thapsus* L.): recent advances in research. Phytother Res 19(9):733–739

Turova AD (1974) Lekarstvennuiye Rasteniya SSSR i ikh Premeneniye (Medicinal plants of the USSR and their uses). Meditisina, Moscow (in Russian)

Turova AD, Nikolskaya BS (1954) Short notes on the medicinal flora of Moscow Province. Farmakol Toksikol 17(1):54–58 (in Russian)

Turova AD, Sapozhnikova EN (1984) Lekarstvennuiye Rasteniya SSSR i ikh Premeneniye (Medicinal plants of the USSR and their uses), 4th edn. Meditisina, Moscow (in Russian)

Tursunova NV, Syrov VN, Khushbaktova ZA, Mukhamatkhanova RF, Sham'yanov ID, Nabiev AN (2002) Effect of the total sesquiterpenoid extract from *Artemisia leucodes* on cholepoietic processes in rat liver in the normal state and during experimental hepatitis. Pharm Chem J 36(2):37–39

Tye A, Baldesberger R, Lapidus JB, Patil PN (1967) Steric aspects of adrenergic drugs. VI. Beta adrenergic effects of ephedrine isomers. J Pharmacol Exp Ther 157:356–362

Ul'chenko NT, Glushenkova AI (2001) Oxygenated fatty acids from neutral lipids of *Artemisia leucodes* seeds. Chem Nat Compd 37(5):406–408

Ul'chenko NT, Gusakova SD, Glushenkova AI (1993) Oxygenated triacylglycerols of the lipids of *Onopordum acanthium* seeds. Chem Nat Compd 29(5):578–581

Ul'chenko NT, Mukhamedova KhS, Glushenkova AI (1999) Lipids from seeds of two species of *Cousinia*. Chem Nat Compd 35(6): 680–681

Ul'chenko NT, Mukhamedova KhS, Glushenkova AI (2000) Lipids of *Helichrysum maracandicum* flowers. Chem Nat Compd 36(1): 93–94

Ulubelen A, Topcu G, Eris C, Sonmez U, Kartal M, Kurucu S, Bozok-Johansson C (1994) Terpenoids from *Salvia sclarea*. Phytochemistry 36(4):971–974

Umarova RU, Gorovits MB, Abubakirov NK (1988) Mussaneoside from *Dodartia orientalis*. Chem Nat Compd 24(2):259

Umralina AR, Lazkov GA (2008) Endemic and rare plant species of Kyrgyzstan (Atlas). Biotechnology Institute, National Academy of Sciences Kyrgyz Republic, Bishkek

UNDP (2007) Kyrgyzstan: Environment and Natural Resources for Sustainable Development. State Agency on Environment Protection and Forestry under the Government of the Kyrgyz Republic and United Nations Development Programme in the Kyrgyz Republic: Bishkek.

Unlu M, Daferera D, Donmez E, Polissiou M, Tepe B, Sokmen A (2002) Compositions and the in vitro antimicrobial activities of the essential oils of *Achillea setacea* and *Achillea teretifolia* (Compositae). J Ethnopharmacol 83(1–2):117–121

Ur-Rehman J, Ahmad M, Saeed MA, Younas M (1991) Antioxidative activity of essential oils of some species of Umbelliferae family of Pakistan: Part-I. J Chem Soc Pak 13:56

Usmanov BZ, Gorovits MB, Abubakirov NK (1975) Phytoecdysones of *Ajuga turkestanica*. III. The structure of turkesterone. Chem Nat Compd 11(4):484–487

Usmanov BZ, Saatov Z, Abubakirov NK (1977) Phytoecdysones of *Ajuga turkestanica*. V. Chem Nat Compd 13(5):595

Utkin LA (1931) Traditional medicinal plants of Siberia. Trudy Nauchno-Issledovatelskogo Khimiko-farmatsevticheskogo Instituta 24:1–133 (in Russian)

Vaisov ZM, Yunusov MS (1987) Methylisotalatisidine – a new alkaloid from *Delphinium confusum*. Chem Nat Compd 23(6):725–727

Vandisheva BI, Yusupova AA, Alimbaeva PK, Nuralieva ZhC (1977) Lekarstvennuiye Rasteniya Kirgizii i Perspektivui Ikh Ispolzovaniya (Medicinal plants of Kyrgyzstan and perspectives on their uses). Ilim, Frunze (in Russian)

Varamini P, Doroudchi M, Mohagheghzadeh A, Soltani M, Ghaderi A (2007) Cytotoxic evaluation of four *Haplophyllum* species with various tumor cell lines. Pharm Biol 45(4):299–302

Venkateswarlu S, Satyanarayana B, Sureshbabu CV, Subbaraju GV (2003) Synthesis and antioxidant activity of 4-[2-(3,5-dimethoxyphenyl)ethenyl]-1,2-benzenediol, a metabolite of *Sphaerophysa salsula*. Biosci Biotechnol Biochem 67(11):2463–2466

Vilner AM (1974) Kormoviye otravleniya (Poisonous fodder plants), 5th edn. Kolos, Leningrad (in Russian)

Volinsky BT, Bender BT, Friedman CL (1983) Rasteniya v Meditsinye (Plants in medicine). Izdatelstvo Saratovskogo Universiteta (Saratov University), Saratov (in Russian)

Vonhoff C, Baumgartner A, Hegger M, Korte B, Biller A, Winterhoff H (2006) Extract of *Lycopus europaeus* L. reduces cardiac signs of hyperthyroidism in rats. Life Sci 78(10):1063–1070

Voynova E, Dimitrova S, Naydenova E, Karadjov P (1991) Inhibitory action of extracts of *Maclura aurantiaca* and *Epilobium hirsutum* on tumour models in mice. Acta Physiol Pharmacol Bulg 17(4):50–52

Vyazovskaya RD (1963) Depressive actions of plant substances on some endocrine functions. Uspekhi sovremennoi biologii 56(2):265–273 (in Russian)

Wang Y, Ohtani K, Kasai R, Yamasaki K (1997) Steroidal saponins from fruits of *Tribulus terrestris*. Phytochemistry 45(4):811–817

Wang AQ, Wang XK, Li JL, Cui XY (2004a) Isolation and structure identification of chemical constituents from the seeds of *Descurainia sophia* (L.) Webb ex Prantl. Yao Xue Xue Bao 39(1):46–51 (in Chinese)

Wang QE, Ma S, Fu B, Lee FSC, Wang X (2004b) Development of multi-stage countercurrent extraction technology for the extraction of glycyrrhizic acid (GA) from licorice (*Glycyrrhiza uralensis* Fisch). Biochem Eng J 21(3):285–292

Warholm O, Skaar S, Hedman E, Molmen HM, Eik L (2003) The effects of a standardized herbal remedy made from a subtype of *Rosa canina* in patients with osteoarthritis: a double-blind, randomized, placebo-controlled clinical trial. Curr Ther Res 64(1):21–31

Wegener T, Kraft K (1999) Plantain (*Plantago lanceolata* L.): anti-inflammatory action in upper respiratory tract infections. Wien Med Wochenschr 149(8–10):211–216 (in German)

Widyarini S, Spinks N, Husband AJ, Reeve VE (2001) Isoflavonoid compounds from red clover (*Trifolium pratense*) protect from inflammation and immune suppression induced by UV radiation. Photochem Photobiol 74(3):465–470

Williams CA, Goldstone F, Greenham J (1996) Flavonoids, cinnamic acids and coumarins from the different tissues and medicinal preparations of *Taraxacum officinale*. Phytochemistry 42(1):121–127

Williams CA, Harborne JB, Geiger H, Hoult JR (1999) The flavonoids of *Tanacetum parthenium* and *T. vulgare* and their anti-inflammatory properties. Phytochemistry 51(3):417–423

Wink M (1999) Introduction: biochemistry, role and biotechnology of secondary metabolites. In: Wink M (ed) Biochemistry of plant secondary metabolism, vol 2, Annual plant reviews. CRC Press, Boca Raton, pp 1–16

Winterhoff H, Sourgens H, Kemper FH (1983) Antihormonal effects of plant extracts. Pharmacodynamic effects of *Lithospermum officinale* on the thyroid gland of rats; comparison with the effects of iodide. Horm Metab Res 15(10):503–507

Wittenburg H, Nehring K (1965) Studies on the effect of pure lupinine alkaloids on the animal body. The effect of lupinine on rats. Pharmazie 20:156–158 (in German)

Wolniak M, Tomczykowa M, Tomczyk M, Gudej J, Wawer I (2007) Antioxidant activity of extracts and flavonoids from *Bidens tripartita*. Acta Pol Pharm 64(5):441–447

World Health Organization (2006) Guidelines for the treatment of malaria. ii. http://www.who.int/malaria/docs/TreatmentGuidelines2006.pdf. Accessed 27 Mar 2007

Wu TS, Li CY, Leu YL, Hu CQ (1999a) Limonoids and alkaloids of the root bark of *Dictamnus angustifolius*. Phytochemistry 50(3):509–512

Wu TS, Shi LS, Kuo SC (1999b) Alkaloids and other constituents from *Tribulus terrestris*. Phytochemistry 50(8):1411–1415

Yagudaev MR, Khalmirzaev MM, Yunusov SYu (1983) NMR study of alkaloids. V. ^{13}C NMR spectra and reconsideration of the structures of 11- and 10-hydroxypleiocarpamines and the structures of new alkaloids from *Vinca erecta*. Chem Nat Compd 19(4):454–458

Yagura T, Motomiya T, Ito M, Honda G, Iida A, Kiuchi F, Tokuda H, Nishino H (2008) Anticarcinogenic compounds in the Uzbek medicinal plant, *Helichrysum maracandicum*. J Nat Med 62(2):174–178

Yang R, Gao G, Liu T, Liu S, Li G (2007) Enhanced ability of hemoglobin to carry oxygen by salidroside. Electrochem Commun 9(1):94–96

Yasukawa K, Akihisa T, Inoue Y, Tamura T, Yamanouchi S, Takido M (1998) Inhibitory effect of the methanol extracts from compositae plants on 12-O-tetradecanoylphorbol-13-acetate-induced ear oedema in mice. Phytother Res 12(7):484–487

Yesilada E, Sezik E, Fujita T, Tanaka S, Tabata M (1993) Screening of some Turkish medicinal plants for their antiulcerogenic activities. Phytother Res 7(3):263–265

You M, Wickramaratne DB, Silva GL, Chai H, Chagwedera TE, Farnsworth NR, Cordell GA, Kinghorn AD, Pezzuto JM (1995) (−)-Roemerine, an aporphine alkaloid from *Annona senegalensis* that reverses the multidrug-resistance phenotype with cultured cells. J Nat Prod 58(4):598–604

Yue J, Xu J, Chen Y, Chen SN (1994) Diterpenoid alkaloids from *Aconitum talassicum*. Phytochemistry 37(5):1467–1470

Yue J, Xu J, Zhao Q, Sun H, Chen Y (1996) Diterpenoid alkaloids from *Aconitum leucostomum*. J Nat Prod 59(3):277–279

Yue ME, Jiang TF, Shi YP (2004) Fast determination of flavonoids in *Hippophae rhamnoides* and its medicinal preparation by capillary zone electrophoresis using dimethyl-β-cyclodextrin as modifier. Talanta 62(4):695–699

Yukhananov DKh, Yatsyno AI, Turova AD, Kuvaev VB (1972) Saponins of the Caryophyllaceae family and their hemolytic activity. Rastitel'nye Resursy 8(3):374–377 (in Russian)

Yuldasheva NP, Rakhimov DA, Turakhozhaev MT (1993) Polysaccharides of *Eremurus*. XXVI. Study of the structure of the pectin from the leaves of *Eremurus regelii*. Chem Nat Compd 29(2):154–156

Yunusov SYu (1974) Alkaloidy izd. 2 (Alkaloids), 2nd edn. FAN, Tashkent (in Russian)

Yunusov SYu (1981) Alkaloidy, izd. 3 (Alkaloids), 3rd edn. FAN, Tashkent (in Russian)

Yunusov AKh, Israilov IA (1974) A study of the alkaloids of *Glaucium fimbrilligerum*. Chem Nat Compd 10(4):555

Yunusov SYu, Sichkova EV, Potemkin GF (1954) Alkaloids of *Aconitum talassicum* Pop. Zh Obshch Khim 24(12):2237

Yunusov AKh, Israilov IA, Yunusov MS (1973) Alkaloids of *Glaucium fimbrilligerum*. Chem Nat Compd 9(5):658

Yusupova SS, Batirov EKh, Abdullaev ShV, Malikov VM (1984) Flavonoids of *Vexibia alopecuroides*. Chem Nat Compd 20(2):233

Zabirov ISh, Kasmaliev DK (1962) About the pharmacology of the alkaloids thalicmine and thalimidine. In: Materialui X Vsesoyuznoi Konferentsii Farmakologov (Proccedings of the tenth All-Union conference of pharmacologists), Volgograd, pp 116–118 (in Russian)

Zafrilla P, Ferreres F, Tomás-Barberán FA (2001) Effect of processing and storage on the antioxidant ellagic acid derivatives and flavonoids of red raspberry (*Rubus idaeus*) jams. J Agric Food Chem 49(8):3651–3655

Zaidi MA, Crow SA Jr (2005) Biologically active traditional medicinal herbs from Balochistan, Pakistan. J Ethnopharmacol 96(1–2):331–334

Zainutdinova UN, Pulatova MP, Badalbaeva TA, Umarova RU, Mavlyankulova ZI, Pulatova TP, Aslanova KhA (1994) Diterpene lactones and iridoid glycosides of plants of the genus *Lagochilus*. Chem Nat Compd 30(1):27–30

Zakordonets AI (1953) Kultura Vuicokovitaminuikh Shipovnikov (Culturing *Rosa* with high vitamin content), Kiev (in Russian)

Zapesochnaya GG, Aitbaeva RKh, Bankovsky AI (1973) Flavonoids and chlorogenic acid in *Sorbus tianschanica*. II. Chem Nat Compd 9(1):112–113

Zapryagaeva VI (1964) Dikorastushchiye Plodovuiye Tadzhikistana (Wild growing fruit-bearing species of Tajikistan). Nauka, Moscow-Leningrad (in Russian)

Zatsny IL, Gorovits MB, Abubakirov NK (1971) Ecdysterone from *Serratula sogdiana*. Chem Nat Compd 7(6):822

Zatsny IL, Gorovits MB, Abubakirov NK (1973a) Arbuitin from *Serratula sogdiana*. Chem Nat Compd 9(3):415

Zatsny IL, Gorovits MB, Abubakirov NK (1973b) Phytoecdysones of *Serratula* II. Viticosterone E from Serratula sogdiana and its partial synthesis. Chem Nat Compd 9(2):170–173

Zavrazhanov VI, Kitaeva RI, Khmelev KF (1972) Lekarstvennuiye Rasteniya Tsentralnogo Chernozemya (Medicinal plants of Central Chernozem). Izdatelstvo Voronezh Inst, Voronezh (in Russian)

Zavrazhanov VI, Kitaeva RI, Khmelev KF (1977) Lekarstvennuiye Rasteniya Tsentralnogo Chernozemya (Medicinal plants of Central Chernozem). Izdatelstvo Voronezh Inst, Voronezh (in Russian)

Zemlinsky SYa (1958) Lekarstvennuiye Rasteniya SSSR, 3-ye izd (Medicinal plants of the USSR), 3rd edn. Medgiz, Moscow (in Russian)

Zhang YD, Shen JP, Zhu SH, Huang DK, Ding Y, Zhang XL (1992) Effects of *Astragalus* (ASI, SK) on experimental liver injury. Yao Xue Xue Bao 27(6):401–406 (in Chinese)

Zhang L, Li G, Houghton PJ, Jackson S, Twentyman PR (1997) Alkaloids in *Sophora alopecuroides* seed and relevant tests for activity. Zhongguo Zhong Yao Za Zhi 22(12):740–764 (in Chinese)

Zhang L, Lau YK, Xia W, Hortobagyi GN, Hung MC (1999) Tyrosine kinase inhibitor emodin suppresses growth of HER-2/neu-overexpressing breast cancer cells in athymic mice and sensitizes these cells to the inhibitory effect of paclitaxel. Clin Cancer Res 5(2):343–353

Zhang L, Yu H, Sun Y, Lin X, Chen B, Tan C, Cao G, Wang Z (2007) Protective effects of salidroside on hydrogen peroxide-induced apoptosis in SH-SY5Y human neuroblastoma cells. Eur J Pharmacol 564(1–3):18–25

Zhao XY, Wang Y, Li Y, Chen XQ, Yang HH, Yue JM, Hu GY (2003) Songorine, a diterpenoid alkaloid of the genus *Aconitum*, is a novel GABA(A) receptor antagonist in rat brain. Neurosci Lett 337(1):33–36

Zhao C-C, Shao J-H, Li X, Kang X-D, Zhang Y-W, Meng D-L, Li N (2008) Flavonoids from *Galium verum* L. J Asian Nat Prod Res 10(7):611–615

Zhou WC (2001) Study on chemical constituents of unsaponifiable matter in chlorophyll-carotene paste from needles and twigs of *Picea schrenkiana*. Chem Ind Forest Prod 21(2):53–56 (in Chinese)

Zhou BN, Bai NS, Lin LZ, Cordell GA (1993) Sesquiterpene lactones from *Inula Britannica*. Phytochemistry 34(1):249–252

Zhou P, Takaishi Y, Duan H, Chen B, Honda G, Itoh M, Takeda Y, Kodzhimatov OK, Lee K-H (2000) Coumarins and bicoumarin from *Ferula sumbul*: anti-HIV activity and inhibition of cytokine release. Phytochemistry 53(6):689–697

Zielinska-Jenczylik J, Sypula A, Budko E, Rzadkowska-Bodalska H (1987) Interferonogenic and antiviral effect of extracts from *Erodium cicutarium*. Arch Immunol Ther Exp 35(2):211–220

Zielinska-Jenczylik J, Sypula A, Budko E, Rzadkowska-Bodalska H (1988) Interferonogenic and antiviral effect of extracts from *Erodium cicutarium*. II. Modulatory activity of *Erodium cicutarium* extracts. Arch Immunol Ther Exp 36(5):527–536

Zitterl-Eglseer K, Jurenitsch J, Korhammer S, Haslinger E, Sosa S, Della Loggia R, Kubelka W, Franz C (1991) Sesquiterpenelactones of *Achillea setacea* with antiphlogistic activity. Planta Med 57(5):444–446

Ziyan L, Yongmei Z, Nan Z, Ning T, Baolin L (2007) Evaluation of the anti-inflammatory activity of luteolin in experimental animal models. Planta Med 73(3):221–226

Zolotnitskaya SY (1965) Lekarstvennuiye Rasteniya Florui Armenii T. 2 (Medicinal plants in the flora of Armenia), vol 2. Izdatelstvo Akademii Nauk Arm. SSR, Yerevan (in Russian)

Zotov EP (1975) Phytochemical study of *Hyssopus seravshanicus* and *ferganensis* and production of medicinal forms. Synopsis of dissertation for the degree of Ph.D. in pharmacology, Frunze-Tartu (in Russian)

Zotov EP, Kolodina LK, Khazanovich RL (1977) Antibacterial activities of preparations of *Hyssopus seravshanicus* and *ferganensis*. In: Sovremennuiye Voprosy Mikrobiologii I Immunologii Tuberkuloz (Current issues in the microbiology and immunology of tuberculosis). Kyrgyz Medical Institute, Frunze, pp 101–108 (in Russian)

General Index

S. Eisenman et al. (eds.), *Medicinal Plants of Central Asia: Uzbekistan and Kyrgyzstan*,
DOI 10.1007/978-1-4614-3912-7, © The Editor 2013

T

U

V

Index to Plant Species

S. Eisenman et al. (eds.), *Medicinal Plants of Central Asia: Uzbekistan and Kyrgyzstan*,
DOI 10.1007/978-1-4614-3912-7, © The Editor 2013

MIX
Papier aus verantwortungsvollen Quellen
Paper from responsible sources
FSC® C105338

If you have any concerns about our products,
you can contact us on
ProductSafety@springernature.com

In case Publisher is established outside the EU,
the EU authorized representative is:
Springer Nature Customer Service Center GmbH
Europaplatz 3, 69115 Heidelberg, Germany

Printed by Libri Plureos GmbH
in Hamburg, Germany